This is
How the Solar System Forms
or
Proving How the Titius Bode law works

In a book this size it is impossible to even give an overview about introducing the new concept I bring to science. This will enlighten readers but it falls far short of displaying the entire concept that my view introduces. In this book I concentrate on proving how the Titius Bode law works as it applies in nature. I do emphasize that I also prove how the other three laws namely the Roche limit, the Lagrangian points and the Coanda effect works and how all 4 laws form gravity. However in this book there is only room to introduce one and briefly touch on the other 3. If I touch on the four cosmic laws I have to show briefly how the Universe started when the Universe started from one point and grew into what we now have. This is the Titius Bode law: The Titius Bode law or also known as Bode's law is named after Johann Elert Bode (1747-1826), who in 1772 published the law, that was formulated by Johann Titius in 1766 published in "Anleitung zur Kenntnis des gestirnten Himmels" [Instruction for the Knowledge of the Starry Heavens], which was printed in a number of editions. In this book, he stressed an empirical law on planetary distances, originally found by J.D. Titius (1729-96), now called "Bode's Law" or "Titius-Bode Law" This law is in astronomy an empirical rule relating the distances of the planets as they are positioned from the sun. The sequence is based on the numerical order 0, 3, 6, 12, 24, …. By adding 4 to each number 0, 3, 6, 12, 24 and divide it then by 10 it gives the sequence a numerical value of 0.4, 0.7, 1, 1.6, 2.8 … This numerical order places planets in a near perfect allocated positions where every planet adheres to this order. This is a reasonable representation of distances in astronomical units for most planets if the minor planets are counted as a single entity at 2.8.

I can and this book proves I do explain as much as I can prove why the sequence is how it is. Furthermore I can explain as much as I can prove why the sequence then ads four and no other number than four. Then I can prove with explaining in detail why the answer is divided by ten…no other number will do but ten. To do that I take this Titius Bode law back to when the Universe began with one point forming singularity and then how the Universe from there developed by implementing the four cosmic laws namely the Titius Bode law, the Roche limit, The Coanda effect and the Lagrangian points. These four laws began the Universe as much as they form gravity because these four laws are all that is needed to apply gravity. The book written to substantiate my claims I named **<u>Nature Annihilating Newton</u>** is the complete book in which I prove and explain the Titius Bode law in much detail. This book is a "sort-of" synopsis to introduce my accomplishment in showing how the Titius Bode law develops the solar system. My accomplishment is that I am the first in physics that apparently could add 3 and 4 and after that get a sum total of 7. This is how I solved the Titius Bode law after 250 years! What surprises me is that with all the brilliant mathematicians at work in physics no one in 250 years was able to add 3 and 4 and get 7 as a conclusion!

Everybody gets scared when hearing about the Titius Bode law and every body thinks it is as scary as hell. Science explains this law as technical as they can present it because they don't understand the law. It's a simple part of cosmology when one understands it.

Do not get scared when reading the definition because everyone does when reading the information. Every person shows heels and run away, so don't you do it…this is simple

In a nutshell the picture below explains the Titius Bode law in very specific detail.

WRITTEN BY Peet (P S. J.) SCHUTTE
ISBN 13: 978-1502708991
ISBN 10: 150270899X
©KOSMOLOGIESE EN ASTRONOMIESE TEGNIKA

http://www.titius-bode-law-explain.co.za/index.html

naturescosmicconcept
www.questionablescience.net

This also is
Proving-the-Titius-Bode-law
That is part of the website
naturescosmicconcept

This book introduces the way in which the solar system forms. It is not presenting a theoretical persuasion but proves a natural law by explaining how nature applies gravity to form the solar system and therefore the cosmos. It is called the Titius Bode law.

Should you wish not to read this website your action may panellise your insight into the future because this is an all-time breakthrough discovery. The Titius Bode law proves the Universe forms space by Π. I apply nature as nature works according to Kepler and I present the formula how Kepler and therefore Nature forms the solar system. Nature applies four laws and I prove that fact by publishing many, many books in which I bring facts supporting a very new cosmic concept...What this concept reveals could never be contained in this.

Please be forewarned: You are going to see me introduce myself numerous times during the length of the book because this book is compiled by many articles written to distinct and not-so distinct physicists, physic journals, science magazines, University lectors and professors, articles and entire books sent off to Universities and damn-well any body so that I could get someone and anyone to listen to what I have to say about my work. I still do it. No sooner you tell the person Newton is a lot of bullshit and they refuse to read further. This time it is your turn to put the book down but when doing that you will remain as uninformed as you were before you started this book. Notwithstanding who you are, what your achievements are, what your level of education is or what your academic standing and history is but what this book reveals as far as science you have never come across or experienced. Therefore then keep your opinion clear until you have been through the book.

This book was done with a $25 ⁰⁰ scanner and a $35 ⁰⁰ printer and the reason I explain inside. For the same reason this book was not edited or linguistically checked. I could not because that does not work because I am in the writing business and not the spelling business and while I check spelling the writing gets more and so does the spelling and grammar errors. I had a choice; doing the books with no funding or not doing it at all because while I rubbish Newtonian science and show it is the fake it is, they will never publish my work because I trash Newton. Not having funds and trying to fight science for the truth with the truth was a fight that physically broke my health and still I am not published except in this manner. I apologise for the spelling and language but in poverty that was the best I could do under the prevailing circumstances in which I find myself.... This book is a first in every sense... it unites science and religion because science and religion was separated by human stupidity. However for this book I took out the Bible and referring to it.

Please take note that I sell information and not words or books and therefore the information takes priority and not the spelling or words used to inform the readers. This represents the work of God and not the word of God and so there is no interpretations applying and versus you can learn and sound intellectual but only cold facts you will have to understand. It is science and not Gospel but that should not make it less religious because in my other books I prove the Birth of the Universe was exactly as the Bible says. I apply physics to prove the Bible and I challenge any person in science to use physics or mathematics to prove my statements about the Biblical birth of the cosmos wrong. If you are an atheist this challenge is mainly aimed at you and your type. I use mathematics t prove Genesis 1 verse 1 to the letter. However, there are those who can believe without physical proof needed to believe and then there are those who have to understand to believe. This book aims to satisfy those who have to understand in order to believe. However, understanding is a mental ability and not an emotion as believing the Bible is. This is science and not theology and to understand this requires intellect and not just rehearsing a few verses from the Bible that you learned off by heart as it is taken from the Bible. To understand this concept I introduce will require a fair bit of study on your part. It is not going to be easy but it will be rewarding.

This is to inform you about my latest book aiming too bring justice to nature by showing how nature and not Newtonian principles form the cosmos. There are four cosmic principles that nature applies and Newtonian science ignores. Nature form the cosmos by applying the following four principles found in nature: These are 1) **The Titius Bode law; 2) The Coanda effect; 3) The Roche limit;**
4) The Lagrangian points and these four still form everything that is committed to gravity.

The Sun and Nine Planets   Copyright © Calvin J. Hamilton

Look at the picture. Envisage how planets are allocated in their positions according to random mass. There is only a very random distribution of mass that follows no pattern or order and unlike the Titius Bode law there can be explanation to prove that the planets are mathematically aliened.
There is no order according to mass distribution, therefore without order mass can't apply mathematically and the Newtonian-invented formulas Newtonian science pretended to have proven is a hoax.
The following is as much part of the Titius Bode law what the planetary position sequence is.

Infinity, Definite and Eternity:
The only borders in the Universe

Infinity: The part of the Universe that can never start

Infinity: The part of the Universe that has no inside

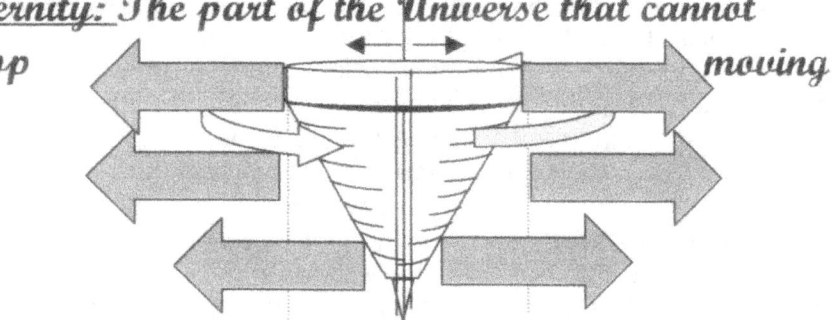

Infinity: The part of the Universe that cannot move

Eternity: The part of the Universe that has no outside

Eternity: The part of the Universe that can never end

Eternity: The part of the Universe that cannot stop   moving

Definite: Forms the part that holds movement and therefore defines space

Definite: which is what name that I gave material

Definite: is the part that converts eternity into infinity or release heat back to eternity

Definite:

Definite: is the part in the Universe that that defines the Universe for what it is by keeping infinity apart from eternity.

Definite: keeps time in infinity apart from time in eternity

There is this phenomenon called the curve ball. When a ball is thrown correctly, the ball follows a curve trajectory where the ball does not follow a straight line but it goes forward by following a curved path as it bends along a designated route. Science has many explanations where the one is as meaningless as the next. This curve ball applies to balls that are hit, thrown and kicked and then the ball travels in a rounded fashion. This action becomes vital in many ball players' careers and executing this with a well-developed technique will bring superiority to the player and the team of the player. It is also a big art of a professional tennis players' offence

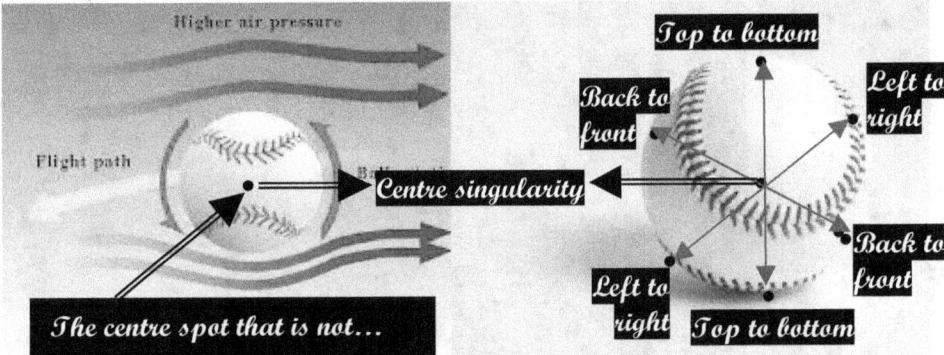

I present my new theory based on this. This is part of singularity and the way singularity affects the movement of round objects. It is called gravity.

In singularity a centre position forms when a ball travels in a circle going in a straight line. This is a cosmic principle by which all gravity applies.

In a sphere there are always a centre point in the very middle of all sphere that holds no space. This point only becomes activated when the sphere turns in a circle.

When this turning starts it activates a centre line that we know as an axis forming. This axis holds no space and in fact is an imaginary line around which the sphere turns.

The spin of the round sphere activates the axis. But everything in the sphere turns around this axis, which means this axis cannot have any space awarded to that sector. If it had space then the one side will have to go left and the direct opposing side will have to go right, which will tear the axis into two parts. This also will apply to the axis of the up and down points and the back to front points This axis line holds no space and yet it controls all that has space within the sphere.

These lines are always in place when a sphere turns and on this principle rides everything that conducts gravity. Around this principle the Universe started when the Universe started with Einstein's first spot. This is the most important feature gravity hold in the Universe. However it is the axis line and that establishes the curve ball but the axis line requires the spin of the circle to get the line exited into existence.

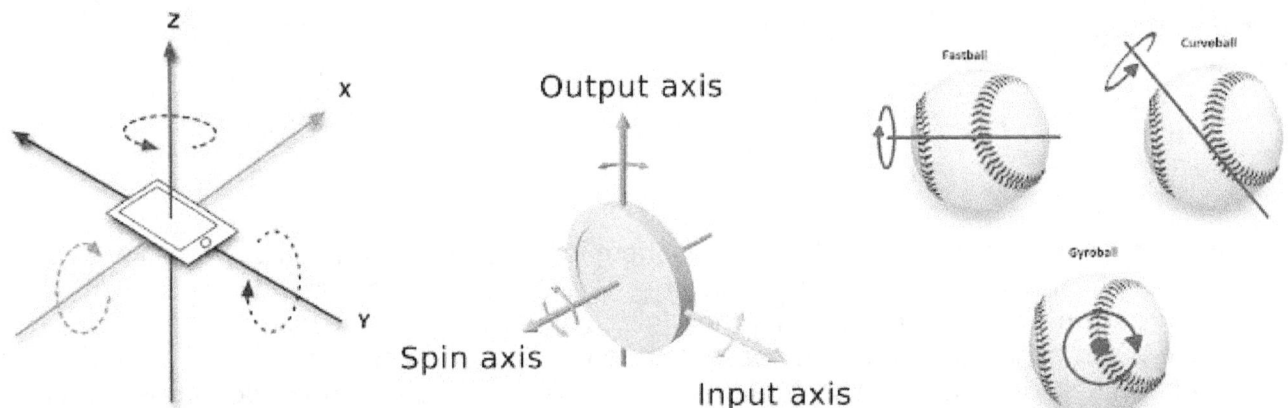

Typical Newtonian science is to give any principle in physics names without understanding the basics of the principle and to see how this principle applies in nature. In rifles it is most crucial that this centre point is established when accuracy is a priority. This centre follows lines in the atmosphere or air or whatever name you wish to give the liquid we walk in and that brings the air resistance that slows down all movement. It follows a trajectory that stabilizes the travel of the bullet by "creating" a path by which the trajectory will travel on one condition; the bullet must spin in a circle while travelling. That is why the grooves are formed in the barrel of

a rifle. Without the grooves there will be no spinning and without the spinning there will be no centre point exciting a point in singularity by which an axis forms and the axis stabilizes the trajectory. Even satellites orbiting the earth have to have a spin in order to ensure stability.

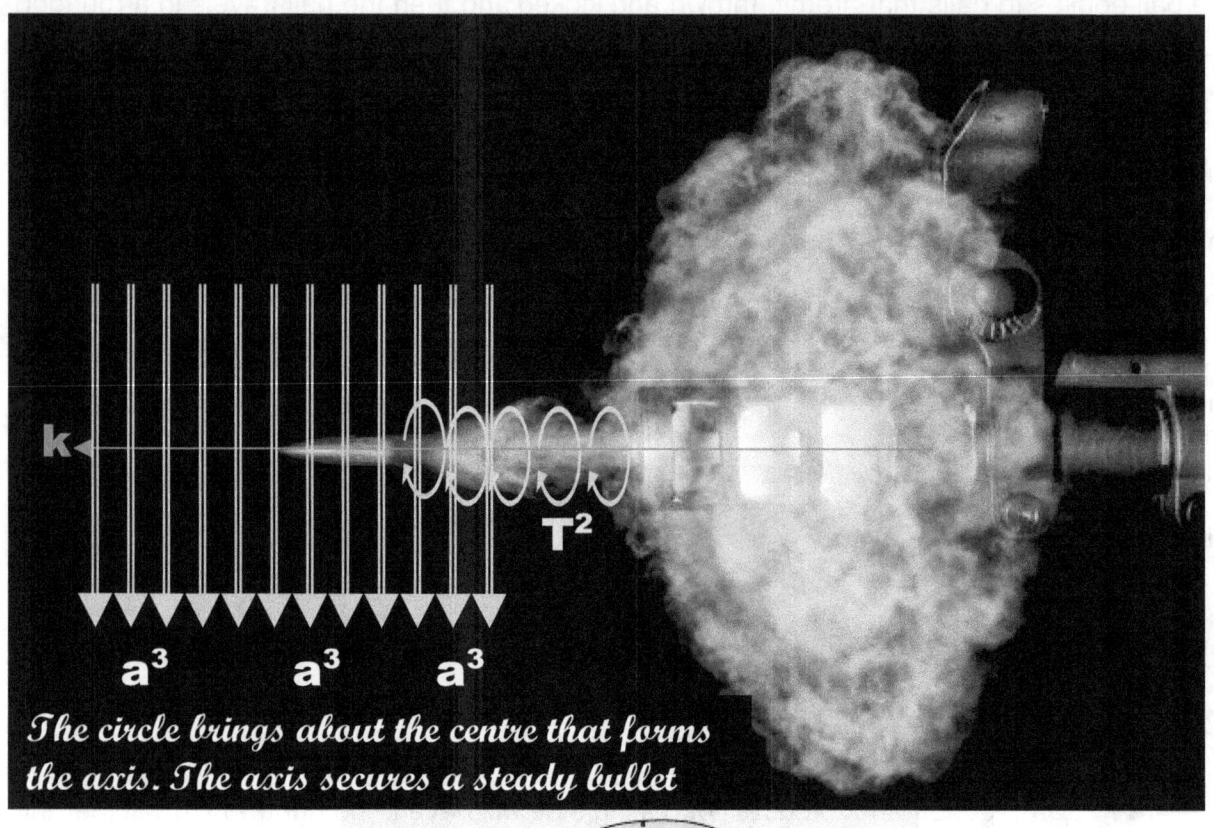

The circle brings about the centre that forms the axis. The axis secures a steady bullet

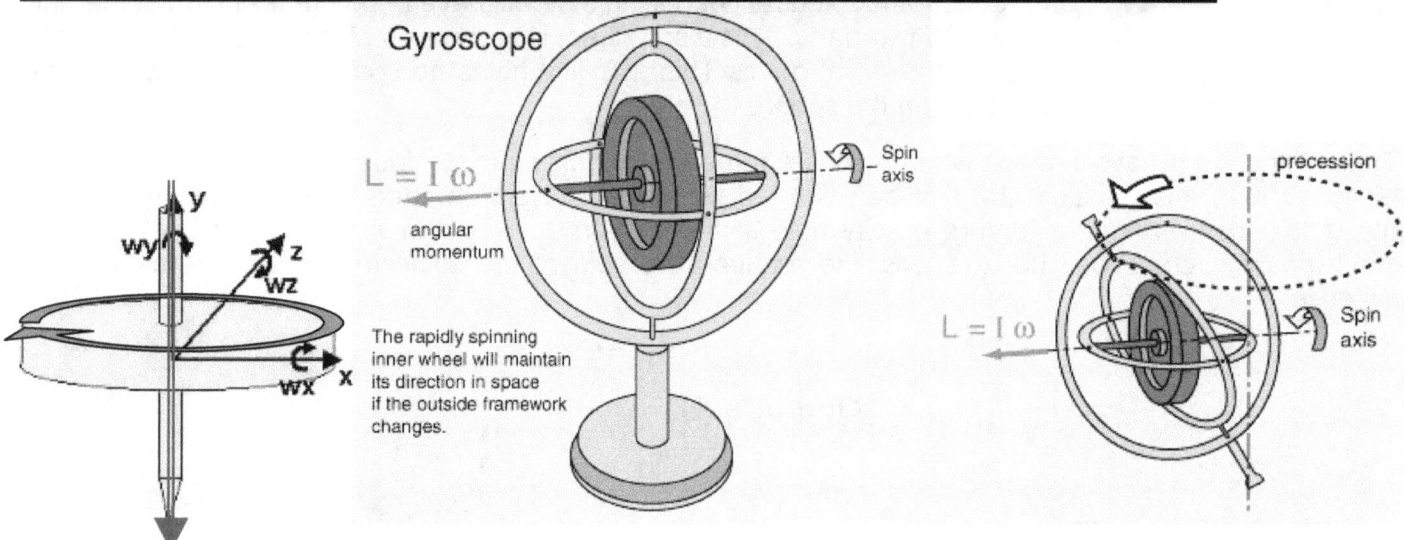

From the centre of the earth or any body that has a sphere in the Universe with gravity runs lines in every direction and all directions there ever could possibly be. These lines are so large they fill the Universe in every dimension and these lines are so small they fall outside the parameters of the Universe. These lines form space when expanding and form heat when compressed. These lines are the heat that forms the Universe and either form material by forming space within material or form space in which material moves. Although these lines are so small they don't truly exist in cosmic standards, yet they fill the Universe by forming everything that the Universe comprises of. They form the Universe as they fill the Universe although not one can be mathematically within the Universe or form part of the Universe.

It is these lines that connect every dot forming the Universe with every other dot forming the Universe. These lines form singularity and from these dots a Universe came about. Science knew about these dots for a long time without ever validating these dots formally because science never came to recognise these dots. It is these dots that are responsible for the four laws forming gravity, which is the Lagrangian point, the Titius Bode law, the Roche limit and the Coanda effect. These lines are gravity and in the Black

Hole we find the Universe disappearing along one of these lines. In 500 years of "modern science" no one came to consider how the sphere works although everything holding gravity is in the shape of a sphere. Science can never take the blame for not knowing. Never is a suggestion put forward that it might be science that holds the shortfall and it is because of science not being adequate that science cannot match what nature proves. I prove how the start came about because I decoded gravity and I did that by finding an explanation about the four cosmic principles. By deciphering the Roche limit, the Lagrangian points, the Titius Bode law and the Coanda effect I am able to show how the very first instant happened when the Universe started the very first point ever formed. These principles are in place and not the principles Newton fabricated... That this book shows. It shows that the cosmos uses other principles than what the Newtonian science promotes. What science says nature uses is not in place or does not hold evidence while what nature does use science deny by just never pressing the issue. I show what is in place and I show why it is in place but first I have to reject what science says is in place because it is not in place.

All Newtonian-thinking mechanical engineers propagate that a car lifts in the air because of the wind flowing underneath. They say the wind beneath the car pushes it up into the air and therefore it lifts from the ground. This is cultured hogwash not well thought through at all and shows very little understanding!

Look at the picture of the car below. The Newtonian engineers teach that at a specific speed the air flowing underneath the car lifts the car from the ground and into the air. This is an argument made by the uninformed to  mislead those with the stupidity of the human race and everyone that has no thought process but still wish to come across as being wise. ...And this mindless dogma has been offered for decades and was never questioned or even challenged by even one person in the world of physics.

It is said that the car lifts into the air because the wind from beneath lifts the car so much it becomes airborne. Again I reiterate this point because I know that any person reading this will be ready to defend this idea forming a logic conclusion as if this argument is his or her personal knowledge.

Giving the situation the minimal thought the truth nails this misconception out of the park. Think of the amount of air pushing the car down versus the amount of air lifting the car up during any particular speed and then reconsider your verdict. There is no match for the ratio of air flowing over the car when it is compared to the ratio flowing underneath the car and the higher the speed the bigger this ratio gets.

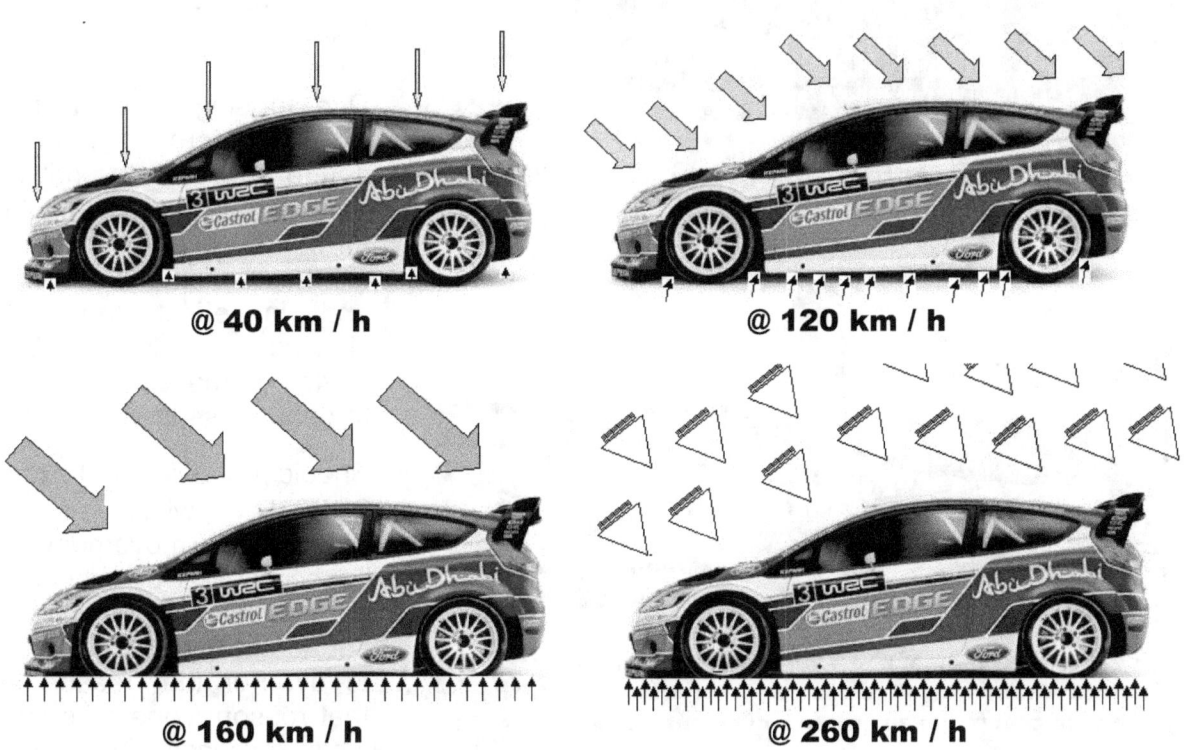

@ 40 km / h   @ 120 km / h

@ 160 km / h   @ 260 km / h

Give this a thought and guard against thought bias because of brainwashing and mind control inflicted on you under the cover of "*learning about science*". There will always be exponentially more air flowing over the car that thrusts the car onto the earth than what there can ever be pushing underneath the car enabling the car to lift into the air. What is even more a daunting fact to keep in mind is that the higher speed that the car is doing the bigger this ratio gets in favour of creating a downward movement in comparison with upward movement. Go sit back and give this a serious thought and think how can the air underneath the car overcome the air pushing the car down. The air pushing the car down will always be completely in a bigger dimension that any air can be when the car is secured to the earth. This way of thinking clashes completely with all logical thought or intellectual mind processing when judging facts.

The article had the purpose whereby I aim to introduce singularity to physics in the way the cosmos uses singularity in the cosmos. Either I fail or everyone else is blind but until now it is clear that I am not projecting my message in a manner that I am reaching the minds of many. If those practising and preaching science would only stop being blinded by Newton's brainwashing then they would see that the entirety in the Universe formed by circles and circles are formed by Π so gravity must be Π. If you remove Π from the cosmos then the cosmos is still left with Π as it is Π^0 located in the Black Hole. Singularity connects to Π by a means that could never be parted and the article proves that above all.

$$7(\Pi\Pi^2)\Pi^0$$

Standing still the car is moving down and moving along with the earth at a relevancy rate of $7(\Pi\Pi^2)\Pi^0$

By turning the rotation of the earth holds a value of $7(\Pi\Pi^2)\Pi^0$. You might think it is eccentric not to just say $7\Pi^3$ but in doing that it would be completely wrong because having a value of Π^3 would refer to space and in the cosmos there is no space. What science thinks of as space is time and what you think of as space is the way time paints a picture on a three-dimensional canvass by using light but this light is only a hallucination of what is time. Space does not exist but to replace time and the history of time in one dot per instant.

Due to singularity Π^0 there are lines forming from the centre of the earth ending in the value of Π. These lines formed by singularity dictates all the movement the earth and in effect the entire Universe holds. When anything moves, it moves through these lines.

The line that follows the earth's rotation and replicated eternity or Π

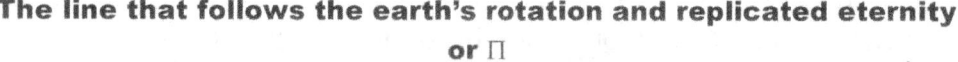

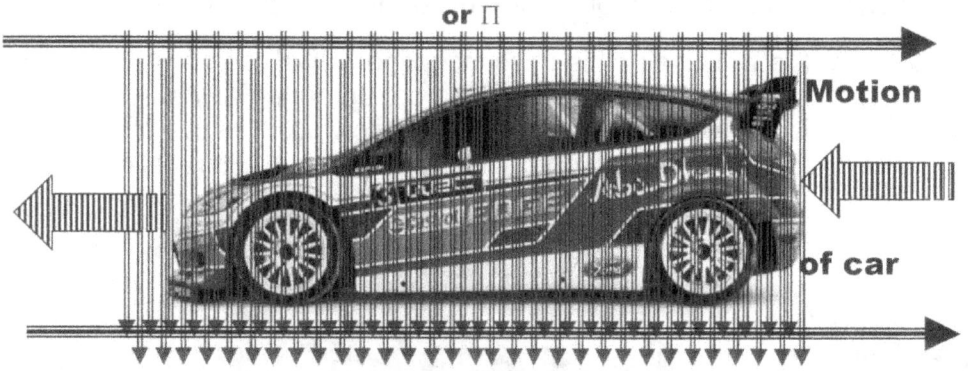

Motion

of car

Innumerable lines representing singularity and leading to infinity Π^0

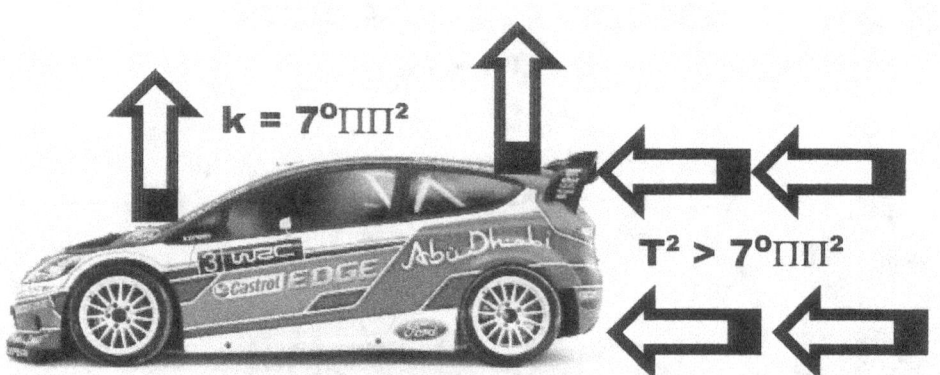

$$k = 7^0\Pi\Pi^2$$

$$T^2 > 7^0\Pi\Pi^2$$

As soon as $T^2 > k$ in the relevance of $a^3 = T^2k$ the car will get airborne. In the earth's gravity the factor k will reverse $k = 7\Pi\Pi^2$ from Π^0 to $4\Pi^2$ because of the density of the car that increases when movement exceeds $T^2 = 7\Pi\Pi^2$

Should the movement exceed $7(\Pi\Pi^2)\Pi^0$, which is the downward movement of space caused by gravity the car will react by leaving the surface of the earth and start to fly. For that reason and that reason alone cars lift off when exceeding the speed limit the earth rotation allows or then gravity put in place, which is $7(\Pi\Pi^2)\Pi^0 = 217$ km / hour.

This is the application of these four laws. These laws are:

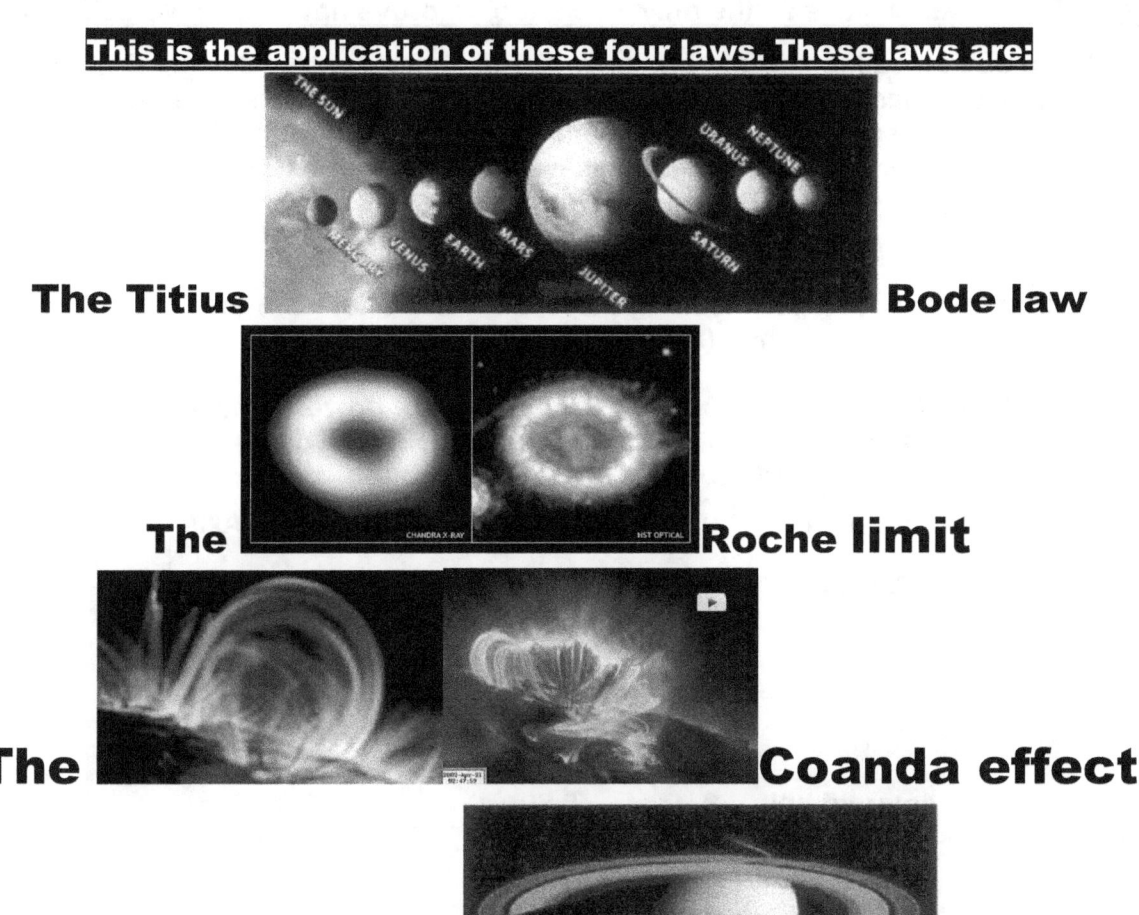

The Titius   **Bode law**

The   **Roche** limit

The   **Coanda effect**

The Lagrangian   **Points**

This is the way one would go about finding the smallest particle in the Universe. This takes material to a mathematical equation even before material begins. This takes material before space forms to where material forms material in space

I explain all four these phenomena because I trash Newtonian gossip, which is what science, are. Nature destroys everything Newton said is true about the cosmos. In the 4 books on the market I explain in various degrees or levels of complexity how the Titius Bode forms. The Titius Bode law is the manner how the solar system forms because it does not form by Newton's science. I presume you that read this never knew about this law?

Nature does not apply mass to allocate planets but use these laws and the Titius Bode law in particular to place plants in the order in which they are. I explain the others and the Titius Bode law in eight other books also on the market but the explaining might not be in the taste of everybody because the level of intensity used is much higher than the four books I promote in this writing. I place these books on the market as cheap as I could to help me as evidence to support me in my fight against their falseness with which science holds the world at ransom.

As you will see I solved the mystery of the four phenomena in cosmic space just because and only because I was the first person in 300 years that tried to add three and four and then reach a conclusion that it is seven. By my ability to be able to add three and four I am the first one (apparently) in cosmology that has the ability to come to a conclusion that three and four is seven

Although I truly believe my work presents an all-time breakthrough in science I am the person that wrote the book on the working of nature applying gravitational density and I am as poor as the poorest can be because I am honest and I tell the truth that nobody in science wants to hear anything about because the world is brainwashed to believe the unbelievable is true. I wish to

present to you an approach to the Universe I think no one before me ever took. It is to prove the insanity of applying mathematics as the only viable tool to prove how the Universe functions.

In 1768, Bode published his popular book, "Anleitung zur Kenntnis des gestirnten Himmels" (*Instruction for the Knowledge of the Starry Heavens*), which was printed in a number of editions. In this book, he stressed an empirical law on planetary distances, originally found by J.D. Titius (1729-96), now called "Bode's Law" or "Titius-Bode Law".>

The original formulation was

$$a = (n + 4) / 10$$

where n=0,3,6,12,24,48 ...

The modern formulation is that the mean distance a of the planet from the Sun is, in astronomical units (AU_{earth} = 147.597 $*10^6$ km):

$$a = 0.4 + 0.3 \times k$$

where "k'=0,1,2,4,8,16,32,64,128 (sequence of powers of two *and* 0)

The following table compares the law's predictions with the actual distances, where the addition of Pluto is a modern modification.

Planet	n	Titius-Bode Law	Semi-Major Axis
Mercury		0.40	0.39
Venus	0	0.70	0.72
Earth	1	1.00	1.00
Mars	2	1.60	1.52
asteroid belt	3	2.80	2.8
Jupiter	4	5.20	5.20
Saturn	5	10.0	9.54
Uranus	6	19.6	19.2
Neptune	-	-	30.1
Pluto	7	38.8	39.4

Titius Bode's law

The number line up appeared logic except that there was a big gap between Mars and Jupiter. Then Titius and Bode decided to skip a number, making Jupiter a particularly good fit. This law was chosen as the correct indicator taken to predict that a planet would be found between Mars and Jupiter. Not once was Newton any choice to consider as an accurate indicator of the planets' layout pattern. Within a few years (1781), Uranus was discovered by Sir William Herschel, and it fit right into the law. This discovery made the law respectable, and the hunt for the missing planet began. In 1801, Giuseppe Piazzi discovered the minor planet Ceres, at just the right distance. Ceres was incredibly tiny for a planet. To date, more than 9000 minor planets (asteroids) have been discovered. At first it was thought that a planet was destroyed by a collision, at that distance from the Sun. Now it is thought that the gravity of Jupiter prevented a planet from forming from the fragments there.

Bode's law (for J. E. Bode), also known as Titius' law or the Titius-Bode law, empirical relationship between the mean distances of the planets from the sun. If each number in the series 0, 3, 6, 12, 24, ... (where a new number is twice the previous number) is increased by 4 and divided by 10 to form the series 0.4, 0.7, 1.0, 1.6, 2.8, 5.2, 10.0, 19.6, 38.8, 77.2, ... , The Titius Bode law holds that this series gives the mean distances of the planets from the sun, expressed in astronomical units. When this relationship was discovered by Titius of Wittenberg in 1766 and published by Bode six years later, it gave good agreement with the actual mean distances of the planets that were then known—Mercury (0.39), Venus (0.72), Earth (1.0), Mars (1.52), Jupiter (5.2), and Saturn (9.55). Uranus, discovered in 1781, has mean orbital distance 19.2, which also agrees. The asteroid Ceres, discovered 1801, has mean orbital distance 2.77, which fills the apparent gap between Mars and Jupiter. However, Neptune, discovered 1846, has mean orbital distance 30.1, and Pluto, discovered 1930 and now regarded as a dwarf planet, has mean orbital distance 39.5; these are large discrepancies from the positions 38.8 and 77.2, respectively, predicted by Bode's law. Some theories of the origin of the solar system have tried to explain the apparent regularity in the mean orbital distances of the planets, arguing that it could not arise by chance, but must be a manifestation of the laws of physics. Some astronomers have argued that the deviation of Neptune from its predicted positions signifies that it is no longer at its original positions in the solar system. However, since Bode's law is not a law in the usual scientific sense, i.e., it is not universal and invariant, and it alone should not be taken as evidence for such a conclusion.

Kepler's Law

The distances and the length of the years of the various planets are as follows:

Planet	Radius of Orbit Relative to that of Earth	Length of Year Relative to Earth's Year
Mercury	0.387	0.2409
Venus	0.723	0.616
Earth	1.0	1.0
Mars	1.524	1.9
Jupiter	5.203	12.0
Saturn	9.539	29.5
Uranus	19.18	84
Neptune	30.06	165
Pluto	39.52	248

If we call T the length of the year and multiply T times itself (T*T) the result is called "T squared" and it is represented as T^2. Call the radius of a planets orbit R and multiply it times itself three times, R*R*R. The result is called "R cubed" and it is represented as R^3. The computations are given below:

Planet	R^3	T^2	T^2/R^3
Mercury	0.05796	0.05803	1.00121
Venus	0.37793	0.37946	1.00405
Earth	1.0	1.0	1.00000
Mars	3.5396	3.61	1.01989
Jupiter	140.85	144.0	1.02236
Saturn	867.98	870.25	1.00262
Uranus	7055.8	7056	1.00003
Neptune	27162.3	27154.7	0.99972

Pluto	61723.5	61504	0.99644

As you notice the ratios of T^2 to R^3 are all close to 1.0. If all of the numbers were precisely correct the ratios would all have been exactly 1.0. This means that for each planet the squared of the length of its year is equal to the cube of the radius of its orbit. This is called Kepler's Law. From Kepler's Law if you know how far a planet is from the sun you can tell how long it takes for that planet to go around the sun.

Titius Bode law

Kepler's Law is exact and can be derived from Newton's laws of the motion of bodies except that Kepler's tables prove that Kepler's laws (named by Newton) is complete hogwash. There is another law concerning the planets and it is a precise indication of how gravity works. The Titius Bode law shows how accurate Kepler's tables are forming and proving gravity. It is often called Bode's Law because it was popularised by Bode, but it was actually discovered by Titius. It is a rule or formula for finding the orbit radiuses of the planets. The Titius Bode- law goes like this:

Take the series of numbers,

0, 1, 2, 4, 8, 16, 32, 64, 128, 256

and multiply each by 0.3 to get:

0, 0.3, 0.6, 1.2, 2.4, 4.8, 9.6, 19.2, 38.4, 76.8

To these numbers add 0.4 to get:

0.4, 0.7, 1.0, 1.6, 2.8, 5.2, 10.0, 19.6, 39, 77.2

Now compare these numbers with the radiuses of the orbits of the planets (relative to the Earth's orbit radius):

Mercury	Venus	Earth	Mars	Asteroid Belt	Jupiter	Saturn	Uranus	Neptune/Pluto
0.4	0.7	1.0	1.6	2.8	5.2	10.0	19.6	39
0.387	0.723	1.0	1.524		5.203	9.539	19.18	30.06 / 39.52

The Titius Bode law gives a pretty accurate approximation view of the radiuses of the orbits of the planets. Where it appears to fail between Mars and Jupiter we find that is where there are many asteroids and where they would have combined to form a planet if Jupiter was not so close by. This I show in other books is the result of the Lagrangian points that pushed a planet too close to Jupiter and with the Trojan system Jupiter placed the Roche limit into effect and destroyed the four other plants. The Law fails to give the right figure for Neptune but Pluto fits the value given by the law quite well. This indicates that some event took place when the solar system formed. At present some intellectuals think that Pluto is not a true planet but an escaped moon of one of the planets but this indicates how little mainstream science knows how the solar system was formed. As remarkable as the Bode-Titius Law is for predicting the orbit radiuses of the planets there is no explanation of the law in terms of other laws of physics.

In April of 2004 a planet-like object of approximately 1000 km in diameter was identified. Its distance is now approximately 86 A.U., not far from the figure of 77 A.U. predicted by the Titius Bode Law. Its orbit is quite eccentric so its distance from the sun may range from 75 A.U. to 100 A.U. over the course of its 10,000 Earth year revolution about the sun. Tentative it is being called *Sedna*, after an Arctic goddess of the sea. It seems like a good choice, in that all planets except Venus and Earth were named after male Roman gods.

If an explanation for the Titius Bode Law were to exist it might be in terms of the size of a radius zone for a planet within which no other planet could form because of the disruptive effects of the gravitational attraction of that planet. There is such a zone for a planet within which a satellite cannot form because of the stresses produced by the gravitational attraction of the planet where the Roche limit defines this zone. Although the Titius Bode Law carries much more credibility than what Newton's misconception ever did, the Titius Bode Law was always regarded by mainstream science with scepticism as a decoy to take the attention away from the obvious misconduct that Newton's views portrayed in misleading science.

This is what those see that is on a mission to find the beginning of the Universe. In this totality of massiveness everybody is looking at and then to try and find the point where it started. Beautiful as it may be it is not where we would go and look for the start of things. To find the start we have to look at where everything starts and believe it or not I have located that very place.

To see where we are we first have to find where we are. In our minds we are the reason why the Universe is in place. Even the atheist has this notion that life is everywhere and that is because the simpleton thinks the Universe is created around the sustaining of life. The atheist wants to construct planets that are plentiful so that life can be abundant so that the Universe holds life as some centre stage from where the lot develops to serve life and more so human life.

The atheist has completely lost the plot in their search to secure life in the Universe. They have cheated and plotted and coerced ideas into frameworks that has lost all validity and that is to convince everyone with no brains that life is   just a belief away but that all forms part of Newtonian ideology and Newtonian religiosity where they can create a Universe by creating a mathematical formula. They look at the large and from that perspective try to allocate a start and that proves their intellect is typical Newtonian, not worth a thought of logic. Therefore lets' locate that smallest place in the Universe.

If you hide from the truth to uphold a fantasy such action contemplate criminality in science. This deed does then pretend there is validity when dreaming about and drumming up a fantasy, which then is explored to be the truth and to give that farce credibility and validity. The mathematical formulating of the impossible then is to make it as too be so complicated that no one could understand and be willing to discard the validity thereof. This pretence then is put to the rest of the world as to seemingly be unquestionably correct. That is the way that those in science guard against questions to surface the truth becoming a realistic and understandable formulation. It is how much science wishes to cheat by creating an untruth that hides the truth in obscurity. If you argue, prove how a space whirl is more than fantasy.

All Scientists that commit only to the truth pretend to know only "the truth" and "only work with facts" **while** all along they have denied to search for the truth and also all the while knowing they promote Newton's ideas. Well knowing they only pretend what they know is correct because Newton in cosmology is one big hoax. That which they use as the basis for all physics is openly false. If you think I am trying to discredit the most honourable men and woman on earth, those you regard as super humans that hold most integrity of all humans on earth they have been deceiving science by backing up Newton.

At school and throughout we all were taught and told that "**mass**" pulls the Universe and "**mass**" holds the Universe together just as "**mass**" is forming the solar system because as Newton said 300 years ago "**mass**" produces gravity by structures pulling one another. That what glue us to the earth as gravity uses "**mass**" to glue the Universe.

All the while nature uses a completely different system that does not even closely resemble Newton and therefore science's version of the Universe. Look at the picture of the planets in the following picture and allow your eyes to explain to you that every planer is in position according to "**mass**". The sizes and the "**mass**" are totally at random and they all are scattered as far as structural size or "**mass**" goes. You see is nature locating planets and Newton's "**mass**" plays no part. Look at the following picture that shows all the planets and see how random "**mass**" is in the order of sizes and the manner in which planets are distributed. There is no order about the "**mass**" distribution.

See the "small planets" with the least **mass** are close to the sun and at the back of the line. The "large planets" are in the centre and in that they are not according to size and mass distribution but scattered. Mercury and Pluto as the "smallest'" are at the start and at the end while Jupiter by far as the biggest is in the centre of the lot. There is no evidence of "**mass**" playing any role in the formation of the planet structure. Science knew about this discrepancy for as long as science knew about planets and yet for three hundred years they have been telling and teaching everyone that "**mass**' controls the solar system and therefore "**mass**" controls gravity which is what controls the Universe.

Everything in the Universe is round. Anything that is round has to apply the value of Π.
This is a fact of mathematics but while Newtonian science forever tells the Universe to have "**mass**" and to use "**mass**" nowhere in science would one find Π used in prominence. Whatever you may study in astrophysics, go where you wish but never would you find Newtonian science taking the fact of Π into any prominence. When you read any of my books you will see that gravity forms by movement applying Π as a value. I have found the four phenomena that put Π in astrophysics. By valuing gravity as Π therefore the Universe consists of gravity that forms by the working of the four phenomena that Newtonian science hardly ever mention but dubbed "a freak of nature".

Science pretends to search for planets far outside our solar system and all the while science has no idea why our solar system functions in the manner that it does. Would it not be much better to first find out how nature applies in forming our solar system before trying to pretend to know about other thought to be planets? Science ignores nature and pretends to know much more while all of science know so little they have not figured out how gravity works…and that I prove! When you only read this insignificant book alone you will know much more about the solar system than what the Super-Educated ever knew.

Infinity and Eternity:
The only borders in the Universe

Infinity: The part of the Universe that can never start

Infinity: The part of the Universe that has no inside

DEFINITE *is the substance spinning faster than light*

ETERNITY *is the substance spinning as fast as the speed or slower than the speed of light*

Infinity: The part of the Universe that cannot move

The Universe are formed by only two substances where I call one definite (better known as material) and eternity (better known as outer space)

Eternity: The part of the Universe that has no outside

Eternity: The part of the Universe that can never end

Eternity: The part of the Universe that cannot stop moving

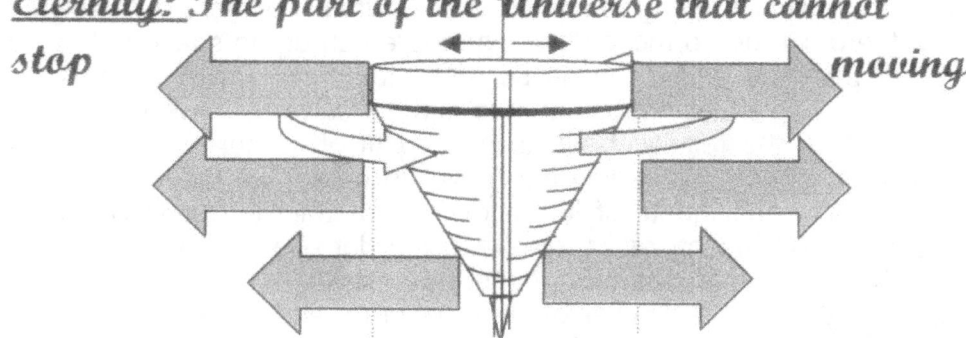

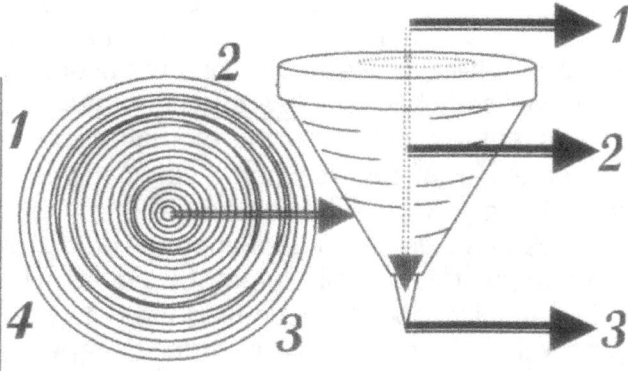

The part that provides space

1
2
3
4

1
2
3

The part that provides time

Where Newton shows that "mass" is the factor that should place planets according to size in their respective positions, which is not true, I prove that "mass" is not used by nature, and for that I am frowned on, not to disprove me but because for the first time in 300 years I confront science by telling the truth about Newton. I show how ridiculous it is to support Newton because Newton has no foundation or support in the cosmos, yet I am ridiculed as the one that is incoherent. I show what nature uses namely the **Titius Bode law**, The **Roche limit**, The **Lagrangian Points** and the **Coanda effect** and how this forms gravity as well as place the positions of the planets in accordance with singularity, not mass. Because I trash Newton's rubbish that does not fit and in that can't apply, no publisher of science books or science magazines will publish my work! I show what goes on in nature while Newton's contribution of mass applying is total rubbish. Because I call it rubbish and I rubbish Newton's ideas I am ignored.

<u>I prove gravity is Π and that four cosmic phenomena forms gravity by producing Π when objects move in relation to each other. The rings you see are in place because gravity is Π. The spherical planet you see is a spare because gravity is Π. The planet spins in relation to Π because gravity is Π</u>

To find validity in my argument one must draw this statement of motion back to the point where singularity is getting sides or said mathematically $Π^0$ is going Π. Π is the **controlling singularity** and Π forming $Π^2$ is in relation to the **governing singularity** $Π^0$. When there is singularity there can be no sides. The one forming singularity $Π^0$ by measure fills no space while form Π develops $Π^2$ into space. The space that even the dot fills being $Πr^0$ does not really exist in the manner we humans see space to exist. It is a spot that is there without being there. It does not visually exist because it is not filling any substance and it cannot be recognised since it is not three-dimensional. The spot and the dot have no dimensional worth of any measure but holds relevance. This Universe I am addressing has never been unveiled by any one since this is the flat Universe. This Universe holds a line in time made up of dots and spots forming no space but holds a Universe relevant.

It is the point forming the very centre that plays the part as the **controlling singularity** within the Universe I have named as **Infinity**, which is better known as the axis. It is where nothing can go smaller and anything within that point can never reduce. That point is where the entirety called the Universe begins and where everything holding substance begins. Once one accepts the fact of singularity being present in that location, that accepting of singularity then is contradicting all the things we know and we can measure and we recognise that point being present by merit of the fact that the point referred to is not being formed by any of the things we can recognise. It is made up of everything we don't know and constitutes of everything we are unable to recognise or visualise. In that spot there is no space. That spot holds <u>Infinity.</u> In that space there can be no motion because there can be no space to have the motion within. It is

declare that

is because of

detect it but

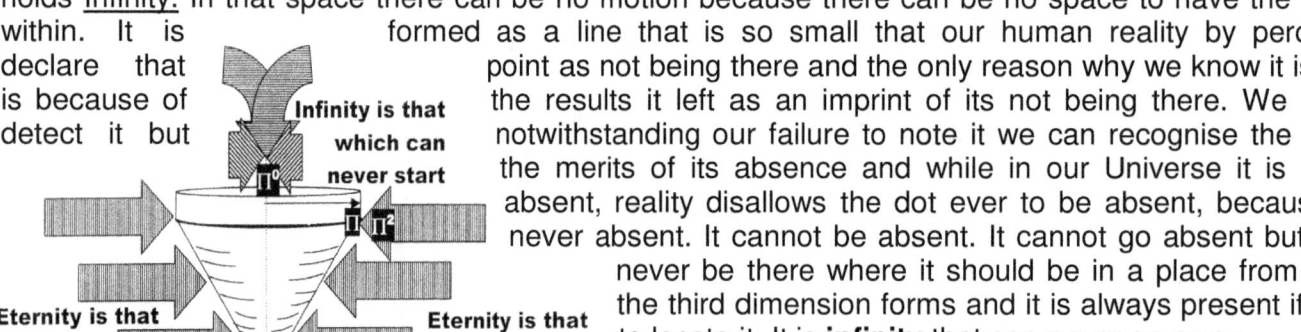

formed as a line that is so small that our human reality by perception point as not being there and the only reason why we know it is there the results it left as an imprint of its not being there. We cannot notwithstanding our failure to note it we can recognise the dot on the merits of its absence and while in our Universe it is always absent, reality disallows the dot ever to be absent, because it is never absent. It cannot be absent. It cannot go absent but it can never be there where it should be in a place from where the third dimension forms and it is always present if I wish to locate it. It is **infinity** that can never go away.

I named the other part of singularity forming space **eternity** because that area never become bigger, or become more or find an end to the outside. Whatever was and is and will ever be is locked in that space I named **eternity** and it is **eternity** that never ends because **eternity** can never end moving. What we think of, as expanding is never ending movement giving eternity the eternal motion that will go on forever. The "so called expanding" of the Universe $T^2 = a^3 ÷ k$ is where singularity is shifting relevance **k** from liquid $k^{-1} = T^2 ÷ a^3$ to solid formulated as $k = a^3 ÷ T^2$ and the process whereby this happens is precisely the same as the Coanda effect. Getting back to my first argument about a line and that no line can start at zero but has to use singularity as a starting point, this is all the proof I require. The line **k** coming from the centre (singularity k^0) forms by forming an initial spot $Π^0$ becoming the dot $Πr^0$. However, I went on to say that whatever the line used to start with has to continue in order to repeat the same that began the line. Therefore the line started with $Π^0$ and it has to continue with $Π^0$ until such a point, as it must end with Π.

Whether the line is Π^0 or is r^0, or uses 1^0 the outcome all refers to singularity being used. By reducing the line we come to the end of the mathematical equation of the circle but the circle does not end there. That is what Newton did not recognise from the figures the cosmos represented to Kepler. The circle only secures the final cosmic figure and the value to singularity where all things have equal value. The movement of the circle splits singularity in two sectors. By forming Π the circle has to form Π^2 due to the movement coming about in securing the space Π^3. Kepler chose to use different symbols to those being valid, but the concept remains the same. Kepler said that $a^3 = T^2k$ while I show that $\Pi^3 = \Pi^2\Pi$. It still confirms that movement $\Pi^2 =$ is the forming of space by three dimensions Π^3 in relation with the movement Π^2 being relevantΠ to singularity Π^0.

To us favouring positions the piston stops at two locations but the fact of the matter is that the crankshaft stops every 7° of rotation and if the crankshaft stops, then so does the piston stop. The stopping is a continuous and is an ongoing process that happen every 7° of rotation. The crankshaft moves in a straight-ahead position going straight and then it stops and redirects by 7° and then it turn by going straight again. It is $a^3 = T^2k$ and then it stops (a^3), it turns (T^2) and then again goes straight again (k) while holding reference with singularity $k^0 = a^3 \div T^2k$ all the time. One cannot part the redirecting and the going straight T^2k because it is the same movement since the space forming a^3 is equal $=$ to the turning T^2 and the going straight k. This is evident when dissecting Kepler's formula $a^3 = T^2k$ that $T^2 = a^3 \div k$ and $k = a^3 \div T^2$ while honouring Newton's 3rd law $k^{-1} = T^2 \div a^3$. Please believe me that this puts movement in such a perspective that it must be the most complicated dimension because this has the material $a^3 = T^2k$ moving $T^2 = a^3 \div k$ in terms of ($k = a^3 \div T^2$ as well as forming $k^{-1} = T^2 \div a^3$) while always referring to singularity.

Kepler gave his formula symbols $a^3 = T^2k$ that do not quite represent gravity in its true symbolic nature and that then was the reason why I came on the idea that gravity has to link to Π more than any other value or symbol. It is because everything holding gravity or representing gravity (not mass) is round. Gravity connects by the use of Π. We have to part what mass does and what gravity does. Mass is where the object connects to one point on Earth and being at that point with mass the Earth does the moving by spinning. The spinning of the Earth then represents the movement or the intention to move because the Earth spins by Π. This movement gives mass its qualities because mass does not possess the influential value of Π since mass is a quantity representative of the amount of atoms and not the spin of the atoms within the mass quantity. If we look at the way the Moon connects to the Earth, committing movement in a circle does it. That represents Π. When we look at the way the solar system connects to the Sun in circles every planet holds an individual symbolic value to Π that circles in relation to the Sun. If we look at the roundness of galactica, the formation represents Π. Every cosmic star holds roundness and roundness only represents one value, which is Π. The connection gravity has is not by mass but it is by Π. When we go in search of a cosmic resolve to find gravity, we better start looking for the influence Π has on the subject or leave the entire subject alone because the gateway in understanding gravity goes by the meaning of Π relating to Π^0.

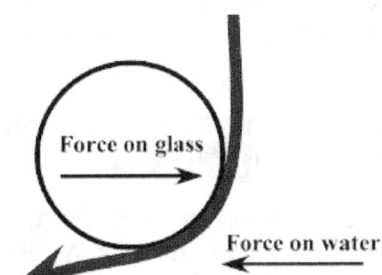

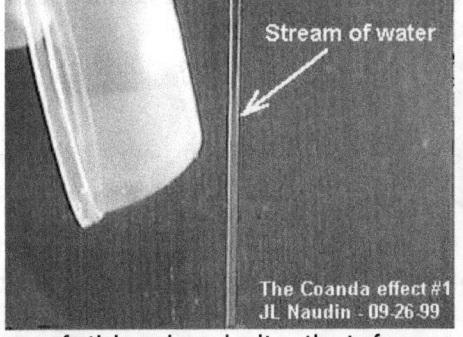

The Coanda effect #1
JL Naudin - 09-26-99

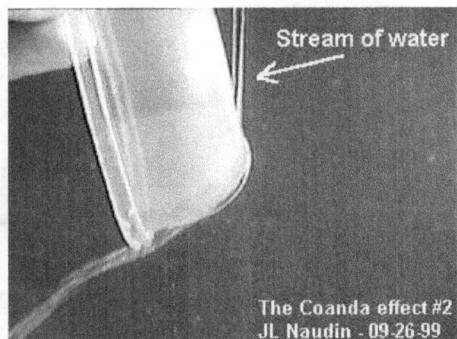

The Coanda effect #2
JL Naudin - 09-26-99

The condition for the presence of this singularity that forms everything, controls everything and is everything is the centralised $k^0 = a^3 / (T^2 k$ singularity that forms by movement $T^2 = a^3 / k$ **of the space** $a^3 = k T^2$ in relevancy $k = a^3 / T^2$ going both ways $k^{-1} = T^2 / a^3$ thereof (Newton's 3rd law). Now put this formula in terms of gravity and we can see the gravitational picture of the Coanda effect come to life.

The condition for the presence of this singularity that forms everything, controls everything and is everything is the centralised $\Pi^0 = \Pi^3 / (\Pi^2 \Pi)$ singularity that forms by movement $\Pi^2 = \Pi^3 / \Pi$ of the space $\Pi^3 = \Pi\Pi^2$ in relevancy $\Pi = \Pi^3 / \Pi^2$ going both ways $\Pi^{-1} = \Pi^2 / \Pi^3$ thereof (Newton's 3rd law).

Reading this mathematically encrypted coded formula of the cosmos given to Kepler and keeping it removed from Newton it reads as being that the space a^3 is equal to = the motion T^2 of the space a^3 in ratio k to a centre k^0, which is relevant to the positioning of k. If we bring in the full equation it will be $k^0 = a^3 \div (T^2k)$ which means half of space is solid $k = a^3 \div T^2$ and half of space is liquid $k^{-1}=T^2 \div a^3$ where liquid is moving. However, it is also true that everything through movement defines a value in relation to one point holding singularity k^0 and that is what the formula $k^0 = a^3 \div (T^2k)$ underwrites. What this proves is that gravity is the motion of space provided by time being the liquid. Please allow me to explain. In the formula $a^3 = T^2 k$ the space forms as the space is in motion.

Newton suggested that $\frac{dJ}{dt} = 0$ where he stopped time to have the motion of the circle demolish the work that the circle does. That means he got time standing still or being T^1 and the motion $T= 0$. Let us ponder on that thought for a while, while we remain with the formula Kepler suggested $a^3 = T^2 k$ and then it will seem that according to Newton $a^3 = T^2$ and in that k then becomes 0. Should that be the case then we have space going flat because $a^3 = T^2k$ where $a^3=T^2 x k=0$ forming a square instead of a cube, and the Universe we have is a three dimensional system in every aspect there is. The concept Newton brought about that $a^3 = T^2$ is putting a person that looks at a mirror equal to the possibility of walking in and out of the mirror by becoming the reflection in the mirror T^2 and then himself a^3 again. It is rediculous to say the very least.

It is quite apparent that Newton saw no difference between the top spinning while the top was standing in an upright position and the top lying down on the Earth. This is a crucial mistake that has such a wide implication that on the one hand it either values the Universe to the value of singularity or on the other side dismisses everything about the Universe to the value of zero.

I am of a very different opinion about Newton's point of view where he declared that forming a circle moving $\frac{dJ}{dt} = 0$, and by doing such the movement then removes Kepler's relevancy factor. This places a value of empty space in which a top would spin and Newton missed the difference there is between a top spinning and a top laying on its side on the Earth. There can be no such a thing as empty space. The fact that space is valid removes an empty connection because space can be anything there is in space except empty space that is filled with nothing. The Universe is time contained in space, which makes it space-time. Space has only one value, and this is to contain time and time provides space with a definite value. I do not disagree for one instant with Newton's calculations whereby he came to the conclusion that $\frac{dJ}{dt} = 0$ and therefore I am not going into repeating the entire calculating process. All of the calculations Newton made are very correct except the eventual and final conclusion Newton came to. Newton never understood the mathematical concept of time playing a part in physics. In the time of Newton singularity and the relevance thereof had no feasibility in any concept regarding physics. Newton had the concept that time could stand still and that is impossible in physics or any other place. Time can never stand still because time is forever moving by establishing space in a three dimensional environment.

This explains the Coanda effect **and** the Coanda effect is gravity **and gravity "glues" the water to the glass! The water forms a value of** $\boxed{\Pi^{-1} = \Pi^2 / \Pi}$ **while the glass forms a value of** $\boxed{\Pi = \Pi^3 / \Pi^2}$.

This process happens to all spinning things and as much as it happens to a piston connected to a crankshaft, just as much this will happen to an atom spinning an electron in a similar manner as the crankshaft is spinning holding a piston connected. This proves that gravity is the Coanda effect and in another book I prove that the Coanda effect has its origins in Π forming a value and that value forms gravity. In order to understand physics applying in cosmology I had to start by dissecting the set-up forming pi. Using this argument I can introduce my theory on the *Absolute Relevancy of Singularity.* At the point in the centre of the circle a line must start. In the beginning when I explained the way I figured how the line starts I said a lot of dots has to continue in order to form a line. It would be 1 + 1 + 1 etc. because the line must form by holding singularity. After that point does mathematics begin but in the line that forms representing space as all other factors, then time holds 1. The line can only form when all the points forming the line have the value of 1 being 1^0. In that conclusion one realises something must separate singularity from all other factors because singularity hosts all other factors but is by own initiativeΠ^0. Only when singularity meets the end value can the end value have Π where the final ring of the spinning circle forms $\Pi\Pi^2$. That will be the spot of origin forming the relevance in Π.

That will hold the eternal spot…the smallest spot ever because all spots that ever can be were secured in a position in the centre of that spot that must continue as a line that forms. Because of the progress singularity follows from the single dimension singularity only allows mathematics a start at Π^0 progressing further onto Πr^0 and from there the line is born as $\Pi^0\Pi^0\Pi^0$ and to $\Pi^0\Pi^0\Pi^0\Pi^0$ etc. where Π^0 then may form the concept and value of r. But the line starts at $\Pi^0 = r^0$.

This forms because cosmology is singularity based and the value is $\Pi\Pi^0$. This line $\Pi^0\Pi^0\Pi^0$ of singularity can only continue because every spinning atom preserves Π^0 in the very centre and since $\Pi^0 = \Pi^0 = \Pi^0$ and is represented in the circle of every atom spinning, the line is the same without finding conclusion except at the end where it forms mass at Π. At the point whereΠ forms, the movement Π^2 of the circle defines the space Π^3 of the circle and it confirms the centre Π^0 of the circle through the rotation going through the atoms. Let's call this the solid forming or if you wish, let's call it Kepler's singularity. After that singularity forms a line $\Pi^0 = \Pi^0 = \Pi^0$ where this forms another line again as Newton stipulated it by $\dfrac{dJ}{dt} = 1^0$. Let's call that the liquid singularity or Newton's singularity and the relevance of singularity having a solid base compared to the singularity holding a liquid base comes about by the movement of gravity. From these conclusions I prove that gravity is the result of four cosmic phenomena interacting to form the value of Π which by movement becomes the value of gravity Π^2 and gravity is equal to cosmic time applying. In order to understand the development of the cosmos and moreover the start of the cosmos and the progress in the cosmos as the cosmos formed, one has to understand the measure of Π.

One has to see that Π is not merely 22 over 7 or that Π is a ratio that no one ever bothered to clarify, but Π is the key that unlocks every lock that hides a secret in origins of the Universe. One has to microscopically dissect the measure of Π to find the cosmos in measure. One has to understand where 7 fit in Π. The fact that Π is 7 at the bottom and that 7 relates to a double value of 10 is a key issue. Furthermore, it is very important to see why Π is 10 times two by adding 1.991 on the top part of the equation. In this measured value is what holds the building blocks of the entirety we call the Universe. It is behind Π that we will find the four phenomena, which I named the four pillars performing as gravity as they form gravity. It is by the actions of Π that the Universe develops. The Hubble expanding goes by implementing gravity as Π in the square through the four pillars on which gravity and time rests. It is behind Π we discover the meaning of singularity and how singularity forms the absolute and only building block as a form that forms the Universe. It is in Π we find the Cosmic Code unlocking the meaning of the Universe. Time is centralised in Π^0 that forms Π as space's limit that becomes space by gravity being Π^2.

Space is time gone to the past in which time confirms its presence it had in the cosmos by moving from the present time into space and then onto the future leaving space behind as the past. The proof of this is again the top standing upright. Time places the governing singularity Π^0 in the centre and then the very next instant the governing singularity of the previous instant moves outward to form Πr^0 which then keeps on moving outward as time goes on the finally establish $\Pi\Pi^2 = \Pi^3$. That is how the top keeps erect and that is how the Coanda principle "glues" the liquid to the round solid surface by guiding the liquid as it flows in relation to the round solid.

The spinning top is the manifestation of the Coanda effect, which is the coming together of the Roche limit, the Titius Bode law and the Lagrangian points. By forming a presentΠ^0, time is in infinity forming singularity that then has to move onΠr^0 and in doing so it leaves a legacy behind being space which will form as $\Pi = \Pi^3 \div \Pi^2$. Time is the movement of everything forming the Universe where in time the movement of time relocates everything in space by moving from the present onto the past leaving behind space as a history of time gone by. That way the top can stay erect. As time becomes the past by going to the future it forms space as it confirms the past, and in that space is what time forms by going to the past leaving space behind. Space becomes what time was at the point where time formed the particular space in relation to Π.

As time becomes the present coming from the past, time has to move on to the future by replacing the past with the present at the same time and as time moved on it left space that represents that instant in time in relation to other space that was in some position at a specific location at such a point in time wherever that point in relevancy might be. The fact of Π not only refers to form but also validates the Universe by splitting infinity from eternity. By forming space when creating Π, time is using Π^0 in establishing movement Π^2. It is in the process of relocating Π to new positions by establishing Π^2 and connecting this as it forms a network consisting of Π^o by forming space Π^3 in relation Π that establishes infinity Π^o that always stays motionless.

If not for movement, the Universe would be one line holding time by repeating singularity Π^o uninterrupted and it is in the diverting of eternity to a position away from infinity that the Universe comes about. This is what happens in a Black Hole where no movement within the Black Hole places eternity that always moves in a standing position to infinity that never moves. Without movement the entire Universe will fall back into and onto one point and everything we thought is real and solid will disappear into that one point holding infinity onto eternity where infinity and eternity then reunites without holding space by any measure thereof. This proves the Universe to be an unreal concept with space being no reality at all but for the movement of space in relation to singularity Π^o whereby Π confirms everything in a location in relevancy to all other things in a specific time slot or space.

When I, as a person forms a part of the Earth by the virtue of having mass that connects me Π to the Earth Π^2, stands on the Earth Π^3, my position in relation to the Earth gives me a specific positional relation to time Π^0 and the Earth. That gives the Moon a future of say one point five seconds being the past in relation to the Earth and that gives the Earth a past in reference to the Moon's future of one point five seconds. Where I am at any specific point in the present, that point I am holding is that which secures my present point in time. The Sun is eight and a half minutes into my past with all the space being in-between the Earth and the Sun and by my view of the Sun I have a present time slot, as it also gives me a past of eight and a half minutes in relation to the Sun since the light travelled eight and a half minutes through space to confirm my past during that present instant.

That secures my past by eight and a half minutes at the point of giving me a present location in time. However, that also secures my future I have from the point I now have in the present by the margin of eight and a half minutes because that establishes a flow of light that would last another eight and a half minutes of filling a presence worth eight and a half minutes while travelling through space by moving with time and every spot filled on the way would secure a position that I will have in a future presence for the next eight and a half minutes, which then becomes my future as it fills my past. Looking at this scenario in a view from Alfa Centauri the allocated position Alfa Centauri holds in space relating to the Earth, gives the Earth a past of say four point six years while this secures the present and having that present secure the Earth to a future of say four point six years by forming time as space between Alfa Centauri and the Earth and this is confirming time to the tune of four point six years. By securing movement it forms time in having a past in relation to the present that by the same margin also secures a future in relation to a definite past. This is how the Universe builds space in establishing time. This applies to all allocated positions of rotating objects throughout the Universe. This means that every point away from Π^o serving as Π, wherever that might be, secures the past the cosmos and I have by giving the cosmos and me a future in terms of the present Π^o.

Take this in relation to Kepler's formula we then find the Earth (a^3), which is in relation as viewed from Alfa Centauri (k) four point six years (T^2). That secures the three dimensional status the Earth has ($a^3=T^2k$) within the space from the Earth to Alfa Centauri (a^3) forming the Universe in terms of a present (k^0) being in the Earth centre which then depends on a location (k) secured by a future (T^2) that will come by movement where the future also doubles as a past ($k = a^3 \div T^2$ and $k^{-1} = T^2 \div a^3$). That is time and that is how time forms space and that is how space-time forms the Universe and that is the _Absolute Relevancy of Singularity_. That then forms time in the centre in infinity in relation to space in eternity in singularity where time that moves forms space by holding time that does not move secured in positions in relevance to where every point that previously formed was in space which is time that has gone by. $\Pi^o\Pi$ divides **infinity**Π^o from **eternity**Π where **infinity**Π^o can't **move** Π^2 and **eternity** Π eternally moves as time $\Pi^o\Pi$ that establishes space Π^3 in motion $\Pi\Pi^2$.

If we put this in terms of singularity (Π^0) we find the Earth (Π^3) is in relation as viewed from Alfa Centauri (Π) four point six years (Π^2) while moving in that space that is time that has gone by. That secures the three dimensional status the Earth has (Π^3) in terms of a present (Π^0) that depends on a location (Π) secured by a future (Π^2) that will come by movement where the future ($\Pi = \Pi^3 \div \Pi^2$) moving forward that also doubles as a past ($\Pi^{-1} = \Pi^2 \div \Pi^3$) by the light coming from and thereby confirming the past. That is space formed three dimensionally by keeping time in infinity apart from time in eternity. The relevance (Π) that forms in relation to the present (Π^0) will relate to movement (Π^2) and the movement is circular which ensures that the relevancy forming is circular (Π) by securing that the movement is circular (Π^2) in terms of one specific point (Π^0) in infinity which then secures a roundness (Π^3) that forms an everlasting eternity ($\Pi\Pi^2$) which validates a never ending circleΠ^3.

In this time in infinity (Π^0) that secures that there is an everlasting eternity ($\Pi\Pi^2$) in space (Π^3), it is not the space that is everlasting but the movement of time by the line ($\Pi\Pi^2$) that is everlasting. The governing singularity (Π^0) holds a positional validity (Π^3) of three dimensions Π^3 =($\Pi\Pi^2$) in terms of any relevance (Π) formed by the controlling singularity ($\Pi\Pi^2$) thus mathematically it equates to $\Pi^0 = \Pi^3 \div (\Pi\Pi^2)$. If a relevance ($\Pi$) did not validate a positional validity (Π^3) securing a governing singularity (Π^0) in terms of movement formed by the gravity (Π^2) that produces the controlling singularity ($\Pi\Pi^2$) in space, with a three dimensional status Π^3, then space (Π^3) would not be obtained and thereby the Universe would not be secured. That is why space-time is $\Pi^0 = \Pi^3 \div (\Pi\Pi^2)$. However this must be seen where it applies. It applies where singularity as time meets space, which means it applies at a point in the Universe where time still grows and that is at the position that predates the Big Bang. It is where material forms before material forms. It is where the visual will never come. It is where singularity Π^0 forms space Π^3 by singularity (Π) moving (Π^2).

Time is the movement of space in relation to any one centralised point not spinning securing such movement. Everything in the Universe moves in relation to any one single point and every one single point that forms in any location everywhere that then has to stand still to form the centre of the Universe wherefrom that point must be motionlessness to allow everything else movement. The point not moving is anywhere and the rest that moves is everything excluding that one specific point that is motionless. In that manner the Universe is constructed and with every point being confirmed only by the movement of all other points around any specific point that means there is no valid solid Universe because the Universe is constructed from singularity (Π^0) that holds no valid space (Π^3) other than being in position (Π) at a specific point ($\Pi^0\Pi$) while having gravity (Π^2) that forms the time (Π^2), which is also the movement (Π^2) which is gravity of space (Π^3).

The flow of time being the present in singularity forms space by moving time in relation to space as much as relocating the present in terms of a past that is determined by the movement of time whereby that action of time moving by the same token is establishing space that confirms the past as it secures the future as time moves on to leave a positional legacy, a footprint of time gone by presented in terms of light which is the presentation of space.

From this we can deduct that the Universe in a three-dimensional form starts at $7/10(\Pi^6) \div 6 = 112$, which is a value forming the start of the element table and that I explain in the Cosmic Code. In the Cosmic Code there are numerous values consisting of Π forming the relevancy by which certain rules comply throughout the cosmos. One is 7/10 which is the Titius Bode law which is the interaction of gravity spinning and by spinning is forming a sphere (Π^6) within a cube ($\div 6$) and that is how the cosmos forms usingΠ. The dimension of $\Pi^0\Pi$ is flat but by spinning Π^3=($\Pi\Pi^2$) the Universe goes in a sphere (Π^6) spinning in a cube 6. In this I prove that for instance amongst so many other things that electricity and gravity is the same thing. By ticking $\Pi^0\Pi$ time forms space by becoming space as time moves into the future leaving the past behind as space. Time is a substance and the only renewable substance with the ability to come into the Universe because from the start it came into the Universe to form the Universe as space. As time moves on space grows by the margin of singularity $\Pi^0\Pi$ leaving spots that form dots. The proof of this is in the value of Π being 3.14159 where 3.14159 -3 = 0.14159 x 7 = 0.9911, which is singularity as the spot (0.9911) becoming singularity 1 as the dot. In other work I explain this in much better detail.

There are two definitions we can use when looking at such a growth. We can look at the space not holding material that grew in size in which the stars finally froze their development to end as Black Holes and the growth was in terms of reducing space by remaining behind in terms of the expanding Universe all because of a lesser developing singularity within material compacting singularity $\Pi^0\Pi$. Or we can focus on the stars growing $\Pi^1 = \Pi^3 \div \Pi^2$ and with that push the outer space much more into expanding by reducing the density of outer space $\Pi^{-1} = \Pi^2 \div \Pi^3$.

As the cosmos grows in space, the cosmos in expanding progresses just as much as the star was reducing in space and the space in the star that became less is the same space as that with which the cosmos expands. This ratio is the ultimate relevancy. This comes from the manner that the star manages to destroy space or dismisses space or compacts space and redirects the space to go from a gas and become more compact and denser by forming a fluid where the fluid is light or heat or the solidity of frozen space as matter really is. In the Black Hole it reduces much further as it claims the singularity, which the object had, and destroys all space and all time there ever was. As the star condenses space on the inside making the star to appear as if it is shrinking away, the space in outer space seemingly becomes more as it seems to be expanding but in real terms this is just a relevancy of one becoming denser and the other losing density. That stars get hotter towards the centre is not the pushing of mass, but is about space condensing.

The entire truth about the cosmos is that the Universe is within the atom that forms a cosmic unit holding singularity as much as it secures singularity and every atom forms a Universe standing apart, parted by time from all other atoms by the spin produced. Every Universe formed by every atom starts in infinity and ends where each atom's spin is forming relevancy between where that Universe starts and ends. All atoms are a Universe formed within the space that time puts between infinity and eternity. All atoms are stitched together by an invisible, unseen singularity - string that is present while also being absent and this invisible string links everything that the Universe is throughout the entirety.

The entirety rests on relevancy. As time moves on forming a line by implementing more dots in relation to the dots that are already there forming the history of time, which is what we call space, the area we call outer space receives many dots that time leaves as a footprint while the dots time leaves within material are less, just because the space is concentrated and thereby is less. The dot that time leaves holds no space but in terms of space moving with time increasing the adding of space-less dots brings about more space which then reduces the concentration of space and the more the dots are, the more the concentration reduces. This is why the top can stand erect when spinning. Its because time forms a governing singularity Π^0 that then shifts in the next instant outwards to form Π as the controlling singularity in terms of the movement Π^2 that then controls the space spinning Π^3. It is time leaving Π^0 that then the next moment forms Π and in the movement of gravity Π^2 the space forms Π^3.

With more dots landing in outer space since there are more space, the space density reduces as the expanding in outer space seems to be more than what is applying to material where space is at a premium, being condensed. With time duplicating to form dots in singularity, every instant that it produces spots forming dots as the present, the space that outer space gains supersedes the space that material gains and that makes material more compact or more and more dense in relation to outer space. The space gained by the space occupied by the moving of material receives fewer dots than the space forming outer space or that part which we see as outer space and the space material holds advances more in density through the loss of density in the space called outer space.

This leaves material more compact in relevancy that seems to hold less space and this is moreover because of the relative loss of density in outer space is there because of outer space gaining space by time leaving more dots. The density in outer space is thereby lost and in that the density in material is gained by the loss of the density in space in outer space being more because it is a greater recipient of time. The dot also leaves one point every time on the dot forming the governing singularity and that confirms the point holding governing singularity in terms of many dots received by the spin of the controlling singularity in terms of the gain of endless space in outer space. In that material always grows as outer space declines in density and that forms the "Hubble Constant" that is no constant. The Hubble constant is gravity expanding, which contradicts Newton's gravity contracting. **In a nutshell that is gravity**. It does not even mention mass because mass has nothing to do with Π while gravity is Π in more forms than what humans are able to imagine. The cosmos grows by gravity which is $\Pi^0 = \Pi^3 \div \Pi^2\Pi$.

That is why the distance between the Earth and the Moon becomes more. That is why the circumference of the Earth becomes bigger. That is why there are Earthquakes and hurricanes. That is why a human grows and heals and that is why hair and nails grow. That also is why there is aging and eventual unavoidable death to material holding life.

The body never stops growing, which brings about the inevitable decline of life's body structure as time becomes more that the body endures. The ever growing of the body makes the body collapse on itself with aging. As time goes by everything on the Earth including the Earth and everything in the Universe around the Earth is gaining in space because that is what time leaves. That is why everything in fossils seems to be bigger the further back the fossil goes in the history of the Earth. Newtonians show millipedes that once roamed the Earth that were one metre wide…and Newtonians not only believe that but also advocate this information as the truth! Everything holding material grows by time leaving space as the history of time that went by. The history or the space of the millipede became bigger as time moved on but the millipede never was one metre wide. That is why we can see galactica so far way. It is through time progressing in space that it carries light to move from there to where we are capable to see where the light came from. Time brings light all the way by progressing in space that carries light through space.

There is and there can be no such a thing as "dark matter" What would make matter "dark"? If the material is "light" it then has a higher concentration of light than where we are at present. This puts the object we see in a denser area than where we are. There is much more movement in that area that concentrates the space in that area and thereby we can see the area because the space released from that area expands as it comes towards us and that light expanding is what we visually see. On the other hand areas that seem dark are more expanded with our light flowing outwards to those areas. Being darker is having light flowing to that area from where we are.

That puts that object in a more expanded environment and in higher expanded surrounding than where we are. If the material is "dark" our light is moving towards that position and that makes that area move slower than what we do. That area is therefore less concentrated and more expanded than where we are. Then again if we see the area as light, the area is more concentrated in density having light poring out towards us by the measure of releasing density. The light flowing towards us will make the region seem as if it is lighter. It again is about relevancies. The part that seems to have brighter light moves faster as the light moves at a greater pace and moves towards us. The area that seems to be darker has light moving much slower because the light is moving away from us as it is the light from our area expanding into that larger area that leaves us with the concept that that area is darker.

As the light moves into an expanded area it will seem to slow down. It is a question of relevancies applying by movement in relation to "standing still" or "moving faster" and "moving slower". If we look at the Earth from the Sun the Earth where we now are would be so dark where the earth is located that Earth would be invisible from the Sun because our space where the Earth is, is so much more expanded than the space is that surrounds the Sun. Again seen from the Earth when looking at Pluto but by only using the naked eye Pluto is so dark it is "invisible" to the normal human perception. It is because "space" is much more expanded out there than it is where we are and if it is more expanded it is moving in relation to the space being available in which it can move making movement seem slower and making that area seem bigger. Then we have Mercury of the approximate same size but are very visible because it is more compressed in that area and therefore more visible than where we are and with the larger density the reflection of the Sun seems to light up the planet.

My question coming from this is why there is this hunt to find dark matter. Dark matter there is because dark matter is only more expanded in terms of denser matter which has light flowing to us which makes us able to see the light coming to us. I am the first to admit that there is no substantiating proof presented in this article alone and I don't even begin to claim that I deliver any proof in this article. There is no room to present even the least bit of proof in any form possible in the space given to this article. With the limited space available to publish information in a journal by way of a small article such as this and having so much information at a premium I decided to release some vital information and the required proof about my claims in other small but comprehensive works that can be obtained.

Have you ever stood outside and thought how is it possible to get that wide area of space that you see into something as small as your eye? If you have never thought of this you have never thought about the cosmos because this is physics and the start of physics. That is how one starts to understand the cosmos and on this thought I base everything I introduce as physics. You are about to experience ideas you have never even thought about and I am introducing to you concept that you are unaware of that it exists.

With my first language not English and the books not linguistically checked by an expert there are bound to be language errors that readers will notice. In the past I tried to check my work myself but after checking say one hundred and fifty pages for language corrections, then after days of toiling instead of having corrected work I ended having four hundred pages of newly written information which is still not linguistically corrected but holds a lot more information. The language and spelling errors compiled instead of reduced. This is because my priorities lie elsewhere. I aim to spend money on correcting the work as far as language goes, as I receive money in the selling of my theses and in the hope that I will receive money. I will have all my work including the one you are reading edited professionally and corrected as I find money to do so...But first I have to get the public aware of the problem to get the academics to appreciate the problem...but after you read this letter you'll know why they have an attitude.

What I am about to explain is not how the cosmos started because that is immeasurably more complex than the depth I am going into with this conversation. My intensions are to show why singularity is so important, why seven is connected to the birth of the Universe and how ten became the liquid of space. Looking at the night sky you see many flickering dots and spots and light coming from afar. The one star you see seems to be a near visible dot in the picture while the dot might be hundreds or might even be many thousands of times the size of the sun...and we think of the sun as big.

The dot is then that much bigger than the sun because the star we think we see could be a galactica hundreds of times the size of our Milky Way galactica but that shows in the sky as one little dot and yet that entire structure as big as it is many times our Milky Way, fit into our eye socket. But that is not all...there are trillions of such light images and they all fit into one eye socket. What we see is immeasurable and yet we see it effortlessly in the space our eye holds...how can that be? It is because we see dots and spots that form a Universe. From studying this concept, it enabled me to realise how the Universe forms and works as gravity applies as time.

Everything we see in the above picture is many, many times bigger than what the Human would be able to see and we have to realise how such vastness in space can reduce by shrinking the space we see into something as small as what may enter our eyes. This might be the size of a photon if it has any size and a photon is so small no one as yet was able to measure the "mass" of one photon. Yet the information such a photon carries is what we see as outer space and one photon conveys that much space into our eyes. Understanding this concept must be very clear should the reader wish to follow the progress of the information that this work contains.

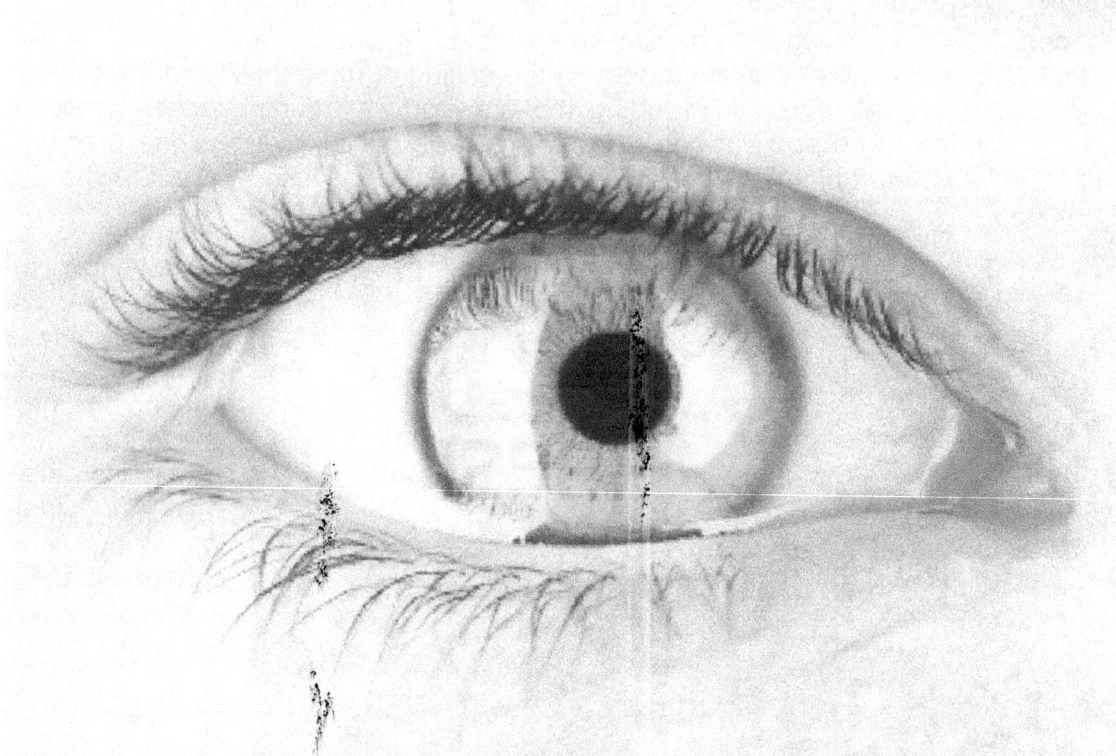

I have asked this many times before but I ask again: Have you, the person reading this, ever thought how it is possible to see that much information that you see at night when looking at the sky with only your eyes? Ever thought about how you are able to see when you see everything in the night sky and how that much light information can fit into such a small space as your eye? If you say it is with light then it shows you have never thought about reality but you depend on simplicity to solve problems. Have you ever sat back and think what vastness in information it is that you see when you see the entirety of the Universe when looking at the Universe at night and what the size is of everything of that which you are able to see?

But that is not all…there are trillions of such light images and they all fit into one eye socket. What we see is immeasurable and yet we see it effortlessly in the space our eye holds…how can that be? How is it possible to fit what we see into the space of our eyes we have? Think how much is the entire information that is visible at night and think about how all of that fit into the space your eye holds? This is what the Universe is and that is what the Universe represents and when I put that into a mathematical equation I wish to see those persons designing space whirls to calculate the volumetric space reduction with physics. Consider how big is what is visible and put that space into the size of what your eye can hold and ask your mathematically educated Professor in physics to find some ratio between what you observe and the size of your eye. This is if not, then certainly one of the most important observations about physics that we can make. This idea defines gravitational physics and proves the understanding of the Universe about physics. This underlines all aspect of physics and place time in singularity and space in a hallucination of non-existence. Realising this places everything in a single dimension and this lays the foundation in realising how the Universe began when it began with one. Moreover this proves that everything returns to singularity formed in the dimension of one, which is where everything began. The entire Universe withdraws to and into singularity and this process helps to explain how the lot started.

One explanation I have to insist on you understanding is that there is no real Universe in space. What you see is light moving forming dimensional inequality in a hologram your imagination creates but what you see is far apart from reality. If you see the moon you have to realise you can't see the moon because what you see and think you see is light that the moon rejects and that rejected light the moon disassociate with is what you associate or accept is the moon. It is not the moon but it is light bouncing off the moon and I see the light, which I accept as information that informs we about a hologram I claim to be the moon. It proves what we observe is not reality but is a figment of our imagination that we validate.

When I look at anything what I see I have to realise it is not the thing I think I see but it is light that bounced off of what I think I see and then by mental association I connect the light that the object rejects with my observation and in that I form an unrealistic association. There is no reality in the Universe I see but only a diagram of time that formed space in line with what the Universe was at that particular instant. I go into this argument much deeper in other books wherein I set room aside to explore this reality much, much further and deeper.

This is the dominant thought one must have when viewing the Universe to form an opinion.

When I look at the moon I see light that bounced off the moon 1.5 seconds ago, but not the moon.

When I look at the sun I see light that left the sun 8.5 seconds ago. I don't see the sun but I see light as an image that the sun rejected and disassociated with 8.5 seconds ago. I see the history of time.

I am in time but the light I use to observe form the history of time, as it was when the light left whatever I see. I am in the very second and space is in some relevancy time left behind as time formed an image of what space was when that light forming that image left the source inn time gone past as history.

This is overwhelming important to accept and understand if you wish to continue reading this book. We have to realise we're in space but we're not part of space but in the time- instant time leaves behind as space. I can't see anything but only a hallucination of the history of time and time forms part of physics.

If I look at the moon, the moon is 1.5 seconds further into the future than the past I now see.

Where I look at the sun the sun developed 8.5 seconds further into the future than the image I now see. What I see is the past. The light of the sun, which I now use to see by, was created in the sun say 10 million years ago and it took that light 10 million years to surface and form light I now use to see everything I see in the daytime. That means the light that sets my present was part of my past the last 10 million years and the light that is now developing secures my future 10 million years to come from now.

What I see is my past and that is what secures my future but in reality it does not exist because it is an image time left behind because there was no longer a need to secure it. When I see what I see it forms an image of what is no longer there to see at that point. The moon moves on. The Sun moved on. The Milky Way moved on. However, for that very reason I can trace the past as the present because the future is light still in development to form a future in time to come. With that I can move to where the Universe started as long as I reject what I see and dismiss space as a hallucination formed as a hologram that is only part of my imagination. The relevancy forming the Universe is that every point in the Universe secures a future by representing a past in the immediate present. That is the line of time and in that physics form a present in which singularity rules the moment. If this is not clear to you then please do not purchase the book for it will be devastating to your self worth. You have to part the past from the present and that secures the future. You see no reality by looking at the Universe but you observe space that repeats time in a flow of space that forever changes instant by instant. This is your introduction to physics and every aspect that brings about information to physics. Physics is not mathematical equating of more rubbish hidden behind a breathtaking mathematical equation. It is ideas mathematics can never prove.

In every one of the books I publish I change science because I prove how nature places the planets in the solar system. However this is a task that will never end while I in my health is very near an end. In every book I manage to approach nature from another angle or show another dimension because I prove in these books the **Titius Bode law**, the **Lagrangian points**, the **Roche limit** and the **Coanda effect**.
These laws allow me (and now you) to trace nature back to where nature started and believe me it is very uncomplicated. Lets quickly go down the road I lead you to where we uncover the start of the Universe. The following few pages will show the cosmos.
Following these four principles and how the four develops I can trace the Universe to a point, one single point from where it started when it started with singularity.
What does these laws prove. They prove gravity forms not by mass but by forming Π as gravity.
All these laws disqualify all of the 300 years of Newtonian rule and thinking about mass forming gravity. The entire solar system disproves Newton while these laws disproves Newton's ideology about mass playing any part in Astrophysics. From determining singularity I can place my finger on the point where the Universe started. I can and will direct you to that precise location, the very point where it all began as the book develops.

$$(r^2 + r_1^2 - 2r_1 r \cos(\pi - \theta))^{\frac{1}{2}}$$

$$m$$

$$(r^2 + r_2^2 - 2r_2 r \cos(\theta))^{\frac{1}{2}}$$

$$M_1$$

$$r$$

$$\theta$$

$$M_2$$

$$r_1 \quad \text{COM} \quad r_2$$

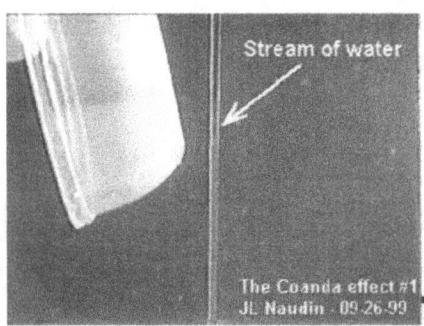
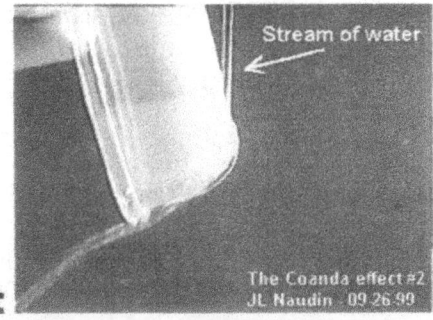

The Coanda effect

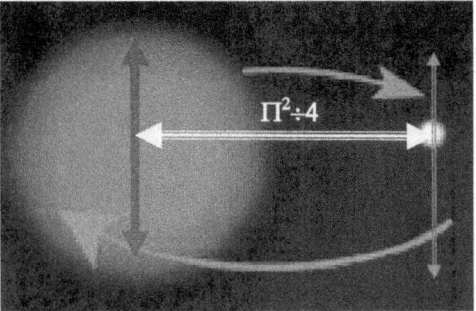

The Coanda effect connects with the previous law. Gravity is a ratio that forms between what is solid space and what is liquid space. This applies when the movement of the solid (star or planet or atom) turns and by turning 7° it reduces the value of the circle of the space surrounding the solid from Π= 21.991/7 to the compressed value Π = 3.142/1 which connects to singularity. There is no pulling of material but only reducing of space when gravity contracts gas into liquid such as what the atmosphere is.

This is the Coanda effect and this works by rotational movement putting liquids in relation to solids. The movement of the solids increases the density of cosmic gas to form cosmic liquids. The Universe consists of materials spinning faster than the speed of light, liquids spinning at or close to the speed of light and gas with a low density spinning not at all. The **Coanda effect** has powered turbine engines and aeroplanes in flight for almost a century and with all the mathematical splendour available to design the most terrific aircraft, not one engineer could mathematically compute one fact to show understanding why this takes place. How sad it is that those claiming to be of much superior intellect in physics remain just no more than having computing power. The understanding is not complex. The entire Universe works on a relevancy that exist between material holding space and material claiming space, in other words the relation there is between occupied and unoccupied space. In short the Universe comprises of solids with a density beyond the speed of light and liquids at or slightly below the speed of light and gas not moving except by gravity. There are two forms in the Universe solids and liquids that can be gas. This relationship is there where movement established Π and in the way Π forms the density of space revalue from 21.991 / 7 to holding singularity at Π being 3.1415 / 1. By changing Π from 21.991 / 7 to 3.1415 / 1 space that surrounds material condense. In that I prove the Titius Bode law. I have published the Titius Bode law in four already published books but in another one I go deeper than the four first published.

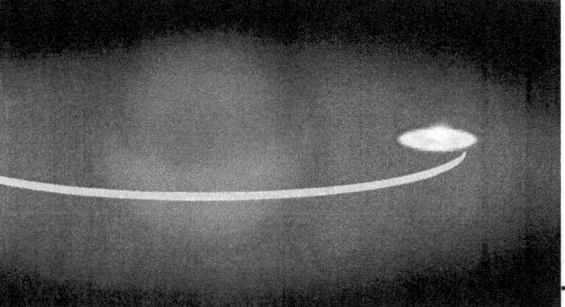

$\Pi^2 \div 4$

The Roche limit

As I start to explain I have to explain beforehand that in singularity there is no dimension positioning but the Universe is a line time forms that develops into space as follows. The Universe is a line and three dots mean the line forms three dots that in space reflect location.

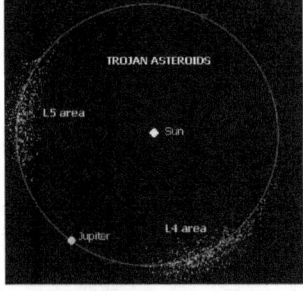
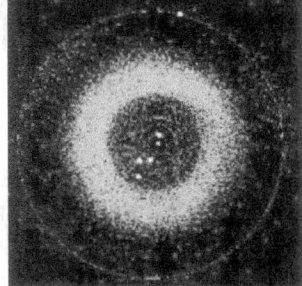
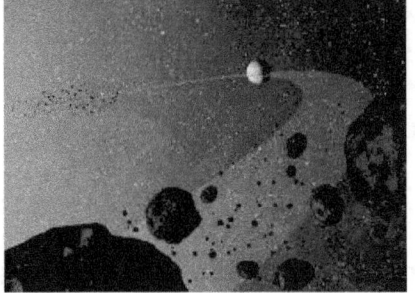

These pictures show the Trojan asteroid and Ceres, asteroid belt. This is how the Roche limit / lobe applies as gravity. This I explain in terms of how the Universe started before material formed as material. The fact that there is the Trojan asteroids and the Ceres / asteroid belt is a direct result of the Roche limit where the Jupiter's gravity crushed planetary remains into fragmented bits by changing the value of gravity applying from Π^2 to $\Pi^2 \div 4$. This was caused by the limitation that the Lagrangian five points bring.

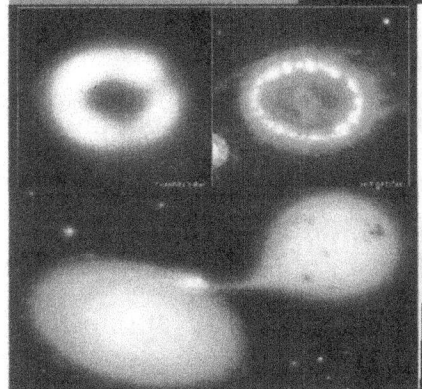

The Roche limit is the law that says stars do not collide ever in spite of Newton's ridiculous idea that stars can or do collide. If two stars are in each other's atmosphere the law reads that when the minor of the two stars is closer than 2.4674 of the diameter of the major star, then the major star will liquefy the minor star into a gas plume that it then treats as more atmosphere. This is most significant in the cosmic principles I now put forward. There is not one instance or any evidence that shows where two stars do collide. When two or more stars are evenly matched by gravity the two stars do not collide either but

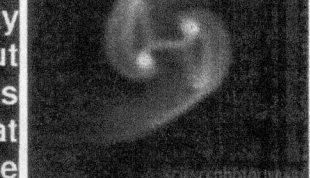

becomes binary stars that spin around each other. Stars or planets never collide and when a meteor enters the atmosphere like at Tunguska in Russia in 1908, the earth vaporises and liquefies the meteor and the meteor becomes fragments as well as more dust clouds in the atmosphere. This totally annihilates Newton's idea that a radius between structures diminishes by the gravitational attraction of mass. **This book shows what really happens in truth in nature.**

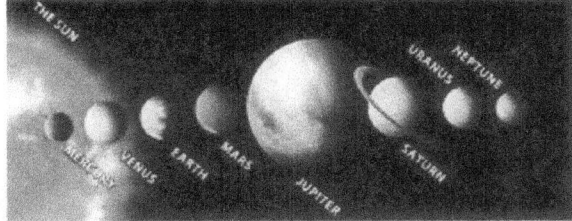

The Titius Bode law is how planets distribute places in the solar system. There is a precise sequence and order in which planets hold places and this annihilates the idea that mass plays any part in this sequence. In contrast the Newtonian system that science promote currently says that gravity "pulls" according to mass. This is fictional. The mass of the planets is totally random and Newton holds no theoretical basis that nature supports. Look at the arrangement and you can visually see using your eyes that the distribution by mass is a hoax and is an invention Newton concocted and science validates for the past 300 years. By explaining the Titius Bode law for the first time ever this book breaks the 300-year-old myth and brings truth

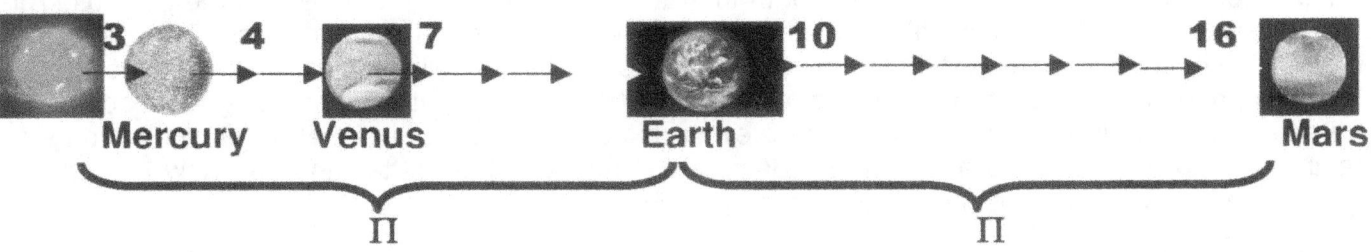

We have the ratio that time turns into space thus forming space-time. There is a specific ratio applying that moves time into space and the Titius Bode law prove that ratio. This is the value of Π and this can be traced back to when time began forming space as the Universe started from a single dot. Also the five points in the Lagrangian system is directly linked to the starting process of the Universe during the very first development process.

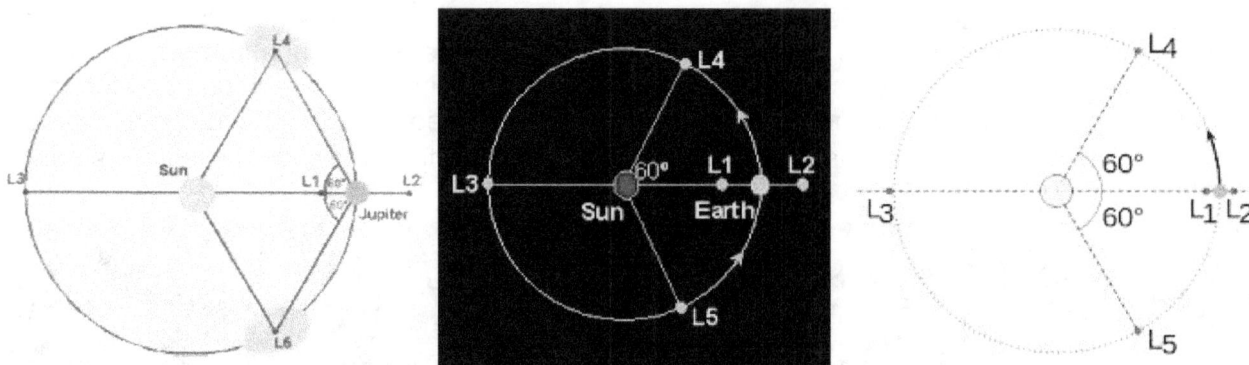

The Lagrangian points show that mass do not attract or come closer but to the contrary stay at a distance predetermined by singularity that controls gravity or movement. There is a ratio applying that has nothing to do with mass

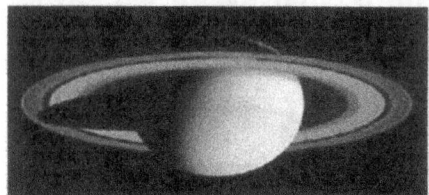

The Lagrangian Points This law proves that science is committed to deception for centuries. It is known that sattelites turn around planets in the same fashion as do planets orbit the sun. Saturn does not "pull" the satelites into its atmosphere or even "pull" the satelight closer and this knowlidge never drove science to question the validity of Newton ot his concept about "mass" pulling "mass". Even the rings formes as debris spins in a circle that always remain constant and science should have rejected newton'ss ideas on the grounds that nature never not once supports Newton and Newtonian science in any way thinkeble.

The Universe form by numbers going into dimensions that ends up as dimensional space. However this is also where we have to find the start of mathematics because mathematics started by starting the Universe. Mathematics is not God Almighty that was before the Universe was but grew as the Universe grew. By enlisting the way mathematics started and comparing that to how the Universe began we can trace the beginning back to a point when space was 1 and mathematics was 1. That formed physics.

To start the synopsis I have to present you with the ratio that apply in the formation of the cosmos. This ratio as presented in the picture above is the only applying way that the solar forms. The Newtonian presentation of how the solar system forms using mass as a force of "pulling power" is the most elaborate hoax ever presented to the world and is completely fake. Where this might surprise you go and do a study and see that this is in place and that proves Newton is not in place and is fake! If this shock you stop reading you are brainwashed and mentally damaged. I resolved the matter and can now present the world with a real solution about how nature forms gravity.

Everything in the Universe is round. Anything that is round has to apply the value of Π. This is a fact of mathematics but while Newtonian science forever tells the Universe to have "mass" and to use "mass" nowhere in science would one find Π used in prominence in cosmic space. Whatever you may study in astrophysics, go where you wish but never would you find Newtonian science taking the fact of Π into any prominence. When you read any of my books you will see that gravity forms by movement applying Π as a value. I have found the four phenomena that put Π in astrophysics.

By valuing gravity as Π therefore the Universe consists of gravity that forms by the working of the four phenomena that Newtonian science hardly ever mention but dubbed "a freak of nature". Science pretends to search for planets far outside our solar system and all the while science has no idea why our solar system functions in the manner that it does. Would it not be much better to first find out how nature applies in forming our solar system before trying to pretend to know about other thought to be planets? Science ignores nature and pretends to know much more while all of science know so little they have not figured out how gravity works…and that I prove! When you read this book you will know much more about the solar system than what the Super-Educated ever knew about science. Ion this book for the first time ever the true nature of gravity is revealed. I repeat the synopses again because I know these principles are totally new to you. They are not new to science because science knows about then for a long time but also science hides these laws for a long time. Again read through the following and be sure these lead to science.

The **Titius Bode law** is about the space every planet holds that becomes bigger the further it is from the sun. Every planet doubles the relevancy in ratio to what the previous plane hold that is more to the inside and then is half the relative ratio of the planet to its immediate outside. That brings about that the space the planet hold is by ratio double every time it is one position further away from the sun. Pluto is 100 times further than Mercury is. This way by doubling space the solar system forms the Titius Bode law and destroys all credibility Newton ever claimed.

The **Roche limit** has been around for centuries and with proving that gravity forms by movement forming Π I show not only that but also why the Roche limit is in place and what influence does the Roche limit have on ordinary gravity applying. The Roche limit is responsible for the "sounds barrier" and it is when movement exceeds the Roche limit in relation to Π that the "sound barrier" becomes a factor. Yet when using the truth about gravity in physics the answer is simple; it is that gravity is Π.

The **Lagrangian points** I prove by proving how Π forms as gravity. Everything about the cosmos forms by the manner in which Π forms and all movement in the solar system adheres to Π allowing movement.

The **Coanda effect** has powered turbine engines and aeroplanes in flight for almost a century and with all the mathematical splendour available to design the most terrific aircraft, not one engineer could mathematically compute one fact to show understanding why this takes place. How sad it is that those claiming of much superior intellect in physics remain just no more than having computing power. The understanding is not complex.

The entire Universe works on a relevancy that exist between material holding space and material claiming space, in other words the relation there is between occupied and unoccupied space. In short the Universe comprises of solids with a density beyond the speed of light and liquids below the speed of light. This relationship is there where movement established Π and in the way Π forms the density of space revalue from 21.991 / 7 to holding singularity at Π being 3.1415 / 1. By changing Π from 21.991 / 7 to 3.1415 / 1 space surrounding material condenses I prove the Titius Bode law. I have published the Titius Bode law in four already published books but in this one I go deeper than the four already published.

What these laws undoubtedly prove is that the entire cosmos is in support of what forms the cosmos. It is a cliché but never was this truer than in cosmology that everything there is also is connected. Numbers forming space-time connects whatever is there in space in time. But in science we follow a route that leads us nowhere because since everybody that practise astrophysics waits for the most bizarre explanation using most impressive equations I now wish to show how simple cosmology is when applying the four cosmic principles in nature.

There is the sun that by gravity connects the earth to the sun and this connection is beyond question. Therefore in relevance we have a factor of two connecting 1 +1 = 3. Every planet holds this ratio with the sun where the sun is 1, the planet is 2 and the distance between the two forms 3. Then the allocated number in the line multiplies with the planet position in the solar system.

Can you accept the fact that since 1776/8 I am the first one with the brainpower to award the sun a position of 1 the planet a position of 2 and the space in between a position of 3 and thus multiplying the planet position number in relation with the space whereby I solve this riddle. For this many decades since 1776 the most brilliant minds failed to explain this? If ever there was a rotten egg, something very stinky that proves science has this conspiracy to hide Newton's blunder then this is it. How simple can any riddle be to solve and in 300 years they all failed?

Space form 3 as a factor

I have shown the relevance of the sun (1) holding whichever planet (1) in space (1), which this forms 3 and that is the value that Mercury has in the Titius Bode law. One planet on and there is Venus with 4. I shall explain this position of four but using the earth as a parameter.

Sun and earth change location when the sun 1 and the earth 1 moves (²) or going = 2

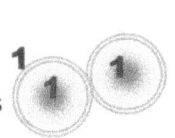

In relation to movement or gravity bringing changes the sun (1) duplicates its position 1^2 and so does the earth 1 duplicate or move

as 1^2. Therefore the sun doubles its position (1+1=2) and the earth also doubles its position the same way (1+1=2)

Sun and earth change location 2+2 or 2x2 =4

However both still forms a relevancy in relation to one another duplicating or changing position by going 2 (the sun) moving in terms of the earth (2) and that brings about (2 + 2=4) by location changing or by gravity or movement or time (2 x 2 = 4)

Sun and earth change location = 4 plus moon = 1 = 5

The sun changes 2 and the earth changes 2 and the moon holds a steady position by remaining in location in terms of the earth forming an earth bound satellite giving it a position in terms of (2x2=4 +1=5) or (2+3=4+1=5). The movement provides stability in terms of the moon that attaches location in terms of the earth and there we have five as a number.

Sun plus space 1 change location = 3

Total change = 6

Earth plus space 1 change location = 3

The sun relocates (2) in terms of the space (2+1=3) and the earth relocates (2) in terms of space (2+1=3) and al the changes amalgamate to six. This is why the sun holds a potion of 1^0 in the Titius Bode law and Mercury in one event holds a value of 3. The table shows that this ratio that holds a value of 3 is multiplied by the number in the line that the planet is. This forms the line 3, 6, 12, 24, 48, 96 and so forth. This is the very simple argument and it took 300 years and someone as simple as I to conclude this riddle. Science stick to mass and there is no and never was any proof of mass forming gravity except in the fantasy world of science make-believe. This stinks of misconduct! However this does not form the best interpretation but this I show to prove that there might be several interpretations where the lot has viability. Newton is the only person in science that never had to prove his work and nature disproves it.

Mercury	Venus	Earth	Mars	Asteroids	Jupiter	Saturn	Uranus	Neptune	Pluto
0.4	0.7	1.0	1.6	2.8	5.2	10.0	19.6	30.1	39.6
	0.3	0.3	0.6	1.2	2.4	4.8	9.6	10.5	9.5

This is the Titus Bode law by which the solar system forms. Notwithstanding Newtonian denial in accordance with these tables it is the only way that the solar develops and there is no other way. These tables present cosmology as nature practise cosmology. Following the Titius Bode law ratio this explanation proves why the order follows this pattern. However there is a second pattern that works on different principles applying different arrangements.

Mercury	Venus	Earth	Mars	Asteroids	Jupiter	Saturn	Uranus	Neptune	Pluto
0.4	0.7	1.0	≈ 1.6	≈ 2.8	5.2	≈ 10.0	≈ 19.6	≈ 30.1	≈ 39.6

Sun plus space form 3 as a factor

Total space relocation $3 + 4 = 7$

Earth rotating $2^2 = 4$

The sun moves by bringing stability to singularity (the sun moving 2 + 1 the space remaining) while the second planet moves by duplicating (2^2) and that places the second planet in a cosmic ratio of 4. Therefore the sun holds 3 and Mercury holds 4 and this translates to Venus being (3 + 4 = 7). However everything (3) moves (3^2) in terms of each other while singularity forms a centre keeping whatever locked ($3^2 = 9 + 1 = 10$) and that is were space ends. Space ends at 10 while material ends at 7 because (3+4=7).

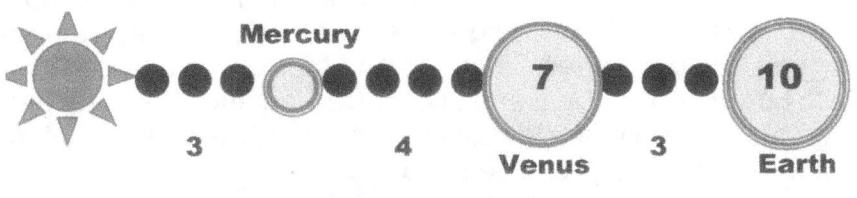

The earth completes the space we have between our centre of the sun and space there is according to the location of the earth. The earth forms ten in the ration but then another ratio begins.

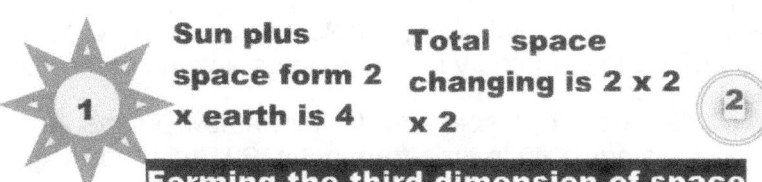

Sun plus space form 2 x earth is 4

Total space changing is 2 x 2 x 2

Forming the third dimension of space

Earth rotating as 2 (the sun) x earth as 2 = 4 x space 2 = 8

Where the sun moves by duplicating (turning 2) in relation to any other object also turning (2) and this creates a ratio of 4 all the while everything spins around the sun by duplicating and this brings the third dimension of depth into the cosmic equation ($2^2 = 4 \times 2 = 8$).

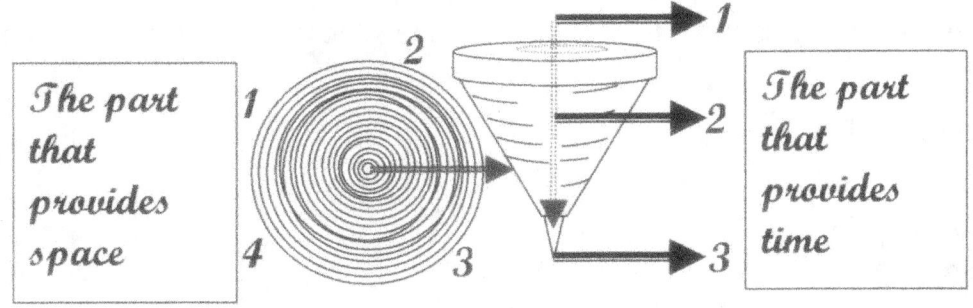

The part that provides space

The part that provides time

The Universe divides into parts forming time. On the very inside of material spinning is a line forms that is singularity in time. On the outside of material is the space part that forms eternity and in the middle as material is the Universe that converts eternity in space to infinity in time.

Then the pattern change and I have written many books in which I explain the forming of the Titus Bode law. There is a ratio by which space forms in the solar system and this is how:

The above explains what no one in science since 1776 could decipher. Not one could explain why 3 + 4 forms 7 or even that 3 + 4 forms 7 and in that there is a difference. This says that if there are three dots formed between the sun and Mercury the ratio will then will form. This law allows space a specific ratio. This is placing what the tables present in a picture to allow better understanding. If there are 3 dots forming between the sun and Mercury another 3 dots will develop between mercury and Venus.

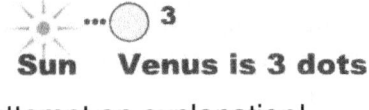

Then according to the Titius bode law Venus holds 3 more dots. In brief I explained that already in less than a page. I wish to remind the reader that all the mighty brainpower since 1776 could not even attempt an explanation!

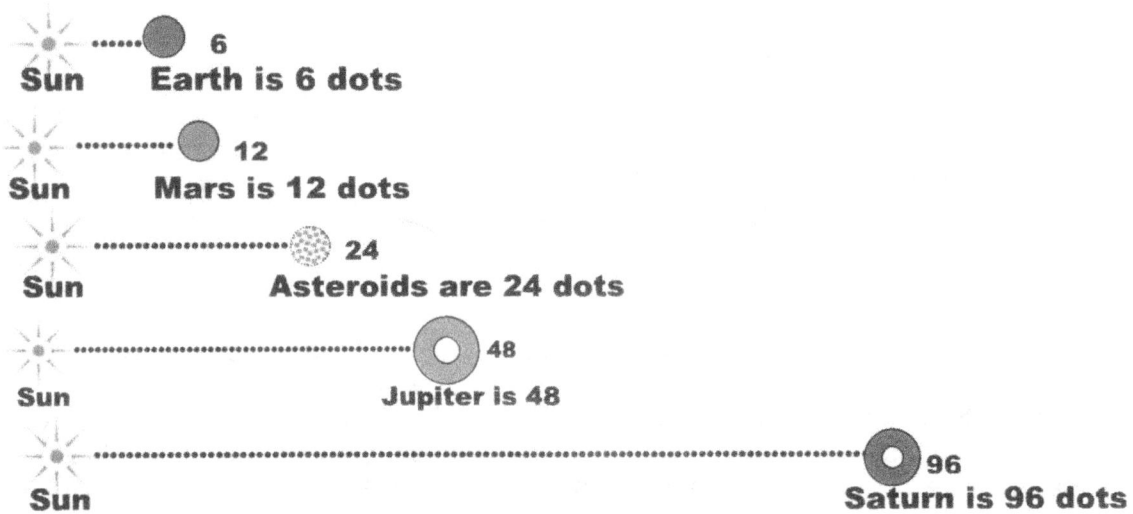

This is why the Titius Bode is so important to understand. This part I can't explain in this part of the book because it will claim too much space but as one can see by reading these laws and how they fit into a ration we can trace this ratio back to where the Universe started. That I do and that is why I can do what no one ever could, I can do it because I wasn't prepared to cheat and accept Newton and I was prepared to add 3 + 4 and get seven. Now I am the first man in 300 years to solve the laws holding physics. Let me again show you how brilliant I am! This is what the law says you have to do.

To find the mean distances of the planets, beginning with the following simple sequence of numbers: Singularity forming time holds a value of three point and singularity in space holds a value in four points.

In that the formation of space is seven points as time develops space and with the flow of time as it forms the future the present and the past it is 7 + 7 + 7 + singularity in time-forming is .991 this forms space 21 in time = Π. Therefore material that forms time represents a mark of 3, which then depends on the allocated position as in the line up that the Titius Bode law presents.

0 3 6 12 24 48 96 192 384

This is a given. This is part of the law that forms the ratio. I did not have to sit for years and figure this out it is presented to me by the cosmos.

Then the law says add 4 to each number:
4 7 10 16 28 52 100 196 388

After which the law says then divide by 10:
0.4 0.7 1.0 1.6 2.8 5.2 10.0 19.6 38.8

I wrote to many academics but never or seldom received any reply concerning my cosmic concept.
When I sent twelve articles to Annalen Der Physics to explain the laws I received the following e-mail in response:

Eternity: The part of the
Universe that has no outside

Eternity: The part of the Universe that can never end

Eternity: The part of the Universe that cannot

stop   moving

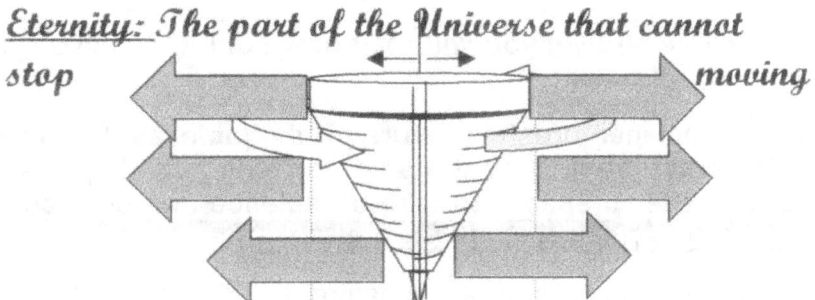

Eternity: The part of the
Universe that has no outside

Eternity: The part of the Universe that can never end

Eternity: The part of the Universe that cannot

stop   moving

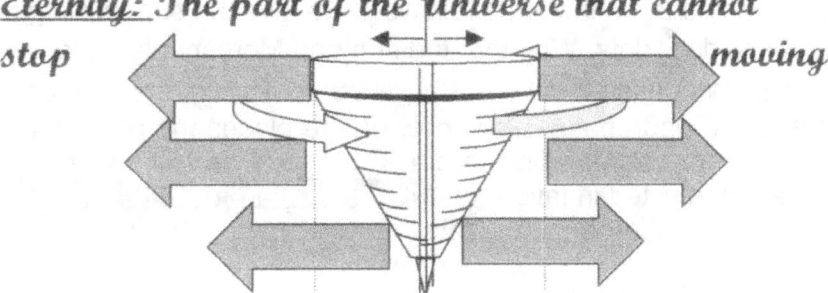

Infinity: The part of   _Infinity:_ The part of
the Universe that can   the Universe that has
never start   no inside

Infinity: The part of the
Universe that cannot move

In the articles that I addressed to Annalen Der Physics I explained the four laws and that the four laws are part of how nature applies gravity and the proof was by my ability to use nature to explain my conclusions. Go look it up before you go on…and why don't science recognize these laws…because these laws brings the entire industry of science to an abrupt end and will stop everything science put in place as science.
Recognizing the importance of these laws will kill an industry worth trillions in books already on the market…
I now introduce you to the Titius Bode law or how the solar system forms.

There are these four laws or phenomenon or principles in the Universe. Newtonian science can't explain it but I can. In doing so I shatter the myth called Newtonian science and I reveal the hoax science portray for three hundred years or more as the truth. I tried to get published but I was unable because then I break the strangle hold physics has.

For the first time in all of human history there is a method deciphered to show how NATURE no less forms the solar system…and in eighty five DVD's sent plus another (about) thirty six or seven e-mails going via sendspace and not one was interested to publish. Go to the Internet and see it is said this code can't be deciphered but I did find a way to decipher.

That science says is impossible and yet you will read how I did it. Science plainly ignores nature while nature is the reality.

Nature is the only reality but science brushes nature off the table, as if nature is madness. To so many publishers I sent the entire book…I sent it as a unit with two chapters more than the book you read and found no publisher prepared to take on science and correct the hoax Newtonian science is. Nature is how things work in practise and therefore it is how theory also have to work!

Now you can find out how to crack the code by which nature (not Newton's fiction) forms the Universe in the manner that it forms the solar system. It is simple; it is adding 3 plus 4 to get 7! Finding the Titius Bode is 7 also turns what you thought was cosmology into an explanation of the truth while the truth turns cosmology into truthful science. This Titius Bode law, its 7 / 10 or 10/7. In this book I explain in detail the layout working process of Titius Bode law.

This book explains the Titius Bode law but what is the Titius Bode law? As I show you later on the solar system does not form as science and Newton try to show but it forms a reality far apart from what those in science try to promote. The Titius Bode law is as follows: if we give the sun forming a circle in a relevancy of **3** say dots then Mercury holds **4** dots. With that the sun and Mercury places a value of **7** on the space between the sun and Venus. With this ratio the earth will be another **3** dots further and will have **10** dots from the sun to the location the earth has. This is how nature places the planets and this totally clashes with what Newton said happens. I am the first to crack the code as to how and why this happens and believe it or not my work is still rejected in favour of Newton's ideas that never applied in nature and in reality. In   a   sketch   this   is   how   nature   works!

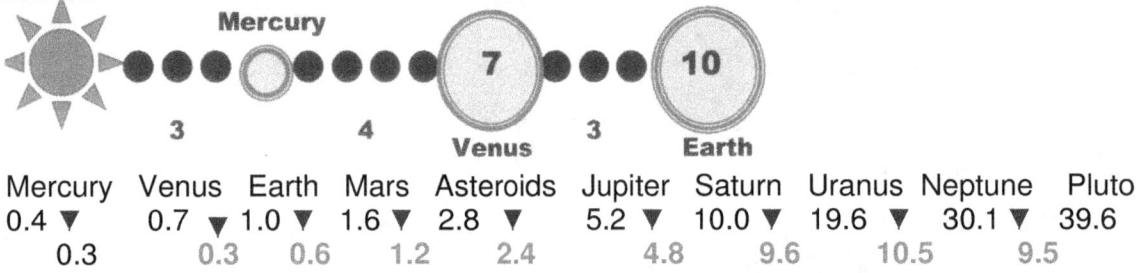

Mercury	Venus	Earth	Mars	Asteroids	Jupiter	Saturn	Uranus	Neptune	Pluto
0.4 ▼	0.7 ▼	1.0 ▼	1.6 ▼	2.8 ▼	5.2 ▼	10.0 ▼	19.6 ▼	30.1 ▼	39.6
0.3	0.3	0.6	1.2	2.4	4.8	9.6	10.5	9.5	

The ratio (not distance) doubles with the placing of every planet except the last two and for that there is a reason. Are you ready to learn the truth about science and see how you were duped? If it is **3** dots from **Mercury to Venus** it is **6** dots from **Venus to the earth** and from **the earth to Mars** it is **12** dots. From **Mars** to the **Asteroids 24** and from the **Asteroids to Jupiter** it is **48** dots with a further **96** dots from **Jupiter to Saturn**. This shows that the distance inexplicably doubles in ratio every time a new planet position is allocated. There is a ratio. If you give the sun **3** dots then the rest is:

3   x 2 = 6   x 2 = 12   x 2 = 24   x 2 =   48   x 2 =   96

Are you familiar that this is the true state of affairs in the solar system and that mass of planets never played a part in the allocating of any position in the planets' layout? Say, if Venus has 3 dots then   x 2 = the earth will have 6 dots

If the earth has   6 dots x 2 = then Mars will have = 12 dots

If Mars has 12 dots x 2 = then the Asteroid belt will have 24 dots

If the Asteroid belt has 24 dots x 2 = then Jupiter will have 48 dots.

If Jupiter has 48 dots   x 2 = then Saturn will have   96 dots.

How aware were you reading this about this law called the Titius Bode law? How much do you know about the presence of this law and were you aware that nature annihilates everything Newton says with this law. This is how nature uses planetary positioning but those in science totally denies this as a reality because nature destroys everything Newton said with the application of this law. The Titius Bode law is reality notwithstanding those in science trying to belittle nature and deny this law and that is a reality you can look up wherever you wish to look. Please take note of my accomplishment. I am the first one in three hundred years to crack this code. Do you even think that I am clever? I surpassed all genius that came before me and this is why: I am clever because I am the 1st in 300 years that could add 3 (+) 4 and get seven (7).

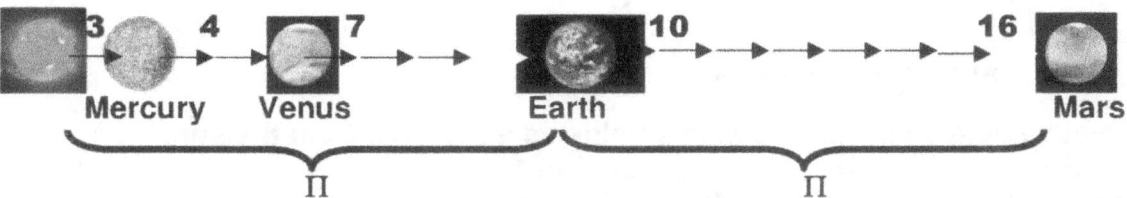

No one in 300 years broke this code because it is so extremely complicated...and you believe that? I am the first person that can add the value of 3 and 4 and get 7!
This is how the sequence work and this sequence have been known to science ever since 1776. When the sun holds an initial value of 3/10 the ratio is as this table predict:

Mercury	Venus	Earth	Mars	Asteroids	Jupiter	Saturn	Uranus	Neptune	Pluto
0.4	0.7	1.0	≈ 1.6	≈ 2.8	5.2	≈ 10.0	≈ 19.6	≈ 30.1	≈ 39.6

Then with Mercury forming the next planet in the order Mercury holds a ratio adding value of 4. Then you divide the total by 10. This is the sequence nobody could unravel.
This is the complicated part. Adding the first 3 and the 2nd value of 4 puts Venus in a ratio position of 7. Then you add 3 to the tally and divide by the 10 the earth has to get 1, which is the earth. The doubling is as simple as I explain inside.
It is so simple as adding 3 plus 4 that then divides by 10. This becomes a mystery no one in 300 years could solve or ever since Newton came up with his idea of mass forming gravity. Do you also smell a rat or is this statement I make going too far? Do you believe this hogwash that nobody could add 3 and 4 to get 7? ...And it took the most brilliant minds 300 years not to conclude this. Those in science, the lot are always going on about telling everybody that this order or planet line –up does not make sense and this is nature's (freak) or failure and this is a reject from the madness that nature is. The question is always about the Titius Bode law or planet ratio that could never be mathematically explained or put into context with gravity. Science always places doubt about nature. It is always putting nature in doubt and then they put Newton's ideas forward as if it is self-explaining and that is untrue because Newton is found nowhere.

There is a conspiracy to prevent the truth about Newtonian science to become unmasked. That statement about corrupt science I prove and that is why the order of science rejects all my work. I ask questions because I give answers they do not wish to discover. If you feel offended or hurt or worried when reading this you are brainwashed and duped at the same time. Please, for God sake then put this book down. Do not read further because with your limited vision and understanding of science this will crack all your confidence. You have been mentally conditioned to believe Newton and to accept science as the only truth there ever was and this will come as a shock and destroy you!
This, the Titius Bode law is there. This is undeniable what is in place. Anything other than this is a daydream and seeing other wise then becomes the King's magic clothes, which is precisely what

Newtonians think they see Newton's magic of gravity that has the ability to allow the mass unexplainably pulling other mass. Newton is not real and nature is reality while this is law and anything ells is a hoax.

3

Sun   Mercury is 3 dots

This discrepancy why Mercury and Venus both form 3 I explain in this book.

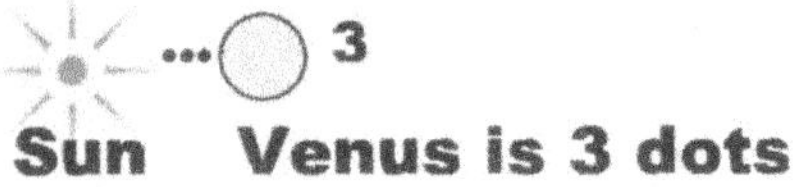

3

Sun   Venus is 3 dots

This is the relative points in ratio that Venus holds as a distance from the sun.

6

Sun   Earth is 6 dots

This is the relative points in ratio that Earth holds as a distance from the sun.

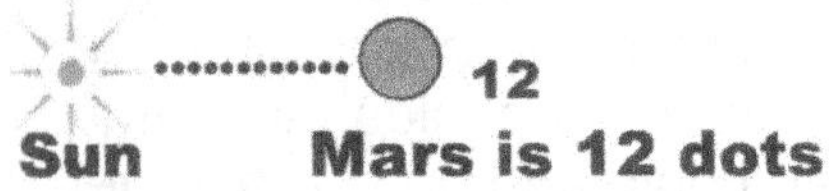

12

Sun   Mars is 12 dots

This is the relative points in ratio that Mars holds as a distance from the sun.

24

Sun   Asteroids are 24 dots

This is the relative points in ratio that Ceres holds as a distance from the sun.

48

Sun   Jupiter is 48

This is the relative points in ratio that Jupiter holds as a distance from the sun.

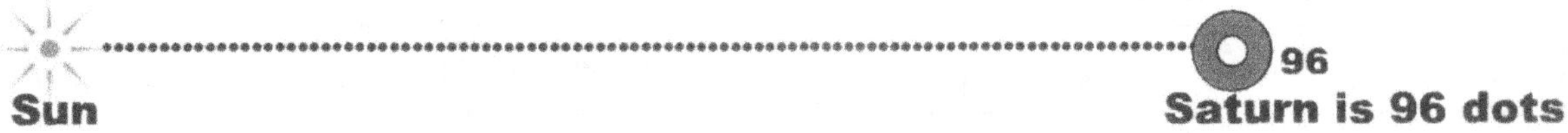

96

Sun   Saturn is 96 dots   **This is the relative points in ratio that Saturn holds as a distance from the sun. There is no space on the paper provided in this book to indicate the others but Pluto is 40 times further than the earth is. Why do the planets' distance double?**

I explain this in detail because what there is in the solar system is this; the Titius Bode law forming in conjunction with the Coanda effect, the Lagrangian points law as well as the Roche lobe / Roche limit that is what forms gravity. This is it!
Newton and his ideology is as absent as the correctness Newtonian science hides and if you do not believe me go and research this by yourself. Giving this truth I contacted (about or more or less) 150 publishers among which there are about fifty or more Universities and not one had any interest in publishing this book. …And I suspect it is because I go against the grain of Newton!

This is the first time in human history that any person had the inclination to explain the forming of the solar system and nobody is interested in this venture? I explain nature and nobody wants to read it!

This is how space forms and out there amongst all not one shows interest in publishing?

Some just believe because they accept and others have to understand to accept truth and you can choose where you are. Please be warned about the following:
Reading this book will intellectually find the reading to be very challenging to any person since what I say was never yet published. Everything you are about to read is new!

That I in principle disagree with science's accepted principles on very basic issues is a fact that is undeniably true. What you read about the principles I propose is new to everyone alike. However, I found that the ordinary persons with a scholastic physics background cope with the difficult explaining much better than does Super-Educated-Masters. The Super-Educated-Masters have information stored by culture and if they can't bring the information to mind by recognition of it they fail to understand new science concepts. You are going to read this in a letter that was sent to me.

The purpose with which I wrote this book is to get around the network of Super-Educated-Masters who strangle any information that forms of science in the form I propose and therefore that does not fit their views or match their liking. If what anyone says does not stroke with what the Brainy Bunch says who controls physics and agree with "Mainstream Science" or echo their thinking, they just smother all intellectual publication on the grounds that it is not fitting their profile on science. If you ignore culture and read with using your logic you will find many accepted norms as ridiculous. With most concepts I disagree most strongly and I disagree because those concepts lack proof but I do also supply detailed proof of my views and that is where Mainstream Science blocks the publishing of my views on science that does not compliment their views.

Read this and wake from the culture you believe in; that which science has lulled you into and made you accept science as the absolute undeniable rock fast truth by instating it as a religiosity then stop reading or get your tranquillising anti depressants next to you with a large bowl of water and a big glass. You will find some mathematical equations, if you are not familiar with it ignore it because it shows the silliness of "Mainstream Science" but if you don't read it you will still understand the explaining by reading the language where I explain it. "Mainstream Science" hides behind maths. I need help to fight their fraud and I need you to help me fight them. What you read I prove to even every last detail and even in this book and therefore I dare anyone to prove otherwise or reprimand me.

You are not going to read a book but you are going to travel a journey. Everything written in this assembly is collected from numerous articles and papers I wrote to individuals, journals, science magazines, Universities, academics in administrative-teaching positions as well as many on line physicists which I tried to interest in my view and my findings on science. I did not remove parts or sections of the articles to disguise this as a book that is ready-written but formed it to be a road on which I travelled and the way I found never-ending rejections.

I never tried to make a name in physics but tried to get some money to make a living from and take care of my children. I always knew there were so many smarter brains out there than what I have and who could see what I saw and take the challenge of correcting science from what I brought to the table. I was stonewalled by a bunch of corrupt conspirators trying to lay claim to rubbish Newton presented as truth and in some cases their efforts to justify Newton's lies became pathetically poor.

I realised there are those that has brains and then you have those that understand Newton and that is why astrophysics remained backwards as it got stuck promoting forces flying all-over pulling to form gravity. I say this straight: no amount of words can describe my utter disgust I have for those in science thought of as flawless performing beyond blame because they are criminals hiding their criminality.

Why would knowing the Titius Bode improve our understanding of physics?

For the first time in all of human history there is a method deciphered to show how NATURE no less forms the solar system…and in eighty five DVD's sent plus another (about) thirty six or seven e-mails going via sendspace and not one was interested to publish. Go to the Internet and see it is said this code can't be deciphered but I did find a way to decipher. That science says is impossible and yet you will read how I did it. Science plainly ignores nature while nature is the reality.

Nature is the only reality but science brushes nature off the table, as if nature is madness. To so many publishers I sent the entire book…I sent it as a unit with two chapters more than the book you read and found no publisher prepared to take on science and correct the hoax Newtonian science is.

Now you can find out how to crack the code by which nature (not Newton's fiction) forms the Universe in the manner that it forms the solar system. It is simple; it is adding 3 plus 4 to get 7! Finding the Titius Bode is 7 also turns what you thought was cosmology into an explanation of the truth while the truth turns cosmology into truthful science.

This Titius Bode law, its 7 / 10 or 10/7. I explain this in detail because what there is in the solar system is this; the Titius Bode law forming in conjunction with the Coanda effect the Lagrangian points law as well as the Roche lobe / Roche limit that is what forms gravity. This is it!

TO WHOM IT MAY CONCERN,

I am P.S.J Schutte, nicknamed Peet. Being a white South African my mother tongue is Afrikaans and my second language is English. I have per suiting a new cosmic theory that I partly present in a six part theses, of which the investigating research began in 1977. First I located what was wrong in physics. I compiled my presentation of The theses called The Absolute Relevancy of Singularity and then six separate thesis parts forming the theses published through LULU.com which I saw as way the only manner whereby I could generate funding by which I would be able to have the twenty seven books I already wrote linguistically edited and then to have the books published on a Print-On-Demand basis. I compiled a new cosmic theory by which I eliminated all the incorrectness that Newton has burdened science with but with this being my opinion I did not find a garage full of academics supporters waiting to applaud me and to uphold my views on the matter. Yet still I was not going to be ambushed by their relentless stonewalling my efforts and blocking my efforts in introducing both the incorrectness and the new cosmic theorem I concluded. This kept me busy for the past going fifteen years on a full time basis whereby I was trying to introduce my findings to many academics without finding much joy from my efforts. This past fifteen years plus saw me go without any income as I tried to get my theorem recognised as well as get my warning noted. Going without a steady income left me almost destitute and in order to find a manner to get my theory across to the attention of influential readers, I decided to publish eight part theses with eight books electronically as to try and get around the stranglehold of Newtonian bias controlling science at present worldwide. I decided to publish electronically which those in power do not control.

With my first language not English and the books not linguistically checked by an expert there are bound to be language errors that readers will notice. In the past I tried to check my work myself but after checking say one hundred and fifty pages for language corrections, then after days of toiling instead of having corrected work I ended having four hundred pages of newly written information which is still not linguistically corrected but holds a lot more information. My motivation to write lies not with language but elsewhere and my work I find most important above and beyond language accuracy.

The language and spelling errors therefore compiled instead of reduced and the problem grows. This is because my priorities lie elsewhere. I aim to spend money on correcting the work as far as language goes, as I receive money in the selling of my theses and in the hope that I will receive money. I will have all my work including the one you are reading edited professionally and corrected as I find money to do so...But first I have to get the public aware of the problem to get the academics to appreciate the problem.

I do find much pride in my status as being Afrikaner and would like to have my names used by pronouncing it in the manner Afrikaans dictates...therefore I would sincerely appreciate the courtesy when readers will take note that my name and last name are pronounced in Afrikaans, which is originally from Dutch and must be pronounced that way. Peet one would pronounce "here" which is the closest English to the pronouncing of the "ee". The "Sch" in Schutte is pronounced exactly as school is where both actually are pronounced Skutte or "skool". By pronouncing my name in Afrikaans you do me the utmost courtesy any one can. Being an Afrikaner is what I am most proud of. If you have been a scholar at school you would know how to pronounce Schutte.

In the books on offer through this web page and in which I am introducing a totally new concept in terms of gravity, the proof I bring is true about gravity being formed as a result of these phenomena. In the past science hardly recognised the existence of such phenomena although they are known to science for centuries.
They are known as

       1) The Lagrangian system
       2) The Roche limit
       3) The Titius Bode law
       4) The Coanda affect.

These laws I prove to be is the foundation of physics and in this is vested all of science coming from mathematics going into Astrophysics as well as physics. Where I differ with Newton and physics has all to do with cosmology and it only concerns cosmology and astrophysics, not general physics. As far as I can tell I am the first person that made an in-depth study on the work of Johannes Kepler.

In a book forming part of the theses *The Absolute relevancy of Singularity* in **Part 8** I prove that the Universe came from one dot and started by developing these four cosmic laws.

Some years ago I was reading about a certain remark that Einstein once made on a realisation or a conclusion that Einstein came to in his younger days while still being a clerk at the patent office. Apparently the idea Einstein came too was concerning the subject of gravity. As I said, this happened while Einstein was still being a patent clerk in his younger days. Apparently Einstein was looking out a window of the multi story patent office, when Einstein suddenly realised that had he, Einstein fall out of the window from the roof to the ground of the patent office where he was working at the time, then he (Einstein) would feel as if he was weightless during the time of his fall. Not only that but also so would all the articles in his office that surrounded him at the time being his office chair, his desk and a pen. By falling with him those articles would feel equally weightless should they accompany his fall down as being part of the falling process in his imagination. As the objects were travelling alongside Einstein down the building to the ground the lot would travel at the same speed from the top to the bottom of the building. That is what Galileo concluded about five hundred years ago. Then I went one step further by supposing the Einstein group's falling was real and no imaginary thoughts were set in the fall, then what was the imaginary factor then? Let's pretend Einstein did fall with his pen, his chair and his desk and Einstein was not imagining his fall. Einstein as a human being can imagine but his falling companions can't. Then during a true fall Einstein may have had an imagination that could tell him about his feeling and in particular about the condition of his weightlessness, but the pen, the chair and the desk had no such imagination and they were travelling at the same speed as he did downwards and therefore had the same weightlessness as he (Einstein) had while they all were falling together being in a downwards fall. If Einstein was imagining his weightlessness, it might be psychological, but in the case of the other travelling companions it was not possible to imagine anything because their participation was one of being mindless in the matter. The falling companions had no such a luxury as having an imagination, however they too had to be weightless as they travelled next to Einstein all the way. There is an immense difference in size between the falling companions and that notwithstanding this they travelled the same speed while descending. If they travelled the same speed as Galileo proved and they all hit the Earth the same time, which then indicated that their weight and mass, that which gravity used to drive and what propelled them downwards and that which was causing the drawing of what the mass was instigating to allow the motion of fall to commence, was equal. Size changed nothing to the equality there was in speed. Einstein should only have thought a little further than he did at the time because that would have made him realise what gravity exactly was and what Kepler found gravity to be. Kepler found space a^3 being equal to the motion thereof T^2 in relevancy to a centre point k. Kepler found space had to move.

When reading this that evening so many years ago, then I came to realise that Einstein could only feel weightless if it was true that he (Einstein) was weightless. He could not feel as if when the as if was part of his imagination because he was truly falling, and in truly falling the falling was then without his imagination doing the pretending. Einstein had to feel his weightlessness as a cosmic fact in the true sense because if he was truly falling, then the part, which was the falling experience, was what he was experiencing in reality by three dimensions with one dimension in time. Then he (Einstein) was feeling weightless through falling and that feeling came as a result of what was happening to him as a cosmic interpretation of reality. That then shows why hid pen and chair and whatever fell with him all the way. The lot had and shared the same weight because the lot were as weightless as he was. He was not pretending to fall whereby he then would feel as if…he was really falling and with that there is no "as ifs". What he then would have experienced came by means of what he was experiencing in reality because of his cosmic state in relation to his relevancy with gravity. If Einstein was experiencing weightless ness, it would be because he was weightless while falling, then Einstein would not imagine the weightless ness because Einstein was truly falling, thus carrying out his cosmic state he was in. His body being in motion ($a^3 = T^2k$) was at that moment truly weightless while experiencing unrestricted gravitational motion. Einstein, the pen, and the chair had the same weight since they were all weighing the same in falling. If there were any mass differences there had to be speed differentiation for the force of the one would generate more motion than the force of the other onto the different mass components but since there is not mass discrepancy amongst the falling while falling the lot is having the same state of weightless ness and therefore they adopt the same speed in the fall. After all it supposedly is the mass that is doing the pulling by creating the force in accordance with the individual mass and more mass does more pulling…except if the mass is not doing the pulling in the first place. With more force applying to different masses there had to be more speed involved and an increase in mass in some participants with more mass has to generate more force.

All four items including Einstein, would be equally weightless during the falling…that was what Galileo found because objects of different sizes and different mass travel at an equal pace (distance over time or space moving divided by time flowing while the object changes position in relation to the Earth ($a^3= T^2k$) while descending. However in reality the bigger objects do not fall quicker than a smaller object and that can only be attributed to one fact; it can only be true if the four weighed the same while falling and no one weighed anything while falling. That means the gravity applied while time flow in relation to the space that was applying the motion, which was what gravity is or numerically $k^{-1} = T^2/ a^3$ according to Kepler. The single line falling is represented by the factor **k** being the relevance of space a^3 that was relocating its cosmic position while all that was happening in relation to the motion of the Earth T^2, which was in relation to the Earth spinning around the sun and that rotation gives us our time T^2. While in motion the four different objects weighed the same since they travelled at equal speed downwards. However, when they stopped moving and came to a standstill, they then weighed different, which then indicated a difference in mass factors amongst them.

By standing still on the earth surface the objects had mass differences and when they were in motion they weighed the same. When the motion became frustrated by being blocked by another occupied space that was also filled with material and that was holding the spot too where the motion was directed, they then had different weight. The two had different levels of frustration with the larger party being more frustrated in the inability to move. Te density of the earth stopped the density of the falling object from falling further. The pushing resulted from the bodies striving to remain independent. It is the independence of the two bodies and the desire the bodies have to remain independent and not to share space that bring about the mass or weight. The two objects were in a fight to claim the position each desired, and that was to fill the centre of the Universe. Being ($a^3= T^2k$) was being in the centre of the Universe because the centre of the Universe was $k^0 = a^3/T^2k$. It may look being a simple mathematical statement but explaining that part to full understanding, requires the reading of *an Open letter to Selected Academics*.

From this one can deduct that gravity is motion or the intent to commit motion and mass is when the motion of gravity is frustrated by some solid structure blocking or preventing the continuing of the motion. Then one may conclude that gravity is motion of space and mass is the restricting of the motion of space. Having mass does not bring about gravity but it does restrict gravity's motion, which is what brings about the mass and weight. Gravity produces mass but mass does not produce gravity or in fact mass produce weight but mass is not responsible for the intended motion. Gravity on the other hand is the intention that the body has to move the very instant the blocking is removed. The intent on moving while being blocked by another object is frustrating the motion of gravity in both cases and the higher the frustration on motion is the more mass there is coming the way of the bigger object who then has the greater desire to move. The reason why it has the desire to move and why space is equal to the moving in time of the space in relevance to the centre of the Universe (which at that point might be the Earth or be the sun) is what the book *an Open letter to Selected Academics* explains.

Mass is the restraining of motion and gravity is material moving about by committing gravity. Mass only comes into the application thereof when two objects filled with space moves into a position where both want to claim the very position in space the other occupy. It is the motion and the independence they show to hold onto their individuality as independent cosmic structures that prevent them the sharing of space which in turn prevent further motion and that prevention of motion causes mass. Gravity is in essence where mass is present, still in a tendency to commit motion but is then in the frustration of motion and gravity at such a point is the commitment to move once the blocking of space is relinquished. Because the one object that has more "mass" would put in a more assertive effort to move in relation to a smaller object and the effort to move will constitute to a greater resisting effort by the blocking object in a fight not to relinquish its position on the space both object claim that the tendency to move and the tendency to block the movement will bring the effect of greater or smaller mass being present during the effort and in line of resisting the effort. However while any space is in motion, the gravity of motion is equal to all and puts everything on an equal basis. Therefore there is no big and small and the big sun does not pull the small Earth closer. The big sun allows the small Earth to glide past in a circle year after year without interfering because the two does not claim the space each other has. Mass forms when the motion is prevented that a differentiation in motion effort becomes part of the picture. The density does not allow further advancing downwards.

Do not be fooled by the seemingly innocent explanation that space is the motion thereof which is what gravity produces because of all things the cosmos creates, motion of space through time is the utmost complex manoeuvre and without bringing a restraining of mathematics into science, it is so complex there is no viable explaining in physics about how the cosmos produce the act of motion of space in time. To get every atom to spin as every atom follows the lead of the atom in front and give direction to follow to the atom just behind while giving coherency to the structure the lot of atoms are holding as an individual unit times the units there are going around in the entire Universe is beyond what the human mind can absorb. While the atom in front is vacating space to fill the space of the atom in front is vacating at that instant, the atom behind is filling the space that the atom in front has vacated in order to vacate and relinquish the previous position in favour of the following position to honour the direction gravity is insisting upon. Times that with every atom there is in the Universe and one may grasp the significance of the calculation. The coordinating of moving one atom from one point to a next point requires the skills that the human mind may never conquer.

We may see the moving of object through space being as simple as merely excepting it as a given fact, as science has done in the past, or we may reason about the complexity as civil person's should do, and come to realise that the complexity of motion of matter is beyond scope of human understanding. Removing material from space by filling material into a position of new space sounds simple because the complexity has never been realised. Reading _an **Open letter to Selected Academics**_ will reveal what the factors are in understanding the commitment of material to move through time. This was all a result of understanding the dynamics of Einstein's arguing about gravity and mass. Then with this information I further realised gravity is motion differentiation between objects. It is the independent motion providing a different speed while sharing a common centre off attracting that allows a discrepancy to establish mass under specific conditions applying between the two in relevancy. While falling the gravity applies as moving of space that is putting time in relation to the distance travelled. That means there is a speed relevancy between particles in motion and synchronised motion would bring about equal orbit around a shared centre. That is the result of gravity functioning. While the object falls the motion confirm gravity. When motion ends mass sets in and becomes the constraining of the object preventing further motion. The motion is still there but now it is reduced to a tendency to move thus establishing the object mass as the limiting of further motion.

Preventing the motion by implementing mass is the resting of objects against each other by resisting the motion to continue, which then is where the mass takes the place of the motion. Where a confronting of objects restricts gravity the action then implements an introducing of the mass as a substituting factor to motion that then replace motion as substitute to the motion that would be and the mass is providing the tendency of gravity being the motion of space. However mass then restricts motion and becomes motion in a tendency to apply motion. While falling gravity applies and motion neutralizes size, mass or weight. Mass counters motion being when the Earth restrains further motion of the falling object and the moving object is stopped from further movement where mass is then preventing or hindering gravity. This is the result of objects claiming an individual and personal claim to space occupied in a dual or in fighting for their individuality and independence of each other while wanting to be in the centre of the Universe. While falling or moving there is no opposition to the body being independent. When the motion seizes the falling object remains individual and still tends to move while Earth individuality resists further movement of the falling body's movement. Further movement is disallowed as other material fill space that falling body wants to lay claim to.

The only manner to remain independent by the falling object will be to relinquish to motion in the securing of mass as a substitute to motion where it then finally comes to rest. Mass then sets in not causing the motion but substituting the motion and from that motion restriction becomes resistance that becomes mass. While falling the object is experiencing gravity because the object is in gravity but when on the soil the object experience mass which is the restricting of gravity or motion by other space filled with material. It is a fight of objects to secure and retain the position they have of being in the centre of the Universe.

Moreover, I came to another conclusion of equal importance. When any person is standing on any place anywhere, while viewing the Universe, that person is filling the centre of the Universe. Let's get more personal. When you, the person that is reading this, are standing at night and is looking at the Universe you are seeing the Universe from the position that one only can have if that person is filling the specific spot in the centre of the Universe. All the light, every single beam that ever left any destiny at any time acknowledges this fact. The light coming to you directly at the spot you fill acknowledges that you are the most important person in the Universe because you are holding the most important position in the Universe. You centralize thee flow of light from anywhere directly to you at the spot you fill.

All the light that come across and travelled all of the vacant space from any and all possible positions in space runs directly towards your position using a straight line towards you where you are filling the centre of the Universe. Not excluding the effort of one photon, all light is heading to meet you where you are in that centre spot and not one photon will pass you by. Not one photon dare miss you because if they do they miss the effort that all light has to accomplish and that is to locate you as the person filling the centre of the Universe. From everywhere and anywhere a straight line forms as the light crosses the Universe to get to the position that you fill and that is then the task of every beam of light filling the Universe. The only task light has, is to reach you in the location you are in and after reaching you the task ends because the flow of light hits you and ends there.

Should you decide to relocate end shift your position to any other place in the Universe, you will shift the centre of the Universe to that location as well. If you install a camera on Mars, the light is obliged to acknowledge your relocating the centre of the Universe at your will to reposition you're being that centre of the Universe. You then by choice relocate where you then choose to meet the light. All the light that ever left its destination crossing the vast spaces of the Universe, excluding no particular light, travelled all the way just to find you filling the centre of the Universe, right where you are. By you're standing anywhere, you fill the centre of the Universe, and the entire Universe admits to that because all the light comes to meet you there. If you shift from the North Pole to the South Pole you will shift the centre of the Universe because all the light travelling throughout the Universe will find you where you then are to where you moved the centre of the Universe. The light left its destination billion years ago as it travelled through space at the speed of light anxious to acknowledge you're being in the very centre of the Universe. No photon will be able to pass you by where you are in the centre of the Universe because all light is heading your way from their starting positions. No wonder every person born has the idea they were born to fill centre of the Universe, which we do fill. The Universe is spinning around you or I, which is filling a centre where all motion is connected.

That is the Coanda effect on the utter-most grandest scale imaginable; nevertheless it is only a manifestation of the Coanda effect. It implicates gravity as wide as can be… Some things mathematics is able to explain but other explaining goes beyond mathematics. Try to explain mathematically the colour of the sky being blue in the sky on a clear sunny day and changing to black when nighttime falls. Use mathematics to explain this to a blind person…first explain the colour blue with the aid of mathematics and only mathematics and then change the entire formula to explain to the blind the change in colour from blue to black by only using mathematics. Use no words but only mathematics. Do the explaining in mathematics to a blind person that had no vision since birth in such perfect mathematical detail that would allow the person afterwards be able to explain the difference between blue and black to other blind persons by using only mathematics. Some aspects of the Universe go beyond mathematics and some even go beyond words. It is our task to find space, to find time and moreover it is our optimal task to find the Universe. We have to see what is solid, what is liquid and what causes gravity.

Gravity is to move or apply the intension to move space a^3 at the distance or relevancy of k while T^2 is the time it is going to take to apply gravity or move the space filled with material space a^3 at the distance of k in the time period of T^2. That confirms Kepler's attribution to gravity where according to Kepler space a^3 is equal to the movement T^2 (time it takes to move) at the distance k from the centre specific. Every aspect of deliberation about the Universe was never discussed in the manner it is discussed in An OPEN LETTER TO SELECTED ACADEMICS written by Peet Schutte.

Apparently an idea concerning the subject of gravity where Einstein came to realise was about him falling off a multi story building and the gravity of being with mass that he then would experience. Remember this was much before flying and parachuting with free fall acrobats showing on TV how a man and a car with a man in the car can descend five or six kilometres while the man and the car were falling side by side. This happened while Einstein was still being a patent clerk in his younger days. Apparently Einstein was looking out a window of the multi story patent office, when Einstein suddenly realised that had he, Einstein fall out of the window from the roof to the ground of the patent office where he was working at the time, then he (Einstein) would feel as if he was weightless during the time of his fall. By falling with him those articles would feel equally weightless should they accompany his fall down as being part of the falling process in his imagination. As the objects were travelling alongside Einstein down the building to the ground the lot would travel at the same speed from the top to the bottom of the building. Then I went one step further by supposing the Einstein group's falling was real and no imaginary thoughts were set in the fall, then what was the imaginary factor then? Let's pretend Einstein did fall with his pen, his chair and his desk and Einstein was not imagining his fall. Einstein as a human being can imagine but his falling companions can't. If Einstein was imagining his weightlessness, it might be psychological, but in the case of the other travelling companions it was not possible to imagine anything. There is an immense difference in size between the falling companions and that notwithstanding they travelled the same speed while descending. If they travelled the same speed as Galileo proved and they all hit the Earth the same time, which then indicated that their weight and mass, that which gravity used to drive and what propelled them downwards and that which was causing the drawing of what the mass was instigating to allow the motion of fall to commence, was equal. Kepler found space a^3 being equal to the motion thereof T^2 in relevancy to a centre point **k**. Kepler found space had to move and when reading this I understood physics for the first time in my life. I realised mass cannot have anything to do with gravity and that **Newton and all other Newtonians were wrong, as I suspected for years.**

When reading this that evening so many years ago, I came to realise that Einstein could only feel weightless if it was true that he (Einstein) was weightless. He could not feel as if when the as if was part of his imagination because he was truly falling, and in truly falling the falling was then without his imagination doing the pretending. Einstein had to feel his weightlessness as a cosmic fact in the true sense because if he was truly falling, then the part, which was the falling experience, was what he was experiencing in reality by three dimensions with one dimension in time. If Einstein was experiencing weightless ness, it would be because he was weightless while falling, then Einstein would not imagine the weightless ness because Einstein was truly falling, thus carrying out his cosmic state he was in. His body being in motion ($a^3 = T^2k$) was at that moment truly weightless while experiencing unrestricted gravitational motion. Einstein, the pen, and the chair had the same weight since they were all weighing the same in falling. If there were any mass differences there had to be speed differentiation for the force of the one would generate more motion than the force of the other onto the different mass components but since there is not mass discrepancy amongst the falling while falling the lot is having the same state of weightless ness, they adopt the same speed in the fall.

After all it supposedly is the mass that is doing the pulling and more mass does more pulling...except if the mass is not doing the pulling in the first place. All four items including Einstein, would be equally weightless during the falling...that was what Galileo found because objects of different size and different mass travel at an equal pace (distance over time or space moving divided by time flowing while the object changes position in relation to the Earth ($a^3 = T^2k$)) while descending. The bigger objects do not fall quicker than a smaller object and that can only be attributed to one fact; it can only be true if the four weighed the same while falling and no one weighed anything while falling. That means the gravity applied while time flow in relation to the space that was applying the motion, which was what gravity is $k = a^3 / T^2$ according to Kepler. The single line falling is represented by the factor k being the relevance of space a^3 that was relocating its cosmic position while all that was happening in relation to the motion of the Earth T^2, which was in relation to the Earth spinning around the sun and that rotation gives us our time T^2. While in motion the four different objects weighed the same since they travelled at equal speed downwards. By standing still the objects had mass differences and when they were in motion they weighed the same. When the motion became frustrated by being blocked by another space that was also filled with material and that was holding the spot too where the motion was directed, they then had different weight. The pushing resulted from the bodies striving to remain independent. The two objects were in a fight to claim the position each desired, and that was to fill the centre of the Universe. Being ($a^3 = T^2k$) was being in the centre of the Universe because the centre of the Universe was $k^0 = a^3/T^2k$. Then one may conclude that gravity is motion of space and mass is the restricting of the motion of space. Having mass does not bring about gravity but it does restrict gravity's motion, which is what brings about the mass and weight.

Gravity produces mass but mass does not produce gravity or in fact mass produce weight but mass is not responsible for the intended motion. The intent on moving while being blocked by another object is frustrating the motion of gravity in both cases and the higher the frustration on motion is the more mass there is coming the way of the bigger object who then has the greater desire to move. The reason why it has the desire to move and why space is equal to the moving in time of the space in relevance to the centre of the Universe (which at that point might be the Earth or be the sun) is what the have the effort to explains. Mass is the restraining of motion and gravity is material moving about by committing gravity. Mass only comes into the application thereof when two objects filled with space moves into a position where both want to claim the very position in space the other occupy. It is the motion and the independence they show to hold onto their individuality as independent cosmic structures that prevent them the sharing of space which in turn prevent further motion that causes mass.

Gravity is in essence where mass is present, still in a tendency to commit motion but is then in the frustration of motion and gravity at such a point is the commitment to move once the blocking of space is relinquished. Because the one object that has more "mass" would put in a more assertive effort to move in relation to a smaller object and the effort to move will constitute to a greater resisting effort by the blocking object in a fight not to relinquish its position on the space both object claim that the tendency to move and the tendency to block the movement will bring the effect of greater or smaller mass being present during the effort and in line of resisting the effort. However while any space is in motion, the gravity of motion is equal to all and puts everything on an equal basis. Therefore there is no big and small and the big sun does not pull the small Earth closer. Mass is when the motion is prevented that a differentiation in motion effort becomes part of the picture.

Again I wish to repeat: Do not be fooled by the seemingly innocent explanation that space is the motion thereof which is what gravity produces because of all things the cosmos creates, motion of space through time is the utmost complex manoeuvre and without bringing a restraining of mathematics into science, it is so complex there is no viable explaining in physics about how the cosmos produce the act of motion of space in time. To get every atom to spin as every atom follow the lead of the atom in front and give direction to follow to the atom just behind while giving coherency to the structure the lot of atoms are holding as an individual unit times the units there are going around in the entire Universe is beyond what the human mind can absorb. While the atom in front is vacating space to fill the space of the atom in front is vacating at that instant, the atom behind is filling the space that the atom in front has vacated in order to vacate and relinquish the previous position in favour of the following position to honour the direction gravity is insisting upon. Times that with every atom there is in the Universe and one may grasp the significance of the calculation.

Removing material from space by filling material into a position of new space sounds simple because the complexity has never been realised. Reading I am in the hope that in this matter I will be able to_will reveal what the factors are in understanding the commitment of material to move through time. This was all a result of understanding the dynamics of Einstein's arguing about gravity and mass. I then with this information further realised gravity is motion differentiation between objects. It is the independent motion providing a different speed while sharing a common centre off attracting that allows a discrepancy to establish mass under specific conditions applying between the two in relevancy. While falling the gravity applies as moving of space that is putting time in relation to the distance travelled. That means there is a speed relevancy between particles in motion and synchronised motion would bring about equal orbit around a shared centre. That is the result of gravity functioning.

While the object falls the motion confirm gravity. When motion ends mass sets in and becomes the constraining of the object preventing further motion. The motion is still there but now it is reduced to a tendency to move thus establishing the object mass as the limiting of further motion. Preventing the motion by implementing mass is the resting of objects against each other by resisting the motion to continue, which then is where the mass takes the place of the motion. Where a confronting of objects restricts gravity the action then implements an introducing of the mass as a substituting factor to motion that then replace motion as substitute to the motion that would be and the mass is providing the tendency of gravity being the motion of space. However mass then restricts motion and becomes motion in a tendency to apply motion. While falling gravity applies and motion neutralizes size, mass or weight.

Mass counters motion being when the Earth restrains further motion of the falling object and the moving object is stopped from further movement where mass is then preventing or hindering gravity. This is the result of objects claiming an individual and personal claim to space occupied in a dual or in fighting for their individuality and independence of each other while wanting to be in the centre of the Universe. While falling or moving there is no opposition to the body being independent. When the motion seizes the falling object remains individual and still tends to move while Earth individuality resists further movement of the falling body's movement. Further movement is disallowed as other material fill space that falling body wants to lay claim to. The only manner to remain independent by the falling object will be to relinquish to motion in the securing of mass as a substitute to motion where it then finally comes to rest. Mass then sets in not causing the motion but substituting the motion and from that motion restriction becomes resistance that becomes mass. While falling the object is experiencing gravity because the object is in gravity but when on the soil the object experience mass which is the restricting of gravity or motion by other space filled with material. I came to realise that gravity depends on gravitational density, which I explain later on. Gravitational density is well proven when comparing a spinning top to a top lying down.

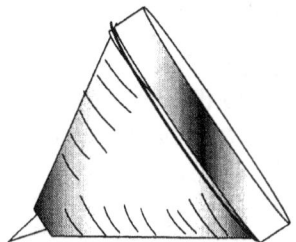 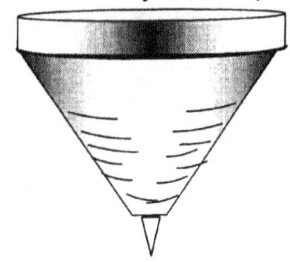

Looking at the top spinning and not spinning brings a question to mind: why would the motion of the top beat the gravity of the top and that of the Earth hands down when the top is spinning. Surely the mass is in effect while the top is spinning just as much as the mass is in effect when it is not spinning and yet when it is spinning the pulling subsides to give way to a situation where the top is apparently free from the mass restriction and that freedom charged top with much excitement allowing the top to stand erect. The excitement is so much it seems to relieve the top of the pulling there is between the top and the Earth. That even strengthened my suspicions more about the fact that gravity is motion and not mass. By spinning the top finds additional motion on the side of the top that brings the side of the top in a more favourite position than was the case before the spinning commenced. The top finds additional gravity in the spinning and the gravity in addition has to bring reconsideration to the position the balance in gravity sets in margins. The top secures a better margin or stretches the parameters of location because the top inherit the Motion of the Earth and in addition to the gravity motion that the Earth provides, the two secures more gravity motion. With more gravity motion in addition to the Earth's gravity motion the top has to have more gravity motion than what the Earth has. That will allow the top the spin while facing such enormous disadvantage there is on the side of the top in mass difference when considering the size disproportions. Let's face it, if it was about mass pulling the top in relation to the top pulling the Earth, then the top had no chance ever to move by the very slimmest of chances there may ever be.

However if it is the gravity motion that the top inherits and with the aid of what motion life can add in addition to the motion already applying then which is the motion the Earth already contributes and considering the mass the earth provides then it will not require that much a bigger effort to get the top going. Singularity charges motion by instigating motion without ever moving. Coming from a spot to a dot and then producing a line running from dot to dot though a spot signify the birth of the cosmos. A line holding time by the square and by the square of ninety degrees announce a birth in the Universe of a birth of the Universe. That which was not there suddenly is there by not being there. The line in singularity was never present and yet by the four rotating it awakes a line of three that then still is not part of the Universe since it has no space but forms a new part of the Universe by allowing the top independence. That which was undetectable suddenly is detectable by being undetectable. It is not my forte to write riddles but in this case there is a Universe within a Universe, which is not in the Universe and does control the Universe from a point no one may ever locate inside the Universe. The Universe is built up by innumerable dots and each dot is charged with being the Universe while being in a representative position since it is not in the Universe. Outer space forms by dots that form no space and yet holds all the space the Universe may have throughout. Outer space forms by a dot that is not present and yet y not being within the Universe is therefore everything that forms a presence we all live in and live by.

The definition of space-time is as follows:

According to the definition of space-time: Space-time is a four dimensional position of the Universe where the position of an object is specified by three coordinates in space and one position in time. According to the theory of special relativity there is no absolute time, which can be measured independently of the observer, so events that are simultaneous as seen from one observer occur at different times when seen from a different place. Time must therefore be measured in a relative manner as are positions in three-dimensional Euclidean space, and this is achieved through the concept of space-time. The trajectory of an object in space-time is called world line. General relativity relates to curvature of space-time to the positions and motions of particles of matter.

In view of the definition of space-time I wish to elaborate on my view of singularity and my deriving of space-time from the likeliness that singularity may produce space-time. In the past singularity was mentioned in the manner one would speak of a ghost hiding in a haunted Black Hole. Let's put singularity in the clear. Singularity is within every sphere due to the natural shape or form the sphere is committed to. The smallest form of material is singularity that is going from Π^0 that then becomes space Π.

Notwithstanding who you are or what you know and putting your achievements in physics aside this is your introduction to physics and every aspect that brings about cosmic information. Physics is not mathematical equating of more rubbish and mindless ideas. It is ideas where physics begin and where physics begin is concepts that mathematics can never prove.

Lets return to Einstein and his falling office equipment. Some years ago when I set out to think in terms of physics that led me to discover the information I put in pen as I was prompted to come to the conclusion I am to reveal when I was reading about a certain remark that Einstein once made on a realisation or a conclusion that Einstein came to in his younger days while still being a clerk at the patent office. Apparently the idea Einstein came to was concerning the subject of gravity. This happened while Einstein was still being a patent clerk in his younger days. Apparently Einstein was looking out a window of the multi story patent office, when Einstein suddenly realised that had he, Einstein fall out of the window from the roof to the ground of the patent office where he was working at the time, then he (Einstein) would feel as if he was weightless during the time of his fall. Not only that but also so would all the articles in his office that surrounded him at the time being his office chair, his desk and a pen. By falling with him those articles would feel equally weightless should they accompany his fall down as being part of the falling process in his imagination. As the objects were travelling alongside Einstein down the building to the ground the lot would travel at the same speed from the top to the bottom of the building. That is what Galileo concluded about five hundred years ago. Then the author of I went one step further by supposing the Einstein group's falling was real and no imaginary thoughts were set in the fall, then what was the imaginary factor then? Let's pretend Einstein did fall with his pen, his chair and his desk and Einstein was not imagining his fall. Einstein as a human being can imagine but his falling companions can't. Then during a true fall Einstein may have had an imagination that could tell him about his feeling and in particular about the condition of his weightlessness, but the pen, the chair and the desk had no such imagination and they were travelling at the same speed as he did downwards and therefore had the same weightlessness as he (Einstein) had while they all were being in a downwards fall. If Einstein was imagining his weightlessness, it might be psychological, but in the case of the other travelling companions it was not possible to imagine anything. The falling companions had no such a luxury as having an imagination, however they too had to be weightless as they travelled next to Einstein all the way. There is an immense difference in size between the falling companions and that notwithstanding they travelled the same speed while descending. If they travelled the same speed as Galileo proved and they all hit the Earth the same time, which then indicated that their weight and mass, that which gravity used to drive and what propelled them downwards and that which was causing the drawing of what the mass was instigating to allow the motion of fall to commence, was equal. Size changed nothing to the equality there was in speed. Einstein should only have thought a little further than he did at the time because that would have made him realise what gravity exactly was and what Kepler found gravity to be. Kepler found space a^3 being equal to the motion thereof T^2 in relevancy to a centre point **k.** Kepler found space had to move.

When reading this that evening so many years ago, I came to understand physics for the first time when I understood why realise Einstein that he could only feel weightless if it was true that he (Einstein) was weightless. If he felt he was weightless it was because he wasn't falling but all space including he Einstein was falling. Therefore I came to one more conclusion than what Einstein could realise at the time. It is not an object falling but it is the space the object holds that falls. And that includes all the space is evenly falling. He could not feel as if when the as if was part of his imagination because he was truly falling, and in truly falling the falling was then without his imagination doing the pretending. Einstein had to feel his weightlessness as a cosmic fact in the true sense because if he was truly falling, then the part, which was the falling experience, was what he was experiencing in reality by three dimensions with one dimension in time.

Then he (Einstein) was feeling weightless through falling and that feeling came as a result of what was happening to him as a cosmic interpretation of reality. He was not pretending to fall whereby he then would feel as if...he was really falling and with that there is no "as ifs". What he then would have experienced came by means of what he was experiencing in reality because of his cosmic state in relation to his relevancy with gravity. If Einstein was experiencing weightless ness, it would be because he was weightless while falling, then Einstein would not imagine the weightless ness because Einstein was truly falling, thus carrying out his cosmic state he was in. His body being in motion ($a^3 = T^2k$) was at that moment truly weightless while experiencing unrestricted gravitational motion. Einstein, the pen, and the chair had the same weight since they were all weighing the same in falling. If there were any mass differences there had to be speed differentiation for the force of the one would generate more motion than the force of the other onto the different mass components but since there is not mass discrepancy amongst the falling while falling the lot is having the same state of weightless ness, they adopt the same speed in the fall.

After all it supposedly is the mass that is doing the pulling and more mass does more pulling...except if the mass is not doing the pulling in the first place. With more force applying to different masses there had to be more speed involved and an increase in mass in some participants has to generate more force. All four items including Einstein, would be equally weightless during the falling...that was what Galileo found because objects of different size and different mass travel at an equal pace (distance over time or space moving divided by time flowing while the object changes position in relation to the Earth ($a^3 = T^2k$)) while descending. The bigger objects do not fall quicker than a smaller object and that can only be attributed to one fact; it can only be true if the four weighed the same while falling and no one weighed anything while falling. That means the gravity applied while time flow in relation to the space that was applying the motion, which was what gravity is $k = a^3/T^2$ according to Kepler. The single line falling is represented by the factor **k** being the relevance of space a^3 that was relocating its cosmic position while all that was happening in relation to the motion of the Earth T^2, which was in relation to the Earth spinning around the sun and that rotation gives us our time T^2. While in motion the four different objects weighed the same since they travelled at equal speed downwards. However, when they stopped moving and came to a standstill, they then weighed different, which then indicated a difference in mass factors amongst them. By standing still the objects had mass differences and when they were in motion they weighed the same. When the motion became frustrated by being blocked by another space that was also filled with material and that was holding the spot too where the motion was directed, they then had different weight. The two had different levels of frustration with the larger party being more frustrated in the inability to move. The pushing resulted from the bodies striving to remain independent. It is the independence of the two bodies and the desire the bodies have to remain independent and not to share space that bring about the mass or weight. The two objects were in a fight to claim the position each desired, and that was to fill the centre of the Universe. Being ($a^3 = T^2k$) was being in the centre of the Universe because the centre of the Universe was $k^0 = a^3/T^2k$. It may look being a simple mathematical statement but explaining that part to full understanding, requires the reading of a few of my books.

From this one can deduct that gravity is motion or the intent to commit motion and mass is when the motion of gravity is frustrated by some solid structure blocking or preventing the continuing of the motion. Then one may conclude that gravity is motion of space and mass is the restricting of the motion of space. Having mass does not bring about gravity but it does restrict gravity's motion, which is what brings about the mass and weight. Gravity produces mass but mass does not produce gravity or in fact mass produce weight but mass is not responsible for the intended motion.

Gravity on the other hand is the intention that the body has to move the very instant the blocking is removed. The intent on moving while being blocked by another object is frustrating the motion of gravity in both cases and the higher the frustration on motion is the more mass there is co0ming the way of the bigger object who then has the greater desire to move. The reason why it has the desire to move and why space is equal to the moving in time of the space in relevance to the centre of the Universe (which at that point might be the Earth or be the sun) is what the information revealed.

Mass is the restraining of motion and gravity is material moving about by committing gravity. Mass only comes into the application thereof when two objects filled with space moves into a position where both want to claim the very position in space the other occupy. It is the motion and the independence they show to hold onto their individuality as independent cosmic structures that prevent them the sharing of space which in turn prevent further motion that causes mass. Gravity is in essence where mass is present, still in a tendency to commit motion but is then in the frustration of motion and gravity at such a point is the commitment to move once the blocking of space is relinquished.

Because the one object that has more "mass" would put in a more assertive effort to move in relation to a smaller object and the effort to move will constitute to a greater resisting effort by the blocking object in a fight not to relinquish its position on the space both object claim that the tendency to move and the tendency to block the movement will bring the effect of greater or smaller mass being present during the effort and in line of resisting the effort. However while any space is in motion, the gravity of motion is equal to all and puts everything on an equal basis. Therefore there is no big and small and the big sun does not pull the small Earth closer. The big sun allows the small Earth to glide past in a circle year after year without interfering because the two does not claim the space each other has. Mass is when the motion is prevented that a differentiation in motion effort becomes part of the picture.

Do not be fooled by the seemingly innocent explanation that space is the motion thereof which is what gravity produces because of all things the cosmos creates, motion of space through time is the utmost complex manoeuvre and without bringing a restraining of mathematics into science, it is so complex there is no viable explaining in physics about how the cosmos produce the act of motion of space in time. To get every atom to spin as every atom follow the lead of the atom in front and give direction to follow to the atom just behind while giving coherency to the structure the lot of atoms are holding as an individual unit times the units there are going around in the entire Universe is beyond what the human mind can absorb. While the atom in front is vacating space to fill the space of the atom in front is vacating at that instant, the atom behind is filling the space that the atom in front has vacated in order to vacate and relinquish the previous position in favour of the following position to honour the direction gravity is insisting upon. Times that with every atom there is in the Universe and one may grasp the significance of the calculation. The coordinating of moving one atom from one point to a next point requires the skills that the human mind may never conquer. We may see the moving of object through space being as simple as merely excepting it as a given fact, as science has done in the past, or we may reason about the complexity as civil person's should do, and come to realise that the complexity of motion of matter is beyond scope of human understanding.

Removing material from space by filling material into a position of new space sounds simple because the complexity has never been realised. Reading the information will reveal what the factors are in understanding the commitment of material to move through time. This was all a result of understanding the dynamics of Einstein's arguing about gravity and mass. When I compiled the information then discovering my new conclusions forming this information I further realised gravity is motion differentiation between objects. It is the independent motion providing a different speed while sharing a common centre off attracting that allows a discrepancy to establish mass under specific conditions applying between the two in relevancy. While falling the gravity applies as moving of space that is putting time in relation to the distance travelled. That means there is a speed relevancy between particles in motion and synchronised motion would bring about equal orbit around a shared centre.

That is the result of gravity functioning. While the object falls the motion confirm gravity. When motion ends mass sets in and becomes the constraining of the object preventing further motion. The motion is still there but now it is reduced to a tendency to move thus establishing the object mass as the limiting of further motion. Preventing the motion by implementing mass is the resting of objects against each other by resisting the motion to continue, which then is where the mass takes the place of the motion. Where a confronting of objects restricts gravity the action then implements an introducing of the mass as a substituting factor to motion that then replace motion as substitute to the motion that would be and the mass is providing the tendency of gravity being the motion of space. However mass then restricts motion and becomes motion in a tendency to apply motion. While falling gravity applies and motion neutralizes size, mass or weight.

Mass counters motion being when the Earth restrains further motion of the falling object and the moving object is stopped from further movement where mass is then preventing or hindering gravity. This is the result of objects claiming an individual and personal claim to space occupied in a dual or in fighting for their individuality and independence of each other while wanting to be in the **centre of the Universe**. While falling or moving there is no opposition to the body being independent. When the motion seizes the falling object remains individual and still tends to move while Earth individuality resists further movement of the falling body's movement. Further movement is disallowed as other material fill space that falling body wants to lay claim to. The only manner to remain independent by the falling object will be to relinquish to motion in the securing of mass as a substitute to motion where it then finally comes to rest. Mass then sets in not causing the motion but substituting the motion and from that motion restriction becomes resistance that becomes mass. While falling the object is experiencing gravity because the object is in gravity but when on the soil the object experience mass which is the restricting of gravity or motion by other space filled with material. It is a fight of objects to secure and retain the position they have of being in the **centre of the Universe**.

Moreover, I came to another conclusion of equal importance. When any person is standing on any place anywhere, while viewing the Universe, that person is filling the **centre of the Universe**. Let's get more personal. When you, the person that is reading this, are standing at night and is looking at the Universe you are seeing the Universe from the position that one only can have if that person is filling the specific spot in the **centre of the Universe**. All the light, every single beam that ever left any destiny at any time acknowledges this fact. You are the most important person in the Universe because you are holding the most important position in the Universe. All the light that come across and travelled all of the vacant space from any and all possible positions in space runs directly towards your position using a straight line towards you where you are filling the **centre of the Universe**. Not excluding the effort of one photon, all light is heading to meet you where you are in that centre spot and not one photon will pass you by. Not one photon dare miss you because if they do they miss the effort that all light has to accomplish and that is to locate you as the person filling the **centre of the Universe**.

Should you decide to shift your position to any other place in the Universe, you will shift the **centre of the Universe** to that location as well. If you install a camera on Mars, the light is obliged to acknowledge your relocating the **centre of the Universe** at your will to reposition you're being that **centre of the Universe**. All the light that ever left its destination crossing the vast spaces of the Universe, excluding no particular light, travelled all the way just to find you filling the **centre of the Universe**, right where you are. By you're standing anywhere, you fill the **centre of the Universe**, and the entire Universe admits to that because all the light comes to meet you there. If you shift from the North Pole to the South Pole you will shift the **centre of the Universe** because all the light travelling throughout the Universe will find you where you then moved the **centre of the Universe**. The light left its destination billion years ago as it travelled through space at the speed of light anxious to acknowledge you're being in the very **centre of the Universe**. No photon will be able to pass you by where you are in the **centre of the Universe** because all light is heading your way from their starting positions. No wonder every person born has the idea they were born to fill **centre of the Universe**, which we do fill. The Universe is spinning around you or I, which is filling a centre where all motion is connected. That is the Coanda effect on the utter-most grandest scale imaginable; nevertheless it is only a manifestation of the Coanda effect. It implicates gravity as wide as can be… Some things mathematics is able to explain but other explaining goes beyond mathematics. Try to explain mathematically the colour of the sky being blue in a clear sunny day and changing to black when nighttime falls. Do the explaining in mathematics to a blind person that had no vision since birth in such perfect mathematical detail that would allow the person afterwards be able to explain the difference between blue and black to other blind persons by using only mathematics. Some aspects of the Universe go beyond mathematics and some even go beyond words. It is our task to find space, to find time and moreover it is our optimal task to find the Universe. We have to see what is solid, what is liquid and what causes gravity.

Gravity **is to move or apply the intension to move** space a^3 **at the** distance or relevancy of k while T^2 is the time it is going to take to **apply gravity** or move the space filled with material space a^3 at the distance of k in the time period of T^2. That confirms Kepler's attribution to gravity where according to Kepler space a^3 is equal to the movement T^2 (time it takes to move) at the distance k from the centre specific. Every aspect of deliberation about the Universe was never discussed in the manner it is discussed by me, Peet Schutte.

Then I reviewed his vision he received from a vision Einstein received and applied such a vision on the findings Kepler received from the Cosmos. It puts all aspects of gravity in the Universe in new dimensions. But the visions formed the beginning because the visions unleashed many new questions. If gravity is motion, what causes motion? What stops motion? That answer is in the Black Hole. In truth the explaining of the Black Hole is as complicated as the Universe may represent and as simple as the cosmos truly is. If a star is about fusing atoms and with such fusing of atoms is thereby growing, what happen when all the atoms fused into one all collective atom in one already all—atom-accumulated star?

What is the gravity if the star has melted all atoms it had into one all-inclusive atom and this all-inclusive atom is providing all the gravity that the star had when the star still had massive volumetric space? If all that space that once filled an entire giant star fused into one specific space less centre holding singularity 1^0 then the enormous gravity is applying to the centre of such a non existing space-less atom and that entire enormous force has been secured in the space less than that which one atom holds. In that case the atom would then show a force that would pull the surrounding Universe flat. The purpose of fusion is to reduce space and magnify space less ness inside the sphere. Where does the gravity of the star end when all the atoms in the star became one giant atom by fusing all atoms into one nucleus? Gravity is smallest where space is least. Where space of an entire massive star is left in the size of one atom the gravity coming from that will pull the Universe flat at that point.

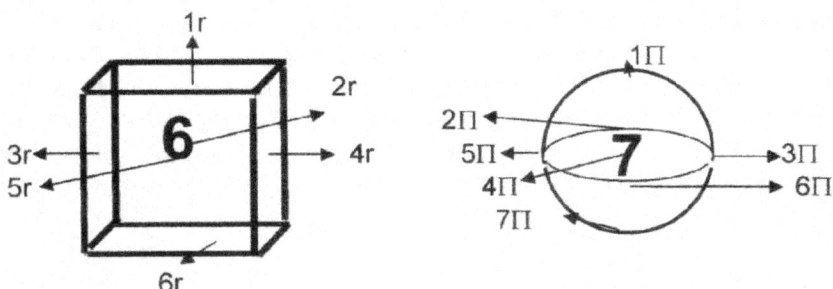

In the sphere there are no radius but only the extending of **k** from the centre **k** in six opposing directions relating to one another by the square but remaining Π because of the unity the matter holds in relating to space. In every sphere there then are the seven Π relating in precise dimensional and positional equality to the centre Π as well as to one another by 90^0 and 180^0 implicating the dimensional positioning. Therefore the sphere holds $_7$ ^{and} the cube holds 6 r^2

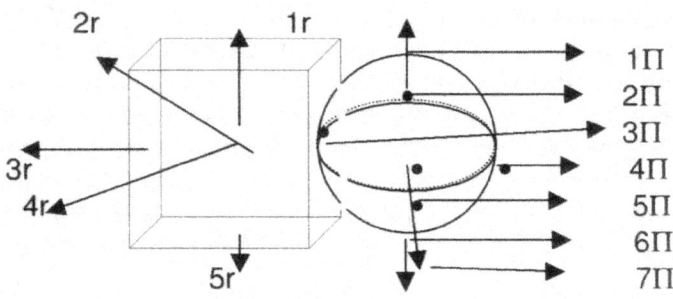

By coming into contact with the sphere the cube loses on dimension to the seven dimensions dominating six bringing about that the cube then has 5 sides to the seven of the cube. That is the Lagrangian system with five cosmic atoms holding relevancy to the centre cosmic atom where the centre cosmic atom stands in for seven and the orbiting cosmic atoms standing in for five positions in space. There is a more explicate explanation about this somewhere else in this book.

Coming to the conclusion about gravity being motion and mass being the restriction of motion was the easy part. The facts that presents the understanding of what produces the motion and what prevents the restriction from overcoming the motion was the part that required thinking. Figuring out why was everything on the move and where did the motion stop, well that was the part that took some figuring and some explaining. What makes gravity move and why does gravity move…the answers are in the four phenomena never yet explained to satisfaction but now turns out to be the cradle of gravity. The answer can only come when the full content of gravity is fully understood as being the unexplained phenomena that produce in conjunction with one another the totality of gravity as we experience it. They are the following:

The Roche limit is:

The region surrounding each star in a binary system, within which any material is gravitationally bound to that particular star. The boundary of the Roche lobes is an equipotential surface, and the lobes touch at the inner Lagrangian point, L_1, through which mass transfer may occur if one of the components expands to fill its lobe. It names after the French mathematician Edouard Albert Roche (1820-83).

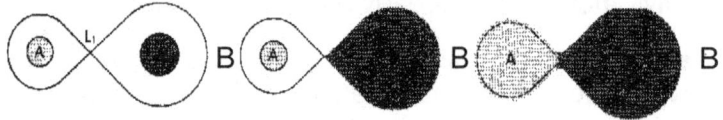

THE ROCHE LOBE: In a binary system, the Roche lobes of components A and B meet at the L_1 Lagrangian point. (a) In a detached system, neither star fills its Roche lobe. (b) In a semidetached system, one massive component, B, fills its Roche lobe. (c) In a contact binary, both components overfill their Roche lobes and share a common envelope.

Gravity is The **Roche limit**,

Gravity is The **Lagrangian system**

Gravity is The **Titius Bode law**

Gravity is The Coanda affect

The TITIUS BODE Principle

5   7   5   = 7 / 10

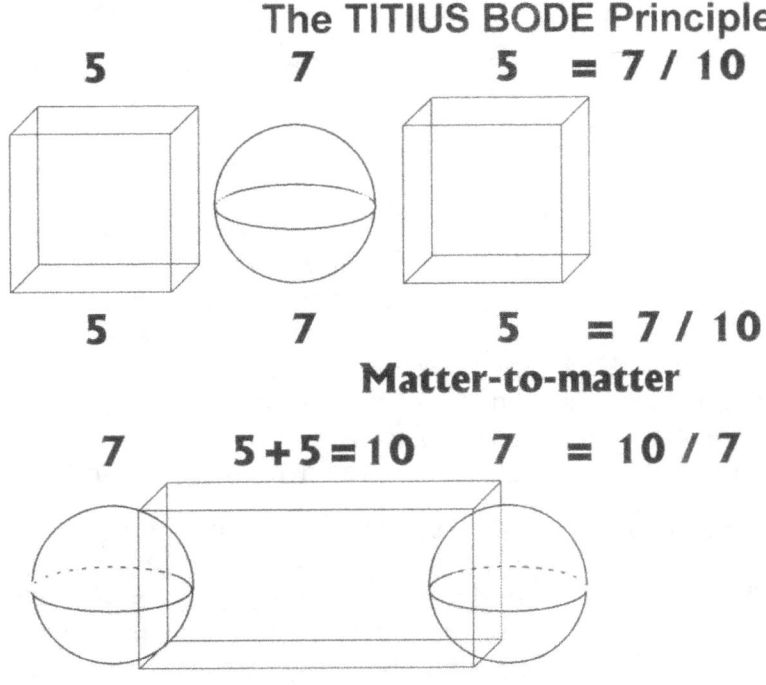

5   7   5   = 7 / 10

Matter-to-matter

7   5 + 5 = 10   7   = 10 / 7

Space-to-matter

In the Roche limit the space factor provides occupied space-time and therefore the value of r is replaced by the value of Π bringing about a square in half of Π.

Gravity is the dimensional changing of heat holding r as reference to the sphere holding Π as the reference. Heat occupying space has the cube that can apply r, as a straight line bringing about the cube with all its other names than may find attachment to specific form but nevertheless still remains only a six-sided cube with angle changing in some cases. The Titius Bode law is an extending dynamic deriving the law from the gravity dimensional factor where the space factor in a square of ten relates to a matter factor in the square by half (half since nothing can be in two places in the universe simultaneously) of the matter factor

The ROCHE LIMIT

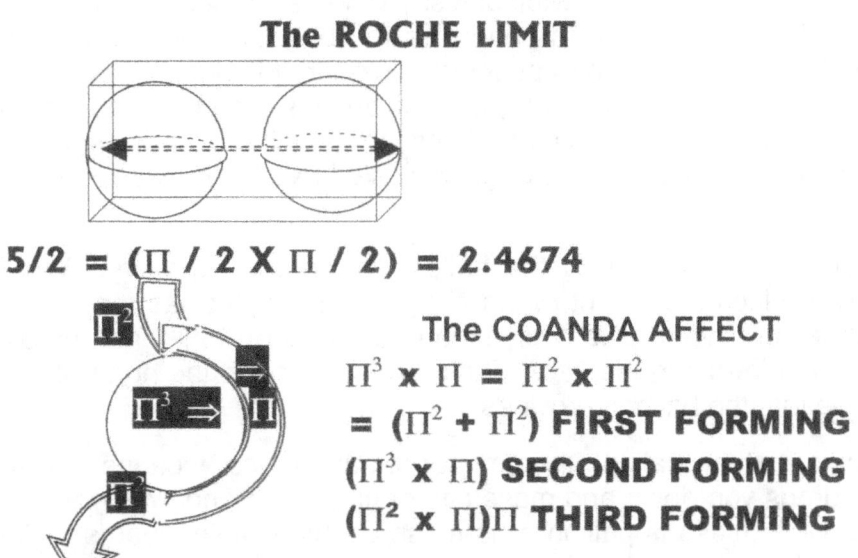

$$5/2 = (\Pi / 2 \times \Pi / 2) = 2.4674$$

The COANDA AFFECT

$$\Pi^3 \times \Pi = \Pi^2 \times \Pi^2$$
$$= (\Pi^2 + \Pi^2) \text{ FIRST FORMING}$$
$$(\Pi^3 \times \Pi) \text{ SECOND FORMING}$$
$$(\Pi^2 \times \Pi)\Pi \text{ THIRD FORMING}$$

Forming the BIG BANG AS THE $(\Pi^2 + \Pi^2)(\Pi^2 \times \Pi)3 = 1836$
ATOMIC RELEVANCY $(\Pi^2 + \Pi^2)(\Pi^2\Pi)3$

Now that you know that you are in the centre of the Universe, and to confirm your status you have in the centre of the Universe, you just have to step outside and see the light flowing to your spot where your gravity is attracting the light from every spot the vastness of space has to offer. By you're filling the most exclusive spot there is in the entire Universe, right in the centre of the known Universe now you'll have to buy the book to know why you are in the centre of the Universe.

It is commonly accepted that Physics demand respect because the general idea going around has the understanding that Physics only work with proven facts that cannot be in dispute or be disproved in any way. Well…I wish to bring to mind some of the facts that physics work with when academics as scientists only work with facts. Remember they are the ones boasting that if facts are not proven then it is fables and those very important academics don't waste time with fables because they only work with facts. Students, it is your liberty to ask them to explain what they say is such correctly proven facts. They maintain it is a fact that we have to have mass in order to produce gravity.

Mass is responsible for gravity. If you don't have mass you're not going to have gravity. Mass is equal to

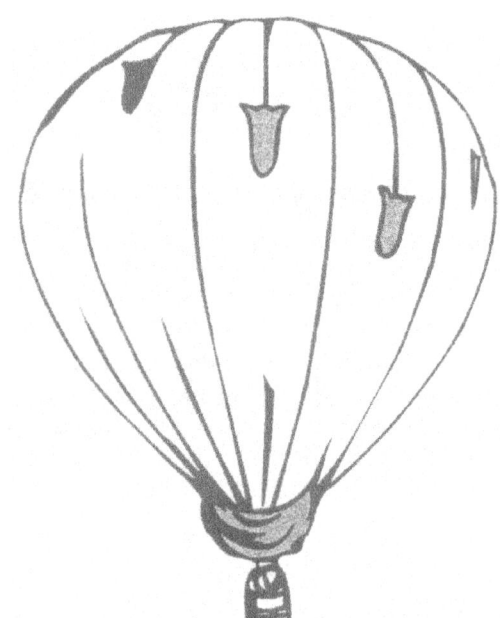

gravity and gravity is only where mass is. If mass is anywhere it should show its presence otherwise mass is absent. If a body falls it is the mass that allows the body to fall because the body receives gravity by ratio of mass and mass that produces gravity in relation to the mass available. It is mass that drags you down because the mass is in charge of the gravity and the gravity finds the value from the mass available. So what happens to balloons? Have they got anti mass or anti gravity? They are moving up when the air is heated. Mass pushes you down by the gravity it forces onto you. If mass drags you down then what are lifting you up in the balloon?

If mass gives the gravity to drag you onto the Earth then why would the hot air lift you up? Is the hot air causing anti gravity or anti mass because gravity and mass drags you down or so Newtonians say. The balloon is lifting the passenger and all that is in the bag plus the bag plus the balloon into the air. So what is then pushing the lot up if it is mass that drags you down. Has the air not got mass because then the air can't have gravity and then the air must escape into the blackness of outer space because by going up it shows a resilience of either mass or gravity. We have seen that it is mass that pulls everything onto the ground.

Why would the air defy mass and allow the balloon to go anti whatever. We find mass being the equivalent of that which brings the object to the ground. The object has mass to produce gravity. Why then would hot air allow the balloon plus everything in the balloon to lift into the air? The balloon lifts in relation to the hot air that blows into the sack. The more hot air and the hotter the air is the more lift and the swifter the lift will be that the balloon provides.

The issue sticking out is that the balloon then must not have mass because with anti gravity it is pulling up. Remember mass drags you down and mass can't pull you up and drag you down at the same time. Then what is pushing while mass is pulling or is mass pushing while what is pulling? The object is not going in the normal direction where it is dragged down by gravity and in all my life I have never heard one Academic mention anything about gravity lifting and that makes the lot very confusing. What is lifting up when the lot should be pushing down and why did everything connected to the balloon lose the mass and if it has mass why is it not dragging down the balloon? If you think this is a little confusing try what is to follow.

They teach you that it is mass that produces gravity and gravity makes you fall because while gravity makes you fall mass drags you down. It is because those mind controllers are lying through their teeth with a menace in which they are the experts, as they know just how to pull cotton wool over you eyes.

Take a truck of 15 tons into an airplane. Put next to the truck a petite little dancer weighing 45 kilograms. Put next to her a frog weighing 150 grams. Then get this lot into the air by airplane and let them jump. Take note that you are told by the wise amongst us that it is mass that produces the gravity that pulls you down. We have just had a lovely debate on how it works and how mass drags you down and wondered if it then is anti mass or anti gravity that lifts you up with the hot air balloon, well take note of this as your airplane reaches 11 thousand meters which is eleven kilometres straight up into the air.

Now we drop the truck and the girl and the frog at the very same time from the airplane. The frog then pretends he drives the truck and the next scene he is dancing with the girl while the truck is falling as fast as a truck can fall. Who do you think is lying? Remember only one group can tell the truth and the other must be lying. Have you thought why one party is lying while the other party has to tell the truth?

The academic Brainy Bunch are telling students all over the world that mass is in charge of gravity and it is mass that's pulling you down. Then the mass is pulling the truck of 15 tons down since the mass produce the gravity and the gravity produces the fall which is three hundred and thirty three times more in a down direction than the mass of 45 kg is pulling the dancer down. The mass providing the gravity that pulls the truck down is doing the pulling down of the truck one million times better than it is pulling down the frog. If the mass is doing the pulling by establishing the gravity the truck must fall 333 times faster than the girl and one million times faster than the frog. It is either that or the three has the same mass because they are falling at the same rate.

Take a truck loaded with cargo totalling of 15 tons into an airplane.

Put next to the truck a petite little dancer weighing 45 kilograms.

Then to keep the dancer on her toes, put a frog of 150 g next to her.

Now we will have the mass of each object "pull" by "gravity" as this lot falls down

If the Brainy Bunch all too wise are correct the frog can fly to America and have a pizza in New York while the truck has a few micro seconds to get down if the girl is going to fall during the normal falling duration of a minute or so. Everyone has seen skydivers jump out of airplanes next to cars and trucks and bags. Every one has seen they all fall at the same rate. The girl can do tap dancing around a jumping frog on top of the truck or below the truck and they can be inside the back of the truck galloping on fresh air inside the truck because the lot is falling at the exact same rate.

The academics wishes to brainwash you by mind control in accepting that it is the mass that the falling takes place and that mass is responsible for the gravity and by mass pulling you down it is gravity that makes you fall. Where is the proof of mass that according to them is that which is producing gravity. They tell you Galileo said all things fall equal and we can see from the TV monitors how all things fall equal. Where is the mass that makes the gravity to let you fall if all things fall equally? They tell you that the truck has a mass of 15 tons and that mass is making the gravity that is having the truck fall while the truck is falling at the same speed and distance than the frog does.

If you take that as proof then they got you. Then they brainwashed you into a zombie. Then if you don't repeat after them and echo every word test after test and exam after exam they will fail your papers and kick you from campus. That is mind control, better than what even the KGB is able to implement. You repeat after them and you live an academic life or you disagree and you go home to play with your toes.

If mass is in the picture then mass must be represented by a factor of more than just one because if mass is not part of the overall picture then mass has a factor of one which proves that mass is not part of the equation since mass can't change the results. With all the objects falling equal mass has no role and if mass has no role then for my money academics in physics can't just go and put everything in as their hearts desire. If it is Galileo that is correct and if all things fall equal then mass has no part in gravity. If mass is the inspiration behind gravity the truck must fall a million times faster than the frog and in fact the frog should almost land in another country because that is how slow it falls.

The fact of the matter is that I don't wish to be near when any of this lot hits the ground because the truck will cause a quarry and the dancer will be a splash of red fluid while the frog might not be that worse for wear if the truck or the dancer doesn't land on the frog. But that is mass. The differentiation of having mass that would differentiate or having equality that would equalise all objects descending and then not having mass and between individual differences in mass by each component that enters the equation when the objects touch the ground. Then every one gets the mass it has.

Only when they touch the ground and land on the soil is mass as a factor awarded. While they fall they all fall equal and there is no distinction between the falling at all. What then is gravity? The gravity is the falling. The gravity is the motion. While the object is in a state of mass it is not moving. The tendency to move and apply gravity is the part that the mass restrains. The mass is preventing the falling from continuing. It is the role of mass to prevent further falling and independent motion to continue. Some of then might even still honestly believe it is mass that produces gravity because they were taught that it is mass that produces gravity and never thought about the matter again afterwards.

They were brainwashed by their tutors as their tutors were brainwashed before them. You don't need the brainwashing because you now can find out what the answer is to gravity. You are the first generation that can receive the light of knowledge about what gravity really is, or you can be the last generation that will live in the lie. You are in a position where you can teach your tutors the truth about gravity if you read what is in the books. The truth is there and the truth is out and the truth will be because the truth is written for all that wishes to read. The academics on the other hand have ignored my work and my being on Earth for the past six years while I was writing them letters about gravity. They ignore me as if I am a rattlesnake because to them I am a rattlesnake. With what I say I will have them tumble down from their pedestals because by accepting my work they suddenly find their position equal to yours as students, and then they will have to learn my work in the same manner as you learn my work because to them everything is as new as it is to you. The Academics of the day have too much to lose to recognise my work and therefore have to protect their interest with all they can muster. For that reason if no other they will rather go on lying to you and cover their corrupt fraud than face up to the truth and admit their work is lost.

The truth will be whether it is recognised by them and they can become the first to admit and repent or they will be the last of the laughing stock that those in the future will refer to as the bunch that couldn't see when things fall equal they cannot have mass and when things do not fall by mass then one can know mass has nothing to do with the falling and the gravity.

It is up to you as students to rattle their cages and make them admit they've been lied to as they are lying to you. Or you can be the last of the fools that couldn't see that when things fall equally they have no mass by which they fall. My book is written and those that read it first will know what gravity is. If you do not accept the role as being zombies that is brainwashed then confront these academics that treat you with disgust and betray your trust. They might tell you the mistake is not that serious and the damage is small but then how will they know how big or small the damage is if they don't even know what damage there is or what the damage is.

Science has stayed so far from the truth that they can't even see the truth any more. If you carry on you will learn about some of it and when you read my books I will entertain you with many more than you ever believed. My books will serve as the light switch that brings the light to you.

I charge your young minds to confront mainstream science about the truth. I wrote to them in the last letter where I informed them that they protect the criminality of their corrupt teachings because when the corruption is removed then nothing remains because they have lived a lie for too long. If you reach the need you may down load it because it is a fair bit of information.

If I come to you with a proposal about something I wish to share with you on condition that you pay me an amount to share with you what I know then I am an academic wishing to teach you. Have you a name for such a person that will force another person to pay him to be brainwashed and be mind controlled because the tutor has absolute control over the life and death of the academic future of the brainwashed being and therefore is willingly forcing this unfortunate creature in accepting what will never amount to the truth? I think they are called Physics professors and rule Universities as draconian authoritarian dictators bent on sadism.

Let's investigate the falling as such and see what happens during the fall. The truck falls at the same pace in which the girl falls, which is the same pace as that which the frog falls. If the truck falls at the same pace as the girl and as the frog there has to be a common denominator in this process and since the common denominator eliminates size form and shape we can eliminate mass. Mass brings distinction and the falling eliminates any form of distinction.

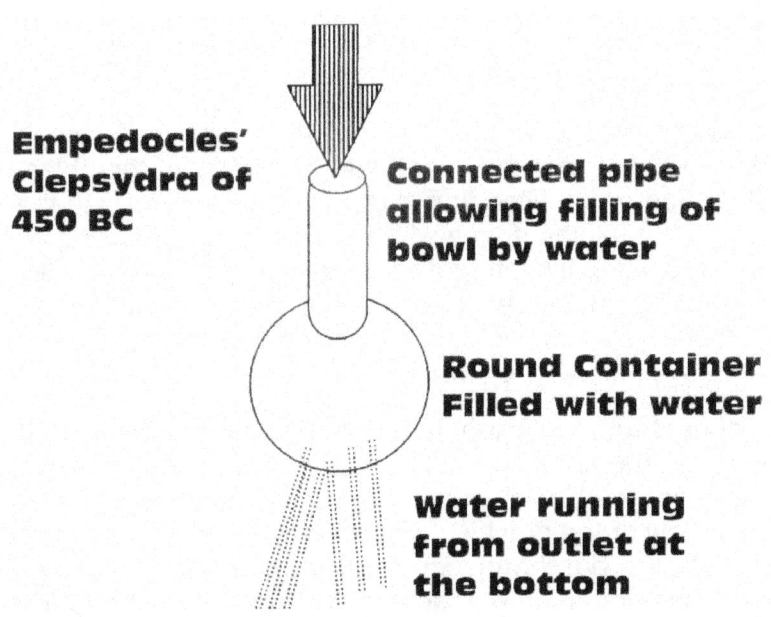

Empedocles' Clepsydra of 450 BC

Connected pipe allowing filling of bowl by water

Round Container Filled with water

Water running from outlet at the bottom

A man by the name of Empedocles had established in 450 BC that when using a tool they called a clepsydra which was a common kitchen utensil at the time water would only run from the water sprouts at the bottom when the inlet pipe at the opening on top was free to let air in. From that science deducted that something fills the clepsydra or the water container because as soon as the pipe was shut by a finger preventing the "something" to fill the container, water would not release from the bottom. From that science concluded two thousand five hundred years ago that "something" and not emptiness flows down into the container and pushes the water out. It is not the mass of the water making the waterfall but the water is pushed out at the bottom by a transparent enveloping and unseen medium that fills the space the water vacates. The process has nothing to do with mass pulling because if it was mass pulling then the water that should be heavier than a human finger must rather try to pull the finger through the pipe as the water mass pulls to the Earth. This was TWO THOUSAND FIVE HUNDRED YEARS AGO AND ALMOST TWO THOUSAND YEARS BEFORE NEWTON, AND STILL

Newtonians have not reached such a high standard of understanding basic principles concerning physics. They would much rather prevent what I write about in order to cover up **scientific misconceptions.**

If it is possible to use the clepsydra to measure time according to measured water flow and the measured water flow works by gravity applying the flow of water at the precise constant pace, then the flow of gravity is also the flow of water and is also the flow of time. When I fall down a waterfall with a boat I travel the same pace, as does the boat. That could be because I am fixed to the boat by sitting in the boat. But my sitting in the boat has certain condition and one is that I can remain sitting because I fall the same pace as the boat is falling.

I fall down with the boat and the boat and me forming a distinctive unit falls at the same pace as the water that forms the waterfall falls. Should I at the time of my falling hold an empty mug in my hand and I wish to fill the mug with water, and then I will have to move the mug against the flow of water streaming down the waterfall. I will have to thrust my mug upwards at a faster pace than my descending is casting the mug down and therefore I accompanying the mug down the waterfall. My mug will not automatically fill with water or if there was water in the mug my mug will not automatically empty with water just because the emptiness filling the mug will be at a different pace than the content that is otherwise the filling of the mug.

The mug being empty falls as fast as the boat and I. The empty space in the mug is falling as fast as the mug will fall when the mug is filled to the brim with what ever can fill a mug to the brim. Notwithstanding the content within the mug or the content within the boat or the content within the water being within the waterfall, the very lot is falling at a similar pace. By lifting the cup while falling the cup will fill with water.

I am not putting the water into the cup but I am exchanging the space that the water holds with space that the empty cup holds and my action in truth has no bearing on the water filling the space, which I then transfer into the cup. I am filling the cup with space that at that point holds water but the holding of water has nothing to do with the transferring of space.

If I leaped from the boat and fell I would fall alongside the boat. The boat will be empty but will fall at the same pace and as the same space as I fall notwithstanding being empty. The mug being empty will fall at the same pace as the boat being empty which will fall at the same pace as the water in the waterfall and I would fall. The space in the boat, which is empty if I do not fill the space, will fall at the same pace as the empty space, which fills the mug, and the mug will fall at the same pace whether the space in the mug contains or doesn't contain whatever can fill a mug. The space filling the mug is falling the same as the water that would fill the space in the mug should the mug be filled with water.

The space in the boat is falling at the same pace as I would fall whether I am filling the vacant space in the boat or otherwise filling the vacant space next to the boat. It is the space that falls and not the object filling the space that are falling. It is the space that is filled or not filled that is dropping down because the space being filled is in decline. If it was not the space that fell the space within the mug would fill first as the mug and the boat fell because the empty space would first fill before it could take anything down. But since the boat falls as fast as whether it is being filled or not we can assume that the space which the boat fills or does not fill is falling as fast as it would fall whether it is holding the boat or I or the boat and I. The space not filled by mass also moves just as fast as space filled by mass.

When the object such as the mug or the boat or I connect with the Earth the Earth disallow the object free motion by taking any more space the object claims through to the centre of the Earth. The object now has to relent the space it claims and take on new space that the object claims to flow by contraction to the centre of the Earth. In forming a blocking it resists the flow or the gravity or space lining up with the centre of the Earth. The flowing of space by contraction is gravity but the object being in the space that flows becomes and obstacle through which the oncoming space must drag in order to flow to the centre of the Earth. It forms resisting of allowing space claimed to release to the normal flow when the object will not relent form in favour of gravity. This resisting such relenting of form and consequently forming a frustrating barrier that blocks the free flow of space towards the centre is time displacement of space and this relenting of space-time flowing freely becomes the mass factor. The density and the resistance that the particles show forms the mass that implicate the degree of the frustrating or preventing or disabling of such free flow of space through time and the displacement of space during time is space-time notwithstanding what ever irrational connection Newtonians wish to add too space-time. Allowing space to displace through time to form time is space-time and that is gravity.

It is not as if I wish to condemn and reject that which is in place without placing something of worth back into the process. All I ask is to read what I bring. Don't be a coward and stop reading as soon as you reach the point where I condemn what is in place! Just move past that to the point where I show what is wrong and how it can be corrected! Just judge me not for condemning what now is so apparently incorrect but for showing why I condemn what now is so apparently incorrect and what I bring to the table and offer as a remedy. See what I have to offer and not only what I am taking away. Don't set your sights on what there is to lose but take a view on what there is to gain! Do not reject me on merits you do not wish to instate because you have the fear you are going to lose what is instated.

Do not judge me by using your double standards that is useless in the face of the truth.

Rather look at the double standards you employ and do not judge me by using your double standards on me. Rather use your mind to detect what is double about your standards and then investigate with me what needs to change. Don't hide the truth. Don't hide from the truth and don't hide behind what you wish to portrait as the truth. Rather come out into the light for the first time in three hundred years and admit to the truth. Follow what I say and see for yourself what there is to gain by trying to detect what is wrong because we all know there is much wrong. The comet does not collide with the Sun and the Moon is not on its way to collide with the Earth in time to come.

Expand science and no the Universe for the Universe is the only aspect that has not the ability to expand.

I challenge all of you Newtonians to prove $F = \frac{M_1 M}{r^2} G$ and not just to declare it proven because it is in use since the Dark ages. Expand your mind and double check the formula you all so vividly underwrite and support. Prove why do you support the formula in a modern and a scientific way. Explore the correctness that this formula $F = \frac{M_1 M}{r^2} G$ underwrites. Be a true exploring scientist and journey with me through the following pages while we venture on the quest to find and vindicate my incorrectness by proving the truth vested in the formula $F = \frac{M_1 M}{r^2} G$ that carries the entire physics everyone uses. Let us start where the lot should start and get two Masters together on one point of argument. Galileo said all things fall equal. That says all things fall alike.

The first thing anyone brings in is the vacuum bit with the feather and the hammer and since we do not live in vacuum there is no chance of finding a feather that will fall as fast as a hammer. Since the feather does not fall as fat as the hammer we immediately jump to the conclusion that there are falling disparities because of the falling discrepancy we find between the hammer falling and the feather falling. Then what would give the feather the time to fall longer than the hammer does. Everyone concludes about mass coming into play and they are correct. But they are half correct while Newton still is completely incorrect by attaching mass to the entire idea of falling. Take away the resisting of the feather and replace it with something far less air resistant and one will come to a different conclusion.

I prove that gravity applies as **gravitational density**, which is a relevancy of movement in space where objects move through space. The Titius Bode is the roof I use to prove my claim that mass only applies when an object contacts and touches the earth. Beyond being in direct contact wit the earth or being in contact and resting on something that is directly in contact with the earth there is no mass applying.

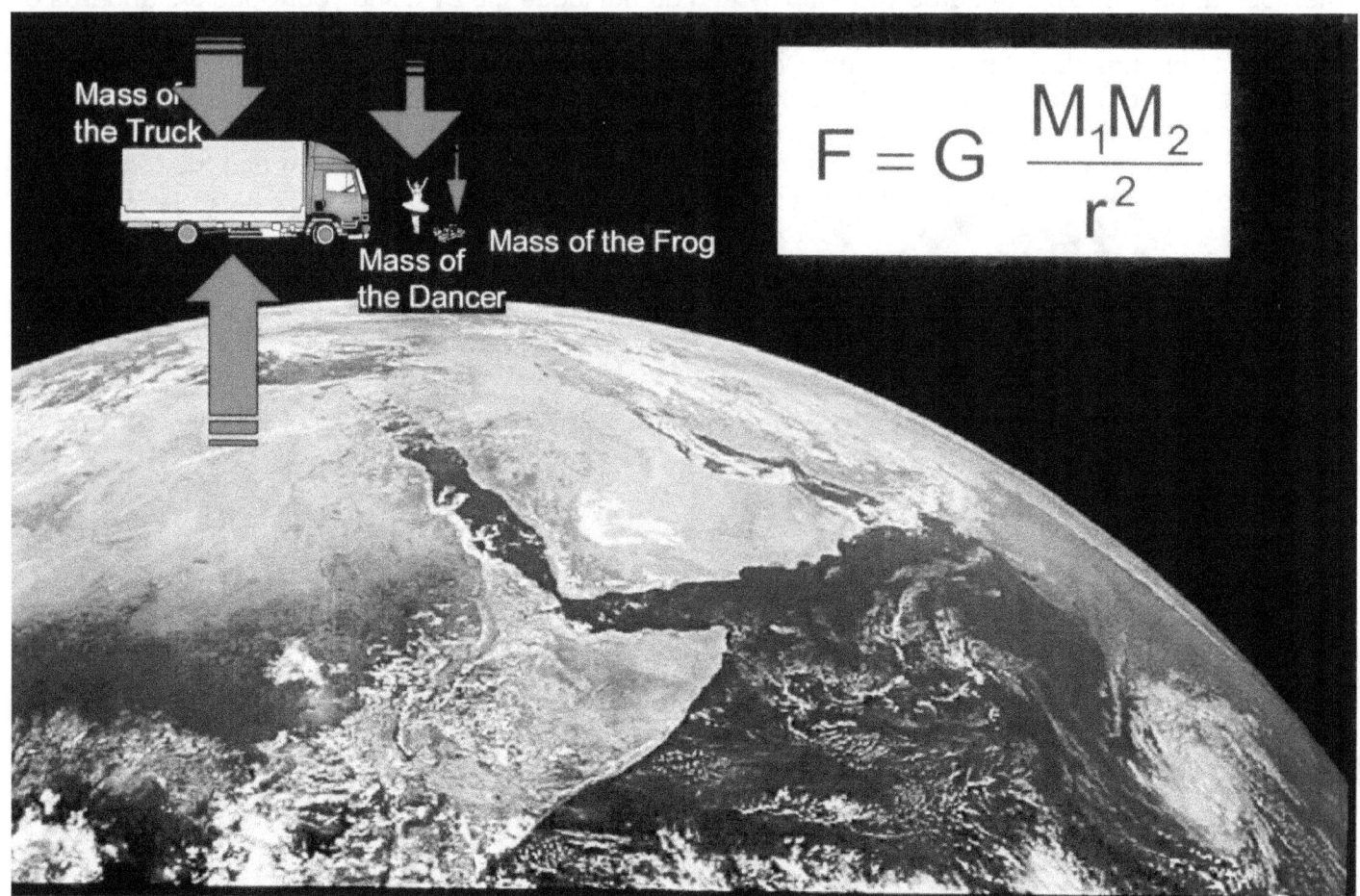

This lot holds obviously very different "mass". Yet, with all the "mass" discrepancy applying this lot fall together at the same pace notwithstanding the "mass" discrepancy and unevenness in size comparison. The only demand to consider is that the conditions by which the lot falls has to be the very same. The heat has to be the same and the air through which the lot travel has to offer the precise same circumstances and conditions. Any variation in uneven ort not matching condition will end in differentiation in the falling tempo. By falling at the same pace although the "mass" varies totally introduces a new concept. "Mass" has nothing to do with the process of gravity but buoyancy plays a non-excluding role in this matter. The buoyancy equalises the "mass" factor by placing everything that falls under the same conditions. If it is the space in which the object are that falls with all the space surrounding the object as far as the space goes, then because it is space and space is the same, then it makes sense. However the "mass" idea is stripped of all common sense and for that reason I am the only person on earth that understands science and do not understand Newton but in my defence I admit that I do understand nature!

Dear Professor,

         I am Petrus Stephanus Jacobus Schutte going by the name of Peet and who is the author of the above-mentioned book(s). I hope you find your reading of this book presented as an open letter a most fruitful experience. I feel I need to warn you, the person reading this letter, that the work contained herein strays widely from mainstream science and for that there is a very good reason. However, in the least, the content will be thought provoking. I researched the work of a man that is most exceptional and therefore should be placed much more prominent in the allocated position his work has in the history of mankind. His contribution in the gathering of information that furthered the entire human species in their accumulation of knowledge as well as the human understanding in cosmic affairs stands second to none in comparison to most others whilst most people are not even aware of the full implication of his work. Whilst recognising the work of Johannes Kepler, Mainstream science bluntly ignores the impact of his work, and in that, they miss the full vastness of the wide influence of his work. Newton shrouded Kepler under a blanket of insignificance and every one since then kept Kepler there. It is therefore almost absolutely realistic to say that what you are about to read in this open letter sent to you for your attention was never yet printed in the near or the far past although the work has been with us for about four hundred years during which time it went unnoticed. It seems to me that any research predating Newton never came into use or into practise. My investigation of Kepler's work brought about a conclusion that no one yet arrived at concerning the findings of Kepler because no one scrutinised Kepler's formula. Kepler found planets rotating around a centre but Newton saw a circle and added what is mathematically required to indicate such a circle. Newton added a mathematical $4\Pi^2$ to the formula of Kepler and removed the distance symbolising measure that Kepler introduced using **k**. On the other side, Newton changed the symbol of **k** by using the mathematical equated symbols $G(m+m_p)$ a formula Einstein could not find any validation for and rejected it. Then to place a cherry on the cake Newton transformed Kepler's work to $a^3 = T^2$ which then was changed to T^2 / a^3 and left out **k** so that the relevancy $T^2 / a^3 = k$ then is not equal to **k** any longer. This is because Newton illuminated **k** as a defining value. This is just an inaccurate manner of not indicating **k** and ignoring the values of the column in Kepler's table. By not defining **k** removes the symbol precisely to the point that eliminates **k** and what **k** stands for does then not exist. Behind **k** I found an entire Universe is hiding the entire process of gravity. **I wish to draw your attention to the matter of Johannes Kepler's findings that Mainstream science considers as resolved and closed for many a century while it is not**. My investigating Kepler helped me too resolve other unresolved matters but it was only possible by using Kepler's work. Therefore only by studying Kepler's work correctly and analysing the work thoroughly can one come to understand the four cosmic pillars as I named the four laws. This means it gives gravity a complete new purpose.

I too am well aware that at first glance you will immediately arrive at the opinion that the theme of the letter has to be considerably below the standard of an intellectual Master such as you must be, due to the position you hold, and because of that, the normal research work you do. Please keep in mind this is a result of studying cosmology from 1977 as time allowed and from 1998 on a permanent basis. The big conclusion I found was that it is not material that moves towards the centre of spinning objects thus material pulling each other but the space in which material turns that is drawn towards the centre of moving objects and that is what Kepler's figures proves. One thing that the Titius Bode law proves is that materials do not move closer to the sun but in fact it moves away from the centre of the sun. The study proves that the space forming outer space and that is identified as the solar system moves towards the sun. The Titius Bode law places material in allocated position according to singularity. Should you find the mathematics too simple in terms of your expectations that is not my fault because that is the level nature applies mathematics? This is how nature uses mathematics and in the simple ness comes the proof.

In my books I explain how four cosmic principles form the solar system and by doing that it forms the Universe as building blocks.

naturescosmicconcept

www.questionablescience.net

ANaturesCosmiConcept NaturesCosmiConcept-E-Z NaturesCosmiConcept-E-Z-R

Titius-Bode-Law-Explained

peet@naturescosmicconcept.co.za mail.naturescosmicconcept.co.za

In order to get an overall in-depth overview of my what my work involves you may visit naturescosmicconcept. To go there click on and naturescosmicconcept **or**

http://www.titius-bode-law-explain.co.za/index.html

Why does the distance from the sun to Mercury double to Venus and that again doubles in distance to the earth and that distance again doubles all the way to Mars and this carries on going throughout the solar system. Tell your physics professor to make sure he or she uses Newton and his idea that the mass every planet has forms gravity and to use only Newtonian gravity principles to explain this.

If you follow the tour and go where I guide you to go you will learn why in the solar system is Mars twice as far from the sun as the earth is and why the earth is twice as far as Venus is away from the sun and Venus is twice as far as Mercury. This principle is called the Titius Bode law and is in place in the solar system instead of the mass factor Newton said is in use by the cosmos.

Let's repeat this again…

The distance that Mercury has from the sun is doubled by that which Venus has from the sun

Then again the distance that Venus has from the sun is doubled by that which the earth has

Then again the distance that the earth has from the sun is doubled by that which Venus has and inexplicably this forms the layout of all planets in the solar system. But why does science never mention this?

Professors in physics never mention the Titius Bode law because the Titius Bode law makes rubbish of Newton's gravitational principles. It proves Newton is a fraud.

This process forming distance between planets carries on throughout the solar system.

There is no room in a room to show this layout in its full compliment where it covers all the nine planets. If mass formed gravity then the layout should be running from the biggest to the smallest.

$$F = G \frac{M_1 M_2}{r^2}$$

The distance should be in terms of size , but it is not, it is according to **the Titius Bode law**, which is some law no one ever hears of because it disproves Newton and his mass concept. It shows Newton has no ground on which to form his concept that is completely wrong! But while the cosmos disproves Newton, science believes Newton in spite of the cosmos using **the Titius Bode law**

This is very typical of science in the way Science prefers to cheat the truth to prove Newton correct.

Planet	Mercury	Venus	Earth	Mars	Ceres	Jupiter	Saturn	Uranus
Bode's Law distance	4	7	10	16	28	52	100	196
Actual distance	3.9	7.2	10	15.2	28	52	95	192

The Titius Bode Law:

A numerical sequence announced by J.E. Bode in 1772, which matches the distances from the Sun of the six planets then known. It is also known as the Titius-Bode law, as it was first pointed out by the German mathematician Johann Daniel Titius (1729-96) in 1766. It is formed from the sequence 0,3,6,12,24,48,96, and 192 by adding 4 to each number. The planets were seen to fit this sequence quite well – as did Uranus, discovered in 1781. However, Neptune and Pluto do not conform to the 'law'. Bode's Law stimulated the search for a planet orbiting between Mars and Jupiter that led to the discovery of the first asteroids. It is often said that the law has no theoretical basis, but it does show how orbital resonance can lead to commensurability. The importance that becomes known is the sequence the Ties – Bode law saw in the number arrangement of 3; 6; 12; 24; 48; 96 etc.

Have you ever thought about where to find the centre of the Universe? Almost my entire lifetime (since I was five) I thought about this question and after decades of studying astrophysics I can prove where the centre of the Universe is. Does science applaud me...hell no because I prove Ptolemy was more correct than what science will ever dare to admit. My finding the facts Kepler introduced put their findings in jeopardy! To find out where you can locate the centre of the Universe then read the "more Information" part and you will be well on your way to locate the centre of the Universe. Look at the cosmic picture that the Universe represent you with and see your role you have within this large concept. Look at every spot of light you can see and see where your position is to where you stand in all of this. Something mighty big is eluding everybody within this picture, something that all the wise in the past missed and the entire mathematical brilliant in the present never came to witness. The largest issue of all issues is to figure out where is the centre of the Universe. Where is the point where everything in the Universe comes together to join in one single spot?

Every light photon come across the sky from all regions there is within the Universe and they all meet at the spot where you are as the viewer of this spectacle. Where you stand in the picture every photon that came from everywhere rushed towards you at the speed of light and not one photon coming from any spot ever missed you at the point where you stand. That places you as the onlooker right in the centre of the Universe. If all the light crosses at the point you fill then all the light puts you in the centre of the Universe. Not one speck of light could dare to miss the point you fill and that makes you standing on Earth the centre of the Universe. Is this wrong to think that way? ...No hell it is the Universe that forces you to think of yourself in terms of filling the centre of the Universe!

Even those not being American has the idea that they are special because they fit bang into the centre of the Universe and all the while it is the Universe telling you and proving to you that you are in the centre of the Universe. Go on and mathematically formulate this truth. Those considering their wits as being superior and above all others go on and mathematically prove how this concept becomes the truth above any criteria they might produce as the truth. These concepts I just mentioned are the realities forming the concepts that rule the Universe and I am able to formulate these concepts mathematically. But as I introduce and formulate the concepts I introduce I show incorrectness in science that all those practising science try to hide because they conceal the anomalies they can't explain ...and so they hide the biggest corruption in science. They tell the Universe what the Universe is but never ask the Universe what the Universe is. In this idea about how you are able to see the entire Universe you will find all the answers to the questions about how physics use time to employ gravity and in forming gravity time produces space as evidence of time that moves on to become the past we so that we are able to see the present coming from the future.

In short the Titus Bode law is the way we find how the solar system forms and this is a fact beyond dispute! There is a sequence in which the planets are spaced and this spacing has no relation whatsoever to mass of any description albeit planets, rocks or distances in formation.

The Titus Bode law named after Johann Elerty Bode (1747- 1826), who in 1772 published the law, formulated by Johann Titius in 1766 as Astronomy an empirical rule relating the distance of planets from the sun, based on the numerical sequence 0, 3, 12, 24,… By adding 4 to the sequence each time after the numbers doubled and afterwards the dividing the resulting number by 10 results in a sequence of 0.4, 0.7, 1, 1.6, 2.8, which is a reasonable representation of the actual distances in astronomical units for most planets but that is if the minor or inner planets are counted as a single entity at 2.8

In a sketch it explains as follows:

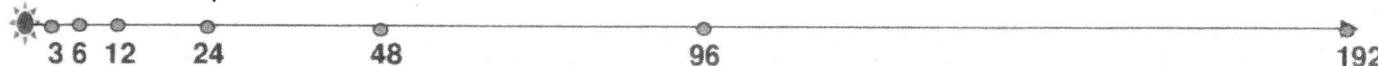

The first planet will adopt a value in ratio of 3. The 3 then adds 4 and the result divides by 10 in order to locate a distance in ratio in order of the Titus Bode law.

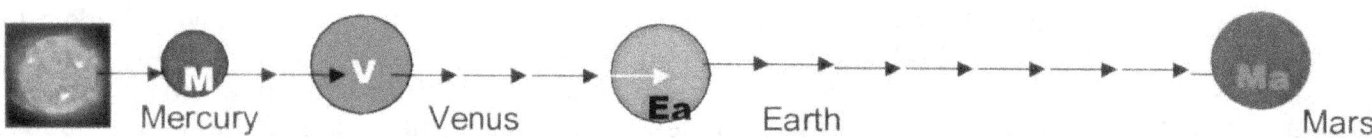

In the first 3 inner planets the ratio does not fit this explanation exactly but the reason why this does not apply has to do with gravitational singularity, which is a major new concept I bring to science. Then to complete this ratio at every planet's worth in distance 4 is added and 10 divide the number. This is how the solar system forms and there is no other way. The reason why it starts with 3 I explain by proving the concept... The reason why 4 gets added I too explain by introducing the correct laws... The reason why 10 divides and the distance doubles every time becomes clear with using very simple mathematics.

Planet	Mass per Earth unit	k^{-1} Movement	a^3 of space volume	T^2 During time units
Mercury	0.06	$T^2 \div a^3 =$ 0.983	$(a^3)=$ 0.059	$(T^2)=$ 0.058
Venus	0.82	$T^2 \div a^3 =$ 0.992	$(a^3)=$ 0.381	$(T^2)=$ 0.378
Earth	1.000	$T^2 \div a^3 =$ 1.000	$(a^3)=$ 1.000	$(T^2)=$ 1.000
Mars	0.11	$T^2 \div a^3 =$ 1.000	$(a^3)=$ 3.54	$(T^2)=$ 3.54
Jupiter	317.89	$T^2 \div a^3 =$ 1.000	$(a^3)=$ 140.6	$(T^2)=$ 140.66
Saturn	95.17	$T^2 \div a^3 =$ 0.999	$(a^3)=$ 868.25	$(T^2)=$ 67.9
Uranus	14.53	$T^2 \div a^3 =$ 1.000	$(a^3)=$ 7067	$(T^2)=$ 7069
Neptune	17.14	$T^2 \div a^3 =$ 0.999	$(a^3)=$ 27189	$(T^2)=$ 27159
Pluto	0.0025	$T^2 \div a^3 =$ 1.004	$(a^3)=$ 61443	$(T^2)=$ 61703

The column showing the **yellow** is **k,** which is exactly what Newton and science ignores because $a^3 = T^2$ To understand the Titus Bode law is to clarify Newton's misunderstanding about Kepler. Kepler's tables prove that there is a movement of space through out the solar system going towards the sun $T^2 / a^3 = k^{-1}$

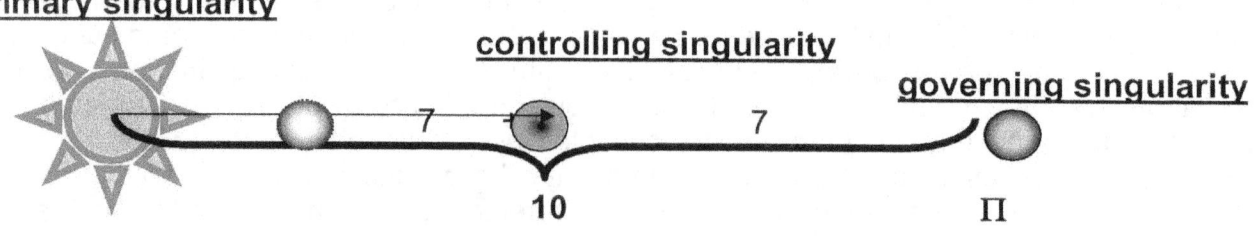

As I explained the relevancies applying the triangle forming gravity places two values in between the immediate inner planet and the planet arranging the outer value of Π at the time.

I am going to show why there is a correlation between the planets and that forming the value of Π positions the planets in relation to position and location.

The correlation forms positional interaction between the Sun (**the primary singularity**) and the very first inner planet (**the controlling singularity**) and the planet forming the allocated position according to singularity (**governing singularity**).

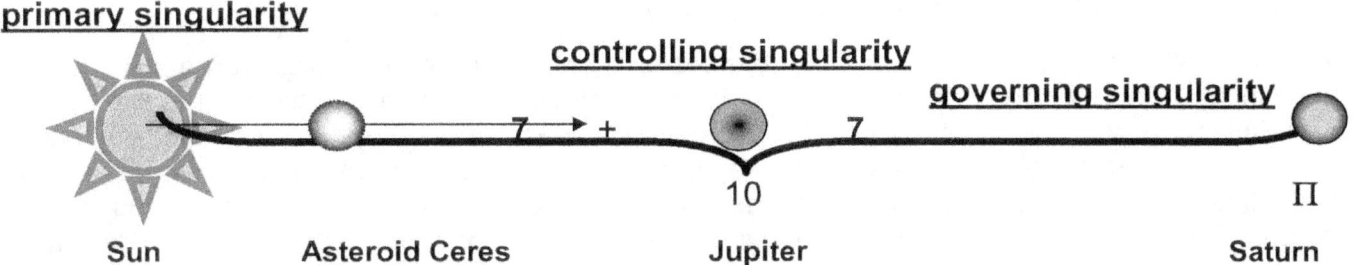

primary singularity
controlling singularity
governing singularity

7 + 7

10   П

Sun   Asteroid Ceres   Jupiter   Saturn

It is the value of forming a double value of seven going square that through the law of Pythagoras forms a value of hundred and the square root of hundred is ten.

The Universe forms space by However I found never to cube (П³)

forming the value of П⁰ПП². work with space in the because space

$$(7)^2 = 49 +1 = 50 \qquad (7)^2 = 49 +1 = 50$$

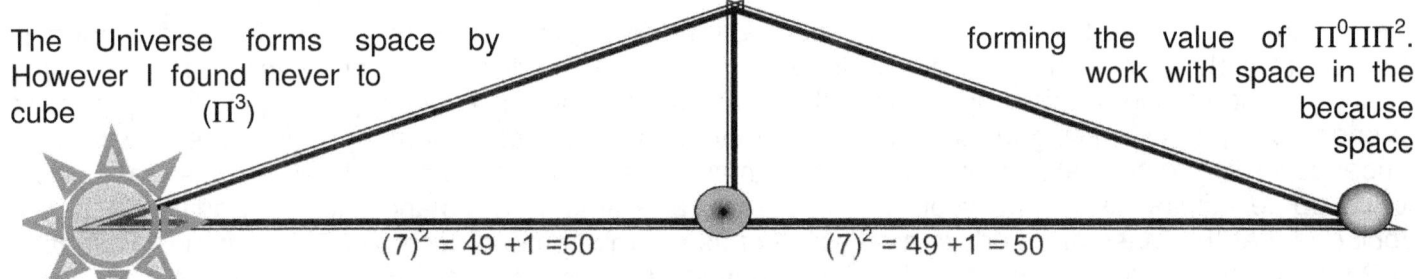

in the cube is just a hallucination created using light and is a hologram forming space, which is also what the Universe is outside singularity.

$$(7)^2 = 49+1 = 50 + (7)^2 = 49+1 = 50$$

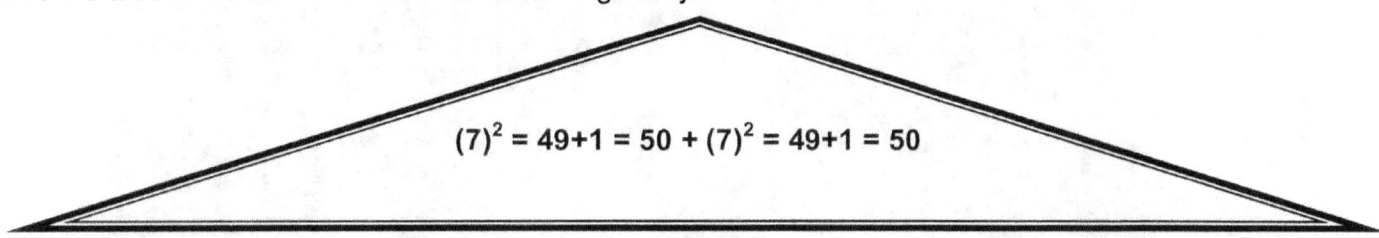

The two point connecting material combine as a sum total of 50 making the square thereof 10
$50 + 50 = 100^{1/2} = 10$

10

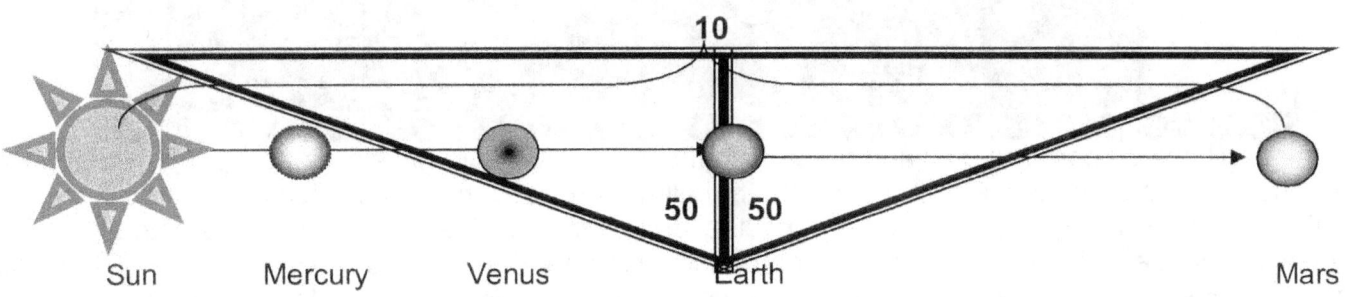

50   50

Sun   Mercury   Venus   Earth   Mars

primary singularity
controlling singularity
governing singularity

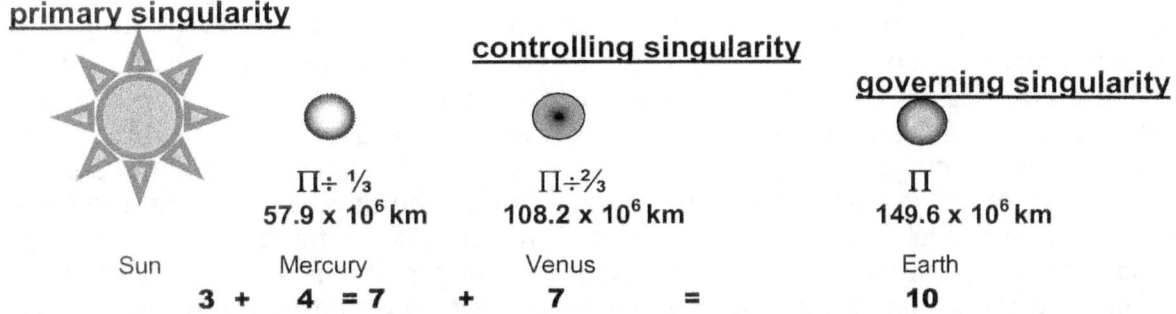

Sun	Mercury	Venus	Earth
	$П \div \frac{1}{3}$	$П \div \frac{2}{3}$	$П$
	57.9×10^6 km	108.2×10^6 km	149.6×10^6 km
3 +	**4 = 7 +**	**7 =**	**10**

The Universe functions in singularity and in singularity we have to locate gravity. I found the Universe is the absolute relevancy of singularity, which controls the entire Universe in all aspects. Therefore the first task is to show where to locate singularity and pinpoint from where does singularity control the Universe.

Singularity is as follows: Singularity is a mathematical point at which certain physical quantities reach infinite values, for example, according to the general relativity, the curvature of space-time becomes infinite in a black hole. In the big bang theory the Universe was born from singularity in which the density and temperature of matter were infinite. From singularity flows space-time.

Space-time is as follows: Space-time is a four dimensional position of the Universe where the position of an object is specified by three coordinates in space and one position in time. According to the theory of special relativity there is no absolute time, which can be measured independently of the observer, so events that are simultaneous as seen from one observer occur at different times when seen from a different place. Time must therefore be measured in a relative manner as are positions in three-dimensional Euclidean space, and this is achieved through the concept of space-time. The trajectory of an object in space-time is called world line. General relativity relates to curvature of space-time to the positions and motions of particles of matter.

SPECIAL THEORY ON RELATIVITY

A theory proposed by A. Einstein in 1905, based on the proposition that the speed of light in a vacuum is constant throughout the Universe, and is independent of the motion of the observer and the emitting body. A consequence of this proposition is that three things happen as an object's velocity approaches the speed of light: its mass goes up, its length shortens in the direction of motion, and time slows down. Hence, according to special relativity, no object can ever reach the speed of light because its mass would then become infinite, its length would become zero, and time would stand still. In addition, Einstein concluded that the mass of a body is a measure of its energy content, according to the famous equation $E = MC^2$, where c is the speed of light. This equation describes the conversion of mass into energy in nuclear reactions within stars. However, there is no functional explanation about the random allocation of planets that form gravitational positions according to size or mass in any argument concerning distribution

The Sun and Nine Planets   Copyright © Calvin J. Hamilton

There is one Universe containing all and preserving the lot. Mainstream physics is accepting this fact. But then by the same margin they accept a principle that allows property that once was part of the Universe to leave the Universe and go somewhere outside the only Universe. Mainstream science places borders within the Universe, which is a concept completely alien to the Universe. They create a loophole whenever it suits them in order to misplace concepts about what they cannot explain readily and logically. In Creation to their and my thinking there can be no hiding or loss of anything but it must remain in the Created Universe. This they admit and confirm although with the same breath those very same intellectuals also admit that there is another place outside of what we are able to find in the Universe. When someone comes up with the marvel where such a person can declare in all honesty that the product of antimatter or singularity escaped from the Universe to God knows where that person should leave the field of science and go for fantasy writing such as fairy tales or reporting about politicians inner deepest chastity and integrity. That is what we can find outside the spectrum of what the Universe can deliver. Since my role model about a New Cosmic Concept falls so far outside the barriers set by mainstream science and is so totally at odds with science I suggest the impossible I wish to suggest we use facts as they are but we have two Universe possibilities, which we call either Newton's solar system and nature's solar system. In Newton's solar system the four cosmic principles does not feature although it is used and refer too when need be and then in nature's solar system mass has no place as we can see looking at the distribution of mass and the random positioning of planets according to mass.

In the very centre of the sphere the form dictates that the shape will relinquish all grounds in space that it can hold and the form will finally be without dimension. Being without dimension means that at a point in the extreme centre of all spheres there is a point that holds singularity because this point with no space has a mathematical position although it is invisible since there is no sides to such a point to give that point any dimensions. When holding the strength of the shape of the sphere in mind as well as taking into account that all cosmos objects of importance is in the form of planets or stars and they are all using the form of a sphere, we therefore may contemplate that it is where gravity originate. We now only have to find the reason why gravity will hold a base in a space less ness as Einstein predicted. It is clear to be seen that gravity is in the centre of the sphere controlling from the centre everything that is outside the space less centre. We can reason with confidence that gravity is the strongest where space is the least. We can further reason that it is gravity that is holding the sphere in true form and since the sphere allows gravity the best working opportunity, gravity can form the sphere in as strong a shape and form as the sphere seems to have. From every point on the surface of the sphere it forms a line running through the centre spot and that is where that point connects with the other side of the surface of the sphere. This point also connects all other possible points forming the sphere as it connect by a line that runs through the space less ness of such a centre of the sphere. Such a line also connects by an angle of 180^0 as well as 90^0 to six other lines running from top to bottom, right to left, and back to front, where all join and cross in the centre of the sphere. There are therefore always no less than six lines crossing and connecting by a centre from any given point on the surface of the sphere. Such points connects in total six surface points on each side of the sphere while they all support one another through the space less centre. In that absolute space less ness in the centre holding singularity we find gravity supporting and controlling all space within the sphere as well as space connected to the sphere. That is where gravity control and guide the space, which falls in the parameters as well as under the influence of the form of the sphere. In the gravity centre space goes singular meaning space becomes space less or flat. That is where Einstein's Universe goes flat because that is where gravity is at its strongest. However my bringing up this statement brings me directly to the point where I get very confrontational about how the brilliant mathematicians treat those they suspect are less inclined to think. This is how to locate singularity.

By dividing the radius r by the half of the value that then reduces r to a point where the left edge of the line reducing will be at the very same place the right hand edge of the line that is reducing will be. At one point the spots that formed the two ends of the line will be at the same spot because the very centre has no space to offer. Any further dividing will land the left hand spot past the right hand spot in the opposing half where it then will grow once again but in the opposing direction. All possible dividing then ends on one spot where such a one spot shares a location with all other possible sides. The centre then physically is in the single dimension applying as one spot to share a location for all sides. At such a point there is no further dividing possible. On several occasions in the past I have been accused of manipulating the argument to produce none-existing or overrate facts. That is not the case. I am not manipulating facts to create an argument as so many accuse me of. What I am talking about is a mathematical fact that any one can prove by calculating following a very simple procedure The formulating value of a circle is $\Pi \times r^2 = CIRCLE$. If you remove r it then is $\Pi \times r^2 / r^2 =$ the CIRCLE Π. The removal is a dimensional concept and not a number in length. At that point reaching Π, all sides share one point $\Pi^0 = 1$, but all sides prevent zero becoming a factor since the sides share one spot. While the different sides are in one place the factor and value is one to all.

Let us find the smallest possible line first. Reducing the line will eventually leave all sides on the same spot. Such a spot must be round in form. The line being the smallest line will start off as a dot. A line so small it has reached a point not dividable any more will have all sides literally on the preside same spot, and I have located singularity in just such a spot. I came to the conclusion that the spot I found had to be singularity purely on the grounds that that spot holds only one side to serve as a start to the starting point of all directions possible. There in that side is only one spot that has only one side applicable and one dimension present. With all the factors given one can only come to one conclusion and that is that there can be only singularity the value of Π^0. In such a case more dividing by two will land further positions on the other side of the divide. That point serves as a position for all points and cannot allow further reducing the radius. It is the smallest line or spot there may ever be. This spot is the result of a most basic process of reduction as the Hubble constant is a most basic process of doubling up during a matter of time. By reducing the line constantly the only value that will eventually remain without dispute from any party arguing about the facts is Π. By only having Π and a radius as one square (the radius effectively becomes one holding any and all sides on one point) of any significant measure as the radius it will be an evenly spaced dot. From the smallest ever possible dot will grow a line in every imaginable direction relating to a prospect of Π not favouring one direction but a growth that puts all directions at equilibrium meaning that any form of what ever might develop from such a spot will have the end and the start being in the same position, which will also have to be a sphere as the flow outward will be equal in all directions. Please think clearly, is that not precisely the commitment we find in gravity, where gravity is flowing from singularity outwards but never favouring any side? This reasoning prompted me to look for singularity in such a spot because if the prime spot from which all came was a spot holding all, then the spot must hold the shortest line but more prominent it will hold the smallest form including the smallest circle or for that matter the smallest sphere. With gravity always being in the centre of a sphere where the space is least available in the entire structure (there is not even space left to fill) one finds a flow of gravity from that centre spot have must grow outwards in all possible direction even-handedly. The fact that the original gravity will begin as a circle or will be a circle is the direction it will take when being the first spot created. All progress will be evenly in all direction because no direction will stand out or be in favour above any other direction at first.

The spot forms a full circle, but the line running through the circle is forever present because that is the future radius of the circle that will one day develop the circle, which is equal to the present diameter. The fact of the presence of such a possible line in such a possible circle dividing the possible circle into two parts makes the centre line equal to the half circle. The line forms the half circle but not only that the line presents the half circle as much as the line is the half circle. The line then is 180^0 and the half circle is 180^0 because in singularity the two factors are the same. The same value is of course $\Pi^0 = 1$.

In this half circle of the future, which is no half circle as yet because of a lack of space there are three future points indicating the space less ness that will go on to become space filled with something. On top of such a circle to form must be a marker indicating an awaiting boundary or future border and at the bottom of the future circle there also must be a similar marker that is no marker as yet. Between the two possible points that are not there yet is a future line running but that is not there yet. Then indicating the possibility of a position to come that will bring about the half circle being a future distance apart from the future line indicating a diameter that will one day be there, a third such a marker must be established for the future. That forms a triangle with two more sides being connected by either a line being one or half pi being one. From singularity comes about that the line is the same as the half circle that is the same as the triangle and all has one value being 180^0. From this come the most basic principles in as much as forming the ground rules of the law of Pythagoras.

When drawing a line, such a line then starts of with a dot serving the spot that holds all sides equal. That means the line serving as the future radius will be equal to the half circle which is then Π. The only aspect of the point that stands in for the end of the single line forming the radius of the circle is that we then mathematically reach the single dimension. We decreased the line to where a circle being Π formed on the single dimension. This dimension also holds the circle dividing line because from there the radius must once again generate a value and by such a gesture that the extending would form the circle that forms the sphere that eventually lead to the formation of particles. This leaves a problem to investigate.

Boys playing games by spinning their tops will never realize the scientific breakthrough their game presents in what cosmic laws it explains and grown ups do not play with toys. In the spinning top we have this little toy played everywhere everyday by almost every boy but whilst it is a game the game holds the answer about gravity and the cosmos that the most brilliant of human Brainpower is seeking. The spinning top presents the answers about all the cosmic riddles no one seems to understand.

Newton said the rotation delivers no work and therefore the effort of the rotation results in a zero. Firstly it bring us back to the zero idea where with all the reasoning in the world and all the leniency I allow I cannot find zero as a value being part of mathematics. Let's move back to the circle to try and find the zero Newton saw in the rotation.

You cannot then say $r^2/r^2 = 0$ and therefore $\Pi \times 0 = 0$. That is nonsense. $\Pi r^2/r^2$ will always be $\Pi \times 1$, and that is the eternal circle. When looking at any rotating object, there has to be a point of no rotation and no rotation means "no rotation", not no existence. No rotation means a factor of 1, not zero. Therefore the notion that $\frac{d}{dt}\left(\frac{1}{2}r^2\theta\right) = 0$ is completely incorrect. The smallest any circle can be is $Circle = \Pi^0 = \left(\dfrac{\Pi^3}{\Pi\Pi^2}\right)$.

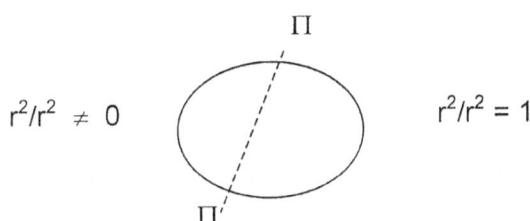

$r^2/r^2 \neq 0$ $r^2/r^2 = 1$

Πr^2 = circle in size but when removing the radius (r^2), only form remains and form is Π, the form of the cosmos. Removing Π by Π/Π what remains is $\Pi^0 = 1^0$.

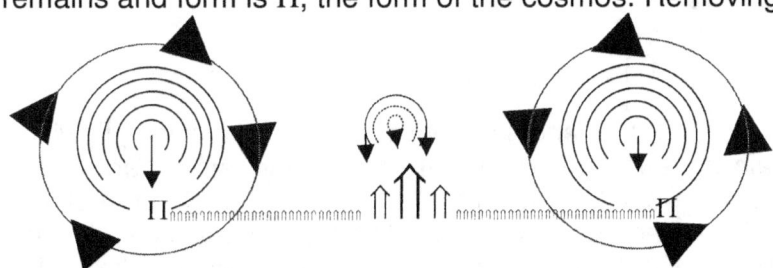

Not only does atomic individual singularity maintain self preservation, but in doing that it also sustain a governing singularity holding structural composition and form within a cluster of matter for example a star. As there is between stars so there are in the same manner a mutual or bonding singularity between atoms in stars, which we see as fusion.

Any object in rotation will have a middle point, a very specific centre point that does not spin. That point once again hypothetical but none the less must be standing still because every line running from that pint in opposing directions are also in opposing directional spin to each other

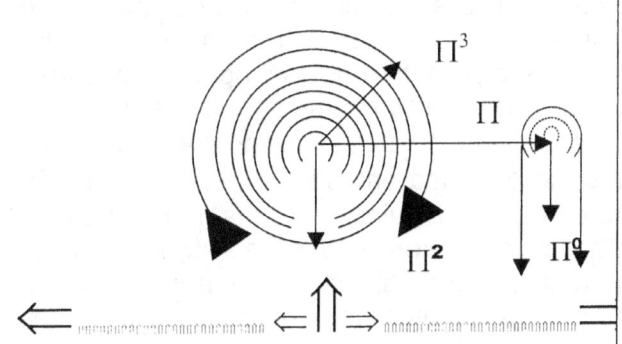

As the stop starts to spin the motion establish the centre line, which activates singularity, which activates space-time that activates gravity at a specific relevancy. Where we locate singularity there is not nothing because gravity cannot come from nothing because only nothing comes from nothing.

After all it is gravity that keeps the top as it is spinning in an upright position while it is spinning because it is gravity that stabilises the cosmos. Moreover, what is actually in progress from the top spinning is the Coanda principle activating gravity and that happens in accordance with Kepler's formula

This means that in the cube at the point of contact between the cube and the sphere the cube experience such a contact point as if the "bottom falls out" of the cube and without a "bottom" to support objects they fall to the sphere as objects does fall to the earth. Remember that a body "floats" in space, but at one specific point it starts to "fall" to the earth. That is gravity and it is a dimension change much more than any force. I shall explain this last remark later on. That too is the Lagrangian system with five cosmic structures holding relevancy to the centre structure where the centre structure stands in for seven positions diverting from singularity and the orbiting structures standing in for five positions in space.

In the centre runs an axis line that forms the division of rotation. No one human will ever be able to indicate the precise line, but such a line must exist because of our logic telling us about such a line. In the centre one will always find one more line smaller than the outside but forever also always bigger as it is towards the inside.

From such a point every other point will be opposing any other point not pointing in the direction to which the first point is pointing, whereby it extends the direction it holds. No matter what the point is or where the point leads, such a point holding a specific direction will be unique in the direction it is rotating because at that or any other specific point wherever, it will be directing not in the direction it spins but in the direction flowing from the centre point outwards.

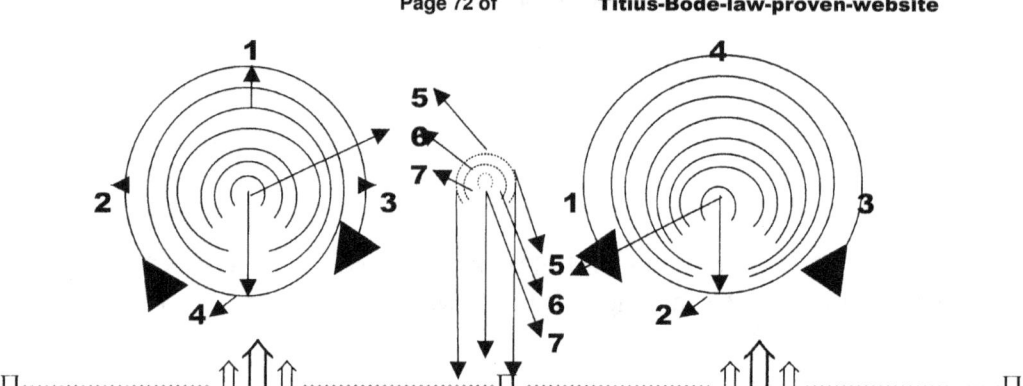

Any point will be it opposing itself within the rotating of 180° changing every aspect of its previous flowing characteristics it previously had or will once again have in 180° from there and in this area there is no space separating this divide. While in rotation from the point of an outside observer all may seem static and never changing but to the object in spin every next second will be a diverting from every aspect it was in very second passing, and the direction it held in relation to the direction it held the previous mille, mille second will totally be incompatible with the direction it holds the very next mille, mille second of rotation. That proves no point can be static or constant, all though it may seem that way to outsiders.

In the very centre of the sphere the form of the sphere dictates that the shape will relinquish space as the line run from the outside towards the very centre. With this natural state of affairs the sphere are naturally inclined to dismiss all space that it can form in the form as the sphere holds space inside and the form will finally be without dimension. All that I attribute to the line shrinking by reducing actually takes pace in every sphere as the diameter reduces to the centre. In the centre where the radius line goes single the form relinquish the three dimensional form it has inside. Being without dimension in the very centre means that at a point in the extreme centre of all spheres there are a point that holds singularity because this point with no space has a mathematical position although it is invisible since there is no sides to such a point to give that point any dimensions. The shape of the sphere is calculated by using the formula 4Π $(r^3) / 3$. By reducing r to a point where r is r^0 singularity steps in because only the form remains as Π. Going even further we find that there then comes a point where Π goes singular Π^0. At that point absolute singularity is present but so is absolute gravity present at that point. When holding the strength of the shape of the sphere in mind as well as taking into account that all cosmos objects of importance is in the form of planets or stars and they are all in the form of a sphere, we therefore may contemplate that it is where gravity originate. We now only have to find the reason why gravity will hold a base in a space less ness as Einstein predicted. It is clear to be seen that gravity is in the centre of the sphere controlling from the centre everything that is outside the space less centre. We can reason with confidence that gravity is the strongest where space is the least. We can further reason that it is gravity that is holding the sphere in true form and since the sphere allow gravity the best working opportunity, gravity can form the sphere in as strong a shape and form as the sphere seems to have. From every point on the surface of the sphere is where that point connects with the other side of the surface of the sphere by a line that runs through the space less ness of such a centre of the sphere. Such a line also connect by an angle of 180° as well as 90° to six other lines running from top to bottom, right to left, and back to front, where all join and cross in the centre of the sphere. There are therefore six lines crossing and connecting by a centre from any given point on the surface of the sphere. Such points connects in total six surface points on each side of the sphere while they all support one another through the space less centre. In that absolute space less ness in the centre holding singularity we find gravity supporting and controlling all space within the sphere as well as space connected to the sphere. That is where gravity control and guide the space, which falls in the parameters as well as under the influence of the form of the sphere. In the gravity centre space goes singular meaning space becomes space less or flat.

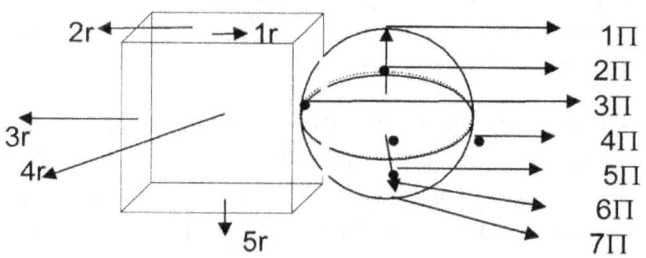

The Coanda effect that joins time by committing material to gravity

Time in eternity forming the motion or liquid that locks space in as material

In the center where we see an axis we find singularity as Π^0. That is r^0 or 1^0 or it cold be $24589421^0 = 1$ and this is all 1, which is the point where singularity is. Where singularity forms space it forms a circle because at that point where singularity Π^0 becomes space it forms a circle Π and where space starts we have Π^0 becoming Π only because movement Π^2 develops $\Pi^0\Pi$ Time in infinity forming the motionless or solid that locks space out as material

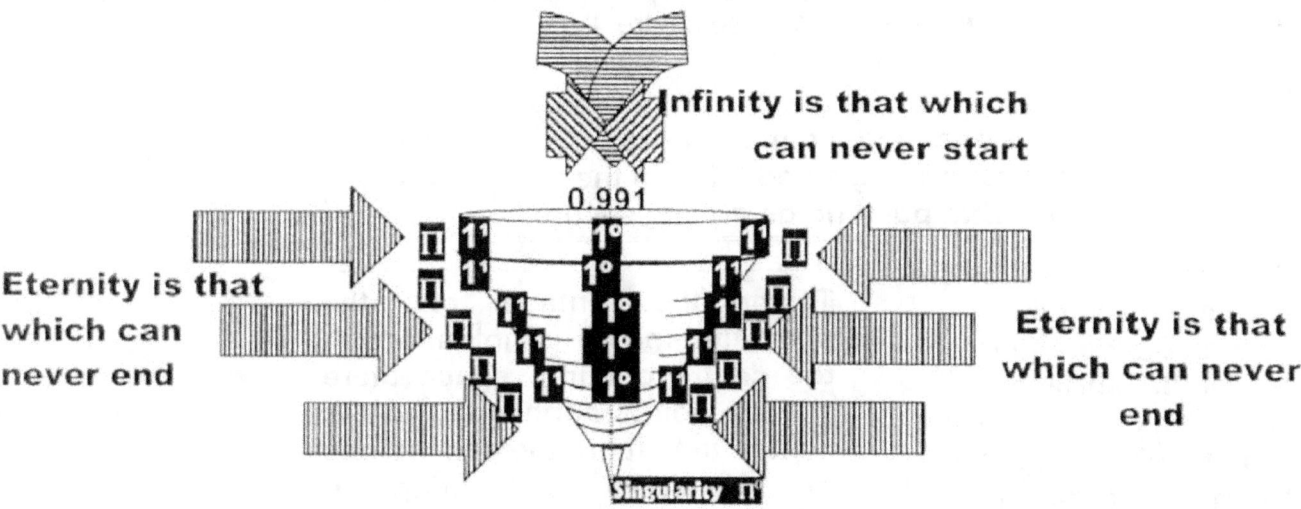

Also it is true that the entire form that is the sphere is controlled from a centre within the sphere. That centre holds the sphere in form and shape. Therefore the strong form is dictated from that space fewer centres where there is no space and no form left. The natural inclining is in the form of the sphere. It is part of the roundness that the overall shape of the sphere represents and this structural strength is carrying down to the very centre. Because the circle is forever reducing that reducing which is inherently part of the form of the sphere becomes a tool in distorting of space in the sphere and is eventually removing all forms of space from within the centre of the sphere. The very centre ends up as having no space because of the reducing that continuous down to become the space less inner centre. The all roundness is the ingredient that forms the backbone of the absolute strength that the sphere has and that is the component that the sphere is so famous for. The form the sphere has allows the sphere to have a control that is coming from the centre deep inside the sphere where the space vanishes and being without space seems to keep the entire structure rigged. From the centre the sphere shape shows strength that the shape as tough as it is. How does it work in its most basic analyses?

It is from the layout that the sphere uses as natural form that we are able to locate singularity. In the case of the sphere the material naturally reduces by measure of the radius becoming smaller to a point where the radius is r^0. At that point the line that will form the radius has gone single dimensional r^0 and that is equal to 1^0, which is singularity.

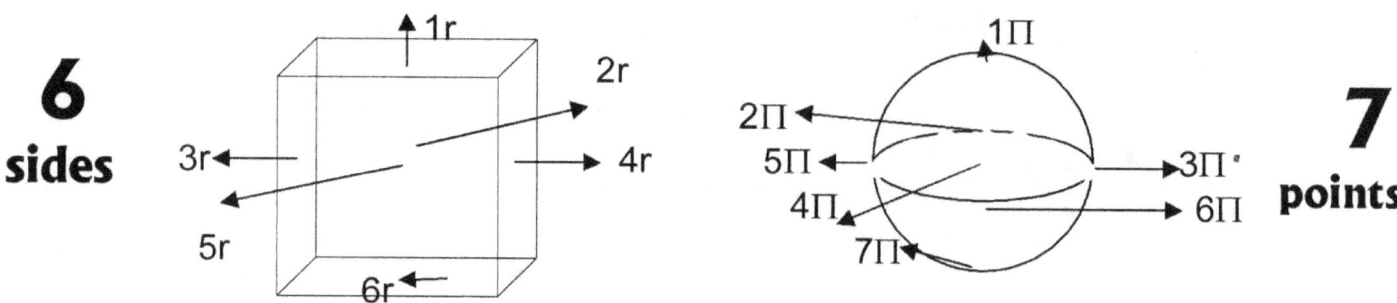

The Universe forms material within the boundaries of gas by turning gas into cosmic liquids.

One such a relevancy is the sphere.

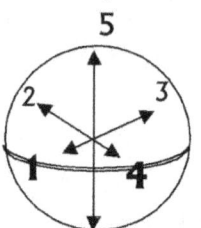

The sphere has six edges relating to one another at all times

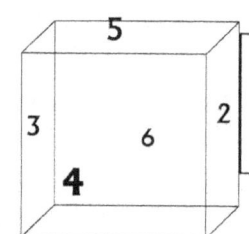

The cube has six sides in three pairs relating to one another at all times

Then connecting the six sides is a centre form where the control comes about that places these edges at specific related points and the points in return puts the centre at the precise centre

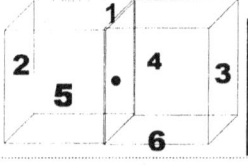

There is one more point in the sphere in the centre forming an addition in the sphere. That point holds gravity secure.

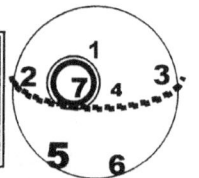

The cube has sides and the sides form a rather weak and flat surface that connects four corners. The flat surface produces a rather indifferent contact point with no special features on the surface. The corners connect to other sets of corners and those corners form a weak structure without any direct support coming from the other five sides. Without material to fill the body of the cube the cube has no direct connecting between any of the sides other than corners connecting at the edges of the sides.

Taking the vantage from the point the sphere is holding from the centre out into space there are ten points connecting to the centre. In that are the dimensions of singularity connecting to space where five connects to space in the second dimension of singularity, and five connects in the third dimension of singularity. On the other hand, the cube does show a very different characteristic, which involves only six sides (at least) connected.

In singularity the triangle is equal to the straight line and the half circle and this only changes when dealing with the cosmos in space.

Since the sphere is so absolutely dominant in form, it converts the liquid of space into material through gravity. Let's look at where to locate all singularity forming gravity within the Universe.

Locating
and
finding
Singularity
1^0

In the precise middle **of all** objects in rotation **is a precise centre dividing the object in sectors that will** start the spinning initiation **from that centre point. Thus, the spinning object** will have a middle point, **a very specific** centre point that does not spin **and only holds** Π **as a specific value. One value such a line** cannot have is zero **because** zero does not start any **line and therefore the** value of the line must be infinite, **just as described in** accordance **and by the** definition of singularity. This line connects al points to all other points. **hypothetical, is also as much a reality none the less and is placed** standing still **because every line** running from that point **in** opposing opposing directional spin the other or opposing side.

That point **albeit where that point** must be directions **are also** in

In considering the spinning detailed instant every every instant of change in had the same before, they oppose the just after the very second in similar points also in rotation. opposing dimensions and

motion in the fraction of time in the aspect of rotation will turn in time. Although the points characteristics only seconds characteristics it had just before which they are and to which they The fact of the graph proves my point in values by opposing the direction it was and

$Π^2$ $Π.$ $Π^2$

and relate by quarterly will be.

From this centre line that is only value always form that becomes theoretical definable, but is still there all the same, a opposing real and distinct when rotating, but even more distinct when not rotating because then the line grows so much it covers all the matter, to a securing spin value of zero, the most original value it had. When not rotating, it is as thick as the material will go. When rotation begins, the line shrinks back to a hypothetical position claiming zero spin that is not less distinct but more distinct because from that point every rotating piece of what ever is then spinning will clearly carry the singularity value of Π implicating rotation which then forms the value of $Π^2$. Everything forming the entirety of the Universe is on the move, which moves in circles and always encircling something of greater importance and prominence. A top can spin but the parameters of its spin are limiting the motion it can apply. By not spinning the top is still spinning as the Earth is doing the spinning on its behalf.

It is the fact that the same affect comes about when spinning too slow that triggers the questions. If the spinning top is all the evidence any one needs to come to such a conclusion what will bring any proof that the singularity governing the top connects too anything anyway. Placing singularity is fair and fine, but what will the evidence be in proving its activeness as part of the creation at large?

The reason why we can be sure it is active is that when spinning it shows borders implicating restraining of further movements outside the set limits. By going faster (past the upward border) the spin goes oblong where it actively tries to change the position the top holds to the earth in relation to the surface of the earth. By going too slow characteristics. attempt to rise confined an effort to shows trying

earth, surmise correctly course the bottoming gauge that to be the normal earth's singularity and we should be just as recognisable normal.

it once again shows identical When going too fast it indicates an into the air, therefore relieve its position according to singularity in part with the earth's singularity. It unmistakable characteristics of to become airborne, securing an independent position from the which holds it down. At the bottom we that it wishes to topple over and fall down. Of out shows the same characteristics whereby we process of falling down. If the bottoming is relative to the recognise the process as normal, then the top of the limits

When looking at the cosmos from whichever angle, it indicates the fact that the cosmos is moving. It is forever spinning and it is all going towards as much where it is coming from another position.

When spinning too fast the top fights something because the alignment keeping it upright starts to tarnish. The same apply when spinning too slowly but that makes sense. The spinning motion controls the axis and determines the critical stance that the axes provide the top's spin. This is very important to recognise. In determining this behaviour as part of a cosmic process where matter interact with matter in an laid down set of rules, we should once more be asking questions and this time it is whether the top will show the same behaviour in outer space as it does on earth. With the reply of "no it would not" this comes as an admitting that the process involves the interacting of singularity of the earth with the singularity of the top where the spinning created independent singularity, as valid as that of the earth because the earth has a role in sustaining it or destroying it at the border ends. Even giving a satellite a rotation secures the stability and by the rotation this provides the satellite or bullet in the case of a rifle much accuracy.

Using the concept that gravity applies Π as the circle factor Π as well as Π^2 replacing r^2 the replacing by Π brings two values as Π and Π^2. That I found is the case with gravity and will be apparent when explaining the sound barrier as well as the Four Cosmic Pillars. In order to create a distinction I remained using r as the indicator of the cube or non-circle that has vacant space and by vacant space I refer to non-solid structures. In the solid structure I use Π as a value for reasons that will become apparent in due time.

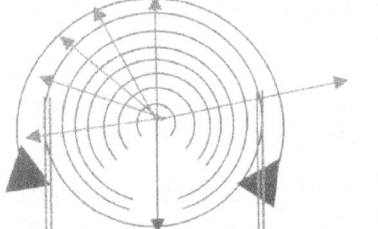

$$\Pi = r$$

Pinpoint positioning of singularity Π^0 with Π positioning space to either side forming the border set by singularity in constant directional change as time flows through rotation.
The new direction pointing to a new location in relation to the previous point will oppose the previous point it had in relation to direction considering the centre point.

Π Π^0 Π

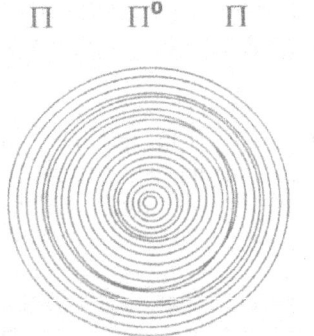

In the sketch the circle to the right would come about from a straight line r growing influencing the appreciation of Π, but to influence Π would lead to a breakdown in r as Π and r are different entities. The circles to the left shows a continuous growth by extending Π every time and since Π is the same part as the previous Π, only extending that billionth of a millimetre each time, the circle will be truly continuous without any signs of a break, making the circle eternal.

Moving in a direction is a straight line that is 180⁰

intermingles without distinction.

Looking at the affect of gravity it shows the precise quality of no distinctive point, as gravity never seems to end at a point but flows all over affecting all that holds a position in its sphere of influence. The gravity coming from China meets the gravity coming from America at no particular spot but

        The value of singularity stems directly from the law of Pythagoras or **Pythagoras** is the result of **the average of singularity. With the shortest line being a dot, all lines must start from a position implicating Π.** A circle is a square without corners implementing Π and a half circle is therefore a triangle without corners. The corners are the factor that confused every one in the past. When replacing the value we normally attach to circle being r with Π, the law of Pythagoras becomes quite meaningful and mathematical.

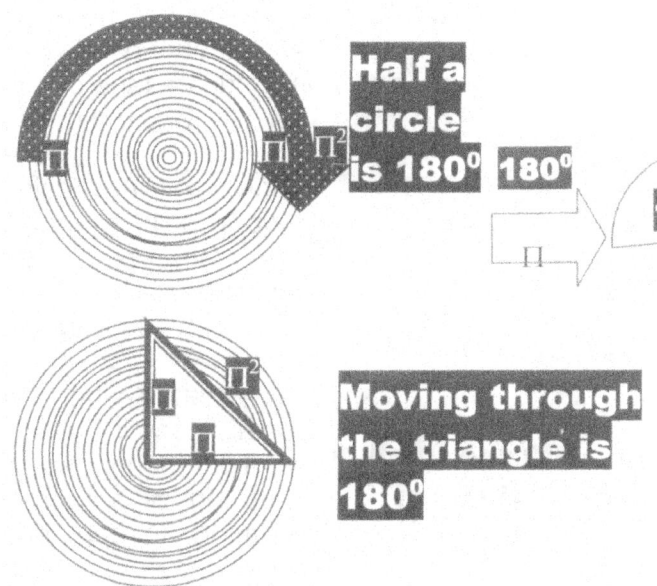

The triangle, the half circle and the straight –line has two things in common, they share 180^0 as a mutual value and they are part of singularity.

By placing a connecting circle on the sides of the triangle half a circle forms. By implicating Π as a relevancy and not the straight-line r, two values of Π applies to each circle, and the straight line is no longer r, but is Π^2. This will bring about that each circle holds half the square value implicated to the allocated conditions applying to Π in that specific instance. By adding the two half squares forming the two half circles and then calculating the square root of the total that then forms the average diameter, an average of Π in the connecting line will come about. As both lines are the straight line forming singularity coming from one line being Π, the connecting line then must be the average of the two lines as Π^2. That is what **the law of Pythagoras says.**

By examining the form of the sphere, we find that there are 6 points on the surface of the sphere that is holding the form at a specific and equal distance from the centre. Lines run from the centre into space at 90° and 180° angles of each other from six opposing sides. There then are six lines at 90° and 180° connecting to the centre from six points on the outside edge of the sphere. As a result of the basic shape that a sphere has, there is a spot in the extreme inner centre of the sphere where the lines in 90° relevance cross each other and others connect by 180°. There is also at that point a spot where all space relinquishes a position and only singularity 1^0 as form remains. At such a point we find the measure of the sphere being Πr^0 with $r^0 = 1^0$. That is where the line that represents the radius as a line disappears, as it becomes singularity r^0. After more reducing continue we get to such a point where we find only Π^0 left. At that extreme point is where space in all form disappears, as the circle providing the sphere the form the sphere has, removes all possible form by going into singularity $\Pi^0 = 1^0$. However, since the Universe is space that formed and this point holding $\Pi^0 = 1^0$ has no space, this point $\Pi^0 = 1^0$ then falls outside the limits of the Universe we know and connects to the Universe outside our Universe.

Then in that area all form of any possible space disappears leaving only the dimensions of singularity 1^0. I cannot delve deeper into the argument since this then connects what is not the Universe with what is the Universe. However, from such a point there runs lines that connect to space on the outside where six points on the outside points connect to the space less point in the inside. In this book I take this argument much further but for now I leave the argument at that. Those lines carry the structural straight the sphere has where the other six support every one of the six by singularity. Where there is no space, there must be singularity 1^0 because the point is present, although space less in singularity 1^0. If zero were a factor where all space finally halted in zero as the value, then zero would be able to remove the space from the centre and such removing would continue to remove the space until all space was removed. This point holding singularity is there whilst being absent of space the point is not absent. It will finally abolish all space in the sphere and it would remove the space but will continue to present the sphere. Zero removes all possibilities of anything coming about and in this case the point cannot be zero because it is present. Since the sphere is there, a zero factor in the centre cannot be present because zero marks the absence of something and the point is present. Only infinity can be a factor from where space may grow because infinity can extend and grow into and up to eternity.

The implication of this is that following the line down to the centre of the sphere we located the centre of the Universe. That is where gravity is. There is a lot more to that but be patient, we are getting there. In every centre we find a point, which is in truth not there but is the mainstay of all that is within the sphere.

The mathematical value of such a point is $\Pi^0 r^0 = 1^0$ and 1^0 is singularity. That is the point where the Universe started and that is where the Universe will finally end. That is the Universe without space-time. That is $k^0 = a^3 / T^2 k$ which proves the Universe is without doubt a sphere…and we just located the centre of the Universe! The centre of the Universe forms inside every point spinning as a point not spinning.

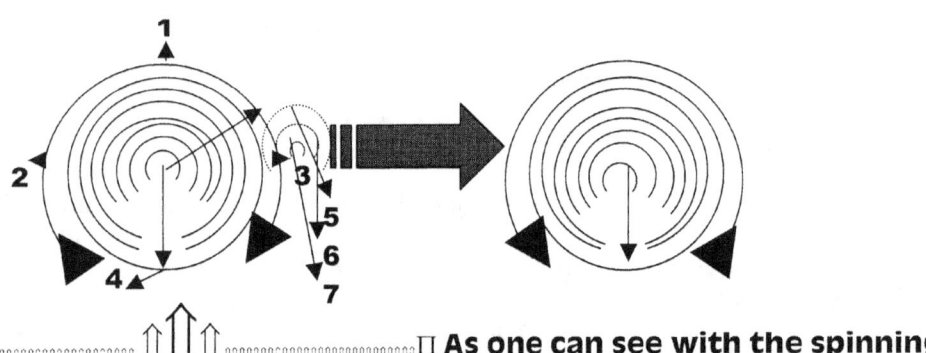

One can see from the top that singularity is established wherever spin occur. The motion generates a position of seven in relation to ten and singularity manifests

As one can see with the spinning top delivering the Coanda principle, every point overheating can spawn space-time by centralising singularity

as 1.9991 as is explained elsewhere. That means any point formed by the sphere spinning can and does start a centre in which no motion holds no space and of which motion surrounds such a point by forming space. Although everything at the time was in the form as a multiple circle, which results in a sphere, the sphere was not the only form present. This too has to do with singularity interpretations. We see a cube, as we know the cube but at first when form came about the cube were not yet forms. .

While the one sphere forms on this point where the dominating sphere secures an edge, the dot may be reserved as an edge marker to the dominating sphere. To the forming sphere in progress of emerging heat gathers at that point because the rotation is a result of duplicating and duplicating is the tendency of naturally growing in space-time $k = a^3 / T^2$. In order to find duplicating coming about there has to be heat in order to duplicate what will form heat. The duplicating process is a process of one factor going softer or less solid and therefore more dynamic that the other. To have singularity is to have gravity but

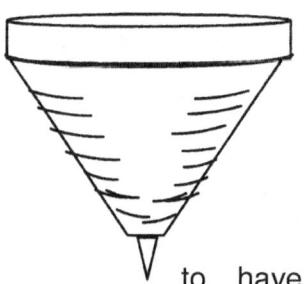

to have gravity there has to be a point of motion and a point of sturdy. The point of sturdy may be in the centre of singularity, but then the solid must be motion. However, even today it still apply: what moves forms liquid in the presence of a solid and at that point singularity presented the solid therefore what we might think of as solid was the liquid because it moved around the solid. Where the one factor is duplicating the other factor is compressing $k^{-1} = T^2 / a^3$

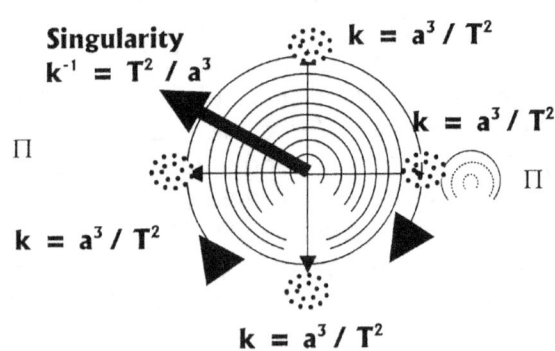

The points duplicating is four moving around a centre by the square of gravity. The motion is the sources of heating because the heat is bringing about the movement. The heat growth therefore provides the action because the action is what energises the points to provide the motion. The motion is purely is space-time duplicating and the duplicating is feeding heat to the centre from the four points overheating thus the points that shows expanding.

But also the duplication leads to the spawning of one point of singularity that provides the installing of the next centre for the next sphere. The Universe forms by lines connecting dots forming singularity.
Therefore the line is of utmost importance in view of understanding the Universe and the concept we form as the Universe.

One possibility that the shortest spot can never have is having a starting point on the zero mark. If the mark of zero holds the start it must also hold the end because the end and the beginning has the same position. If the position of zero then is the beginning, the end will also be zero leaving the line without an end as well as without a beginning. The conclusion from this is that no line can start at zero because that will be a mathematical impossibility. Zero means the absence of anything and starting from zero means starting from the absence of a start and that is not starting at all. The importance of understanding this is to show that in the Universe the number we connect to zero does not exist. If that line that started from zero did start from zero such a line technically would form line or spot starting at a point shorter than any possible line could and would therefore be shorter than the shortest line possible. This we see in evidence looking at a sphere. The radius of the circle forming the sphere has to start where the shortest possible line can start, but it cannot be at zero because zero would remove such a point leaving no line to grow. A line growing or extending from zero can never leave at point zero because of the influence of being zero disqualifies any possibility of growth. If the line then had to grow in all directions at the same pace the line must therefore be a circle. The value of the circle is Π, and that is where creation started. That gave me the clue where to start looking for singularity and therefore the start of the Universe. One would find singularity in the value Π and the value Π will be in all things rotating in a circle. To start my explanation about my cosmic theory I wish to firstly bring some nostalgic and the relevancy will become apparent later on. Such is the importance however that I wish to place this at the very start of the prologue.

When we were boys we played with a top we called the spinning top. I cannot imagine that there is one boy in the western world that did not hold such a devise in his hand. Tying a string securely around the tapered cone started the operation and then with a jerking or pulling throw the devise is launched in a projectile manner and the big knack to success was getting the nail end firmly on the ground and by the realizing jerk the top was rotating. The champion was always the one boy that could throw his top to spin the fastest and that would create a humming sound. The louder the sound produced the bigger champion he was.

When a back braking effort produced a throw of enormity the spinning top would not only produce sound varying in pitch but also create a spin that would seem to have some instability. The top would start to jump up and down as it tried to get airborne. There are very many limitations about the spin, parameters that determine the slowest and the highest spin rate and spinning is within the parameters of such settings. The question arising is why such parameters are there in the first place?

An enormous effort will have the top going oblong while spinning violently and as the pace reduced the top will stabilize by coming to an upright position. In the upright position it will then spin for the remainder of the period where it will in the end start tilting to the side and in a last effort throw a few wild oblong turns and fall over seemingly as if defeated by a more overwhelming power.

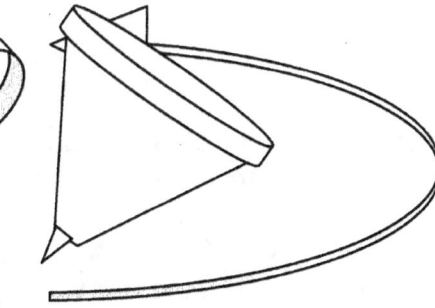

Boys playing games will never realize scientific breakthrough explaining and grown ups do not play with toys. In this little toy played everywhere everyday by almost every one is the answer that the most brilliant of human Brainpower seek answers about. Behind this answer hides all the cosmic riddles no one seems to understand. In the spin as such one may find two vital boundaries in the motion and the boundaries are marked by a wobble coming about as if the top is fighting some other influence. Spinning too fast pulls the centre off the earth as if the centre tries to lift and so does spinning too slow, which then thrusts the centre down to earth. It is the same influence coming about at both ends of the limitation in the spin. There are influences at work, but force…no; it cannot be forces setting such boundaries. From that I started per cuing what sets such limitations because that limitation must be universal as all matter is spinning in one way or the other. In the past these remarks made me the clown in the courtyard and no friends came to my aid because no friends were in support of my statements. A description that would be closer to is that no friend wanted to admit any friendship because such admitting may also reflect on his or her sanity. Everyone thought I am mad but I knew I was onto something, which later became the four pillars of the Universe.

When looking at the cosmos from whichever angle indicates the fact that the cosmos is moving. It is forever spinning and it is going to as much as it is coming from. Everything is on the move and always encircling something of greater importance. A top can spin but the parameters of its spin are limiting the motion it can apply. By not spinning the top is still spinning as the earth are doing the spinning on its behalf. By not spinning the top loses an independent claim on singularity and takes on the earth's point in singularity.

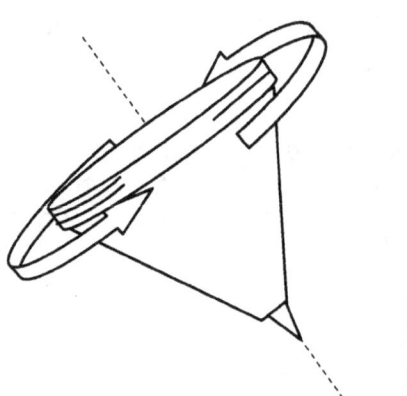

Standing erect one can see the top gained something precious it is fighting to keep. The top is in a struggle of life and death and the top will not subside or relinquish the something it has gained and that important status it is fighting to keep.

Before the top was spinning the top was lying on the Earth surface. The top was not motionless. The top presumed in the status and according to the motion the Earth subscribes. The top presumed in the the Earth dominated the top. The top was be. The top was killed by all the mass the top received the Earth and on condition it presided in the motion and Earth dictated.

position and at the speed by which what the Earth said the top has to from the Earth as being part of with the status to which the

When spinning too fast the top fights because the alignment keeping it upright starts to tarnish. The same apply when

something

Earth

spinning too slowly but that makes sense. It is the fact that the same affect comes about when spinning too slow that triggers the questions.

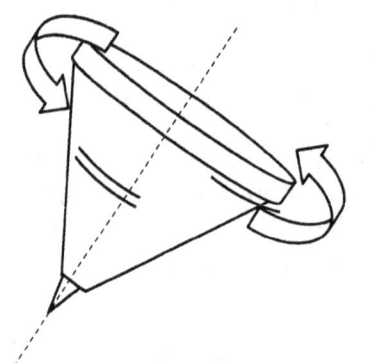

The spinning top is all the evidence any one needs to come to such a conclusion.

Singularity is a mathematical reality. Einstein may be the first to name it and Galileo (unwittingly) may have been the first to define it as Kepler was the first to formulate singularity, but in mathematical terms singularity is the most basic principle. At

this point I wish to establish a fact that seems lost in all other grandeurs of cosmology. A straight line cannot begin at zero or nil but it can only start at infinity Π^0. Such a statement will hardly seem appropriate but the relevancy of this fact has no limits because the Universe that grows from this has no limits.

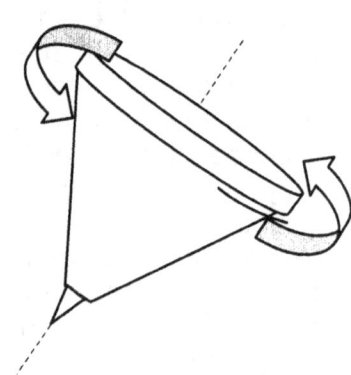

Singularity of the top trying to exceed the boundaries the Earth's singularity dictates to the individual status off the independently spinning top and behind this is a gravitational fight by applying four cosmic laws.

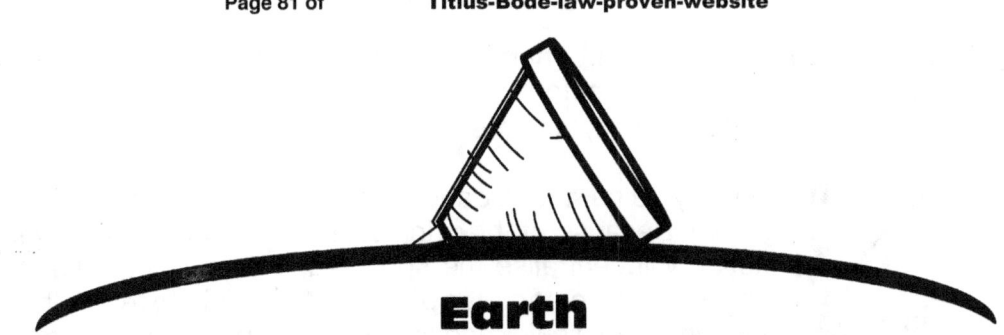

Earth

Earth providing mass by destroying independence

Singularity is in this case exceeding the equilibrium that the top has with the Earth's singularity by the earth providing the mass and the gravity

Singularity of the top trying to exceed the boundaries the Earth's singularity dictates to the individual status off the independently spinning top

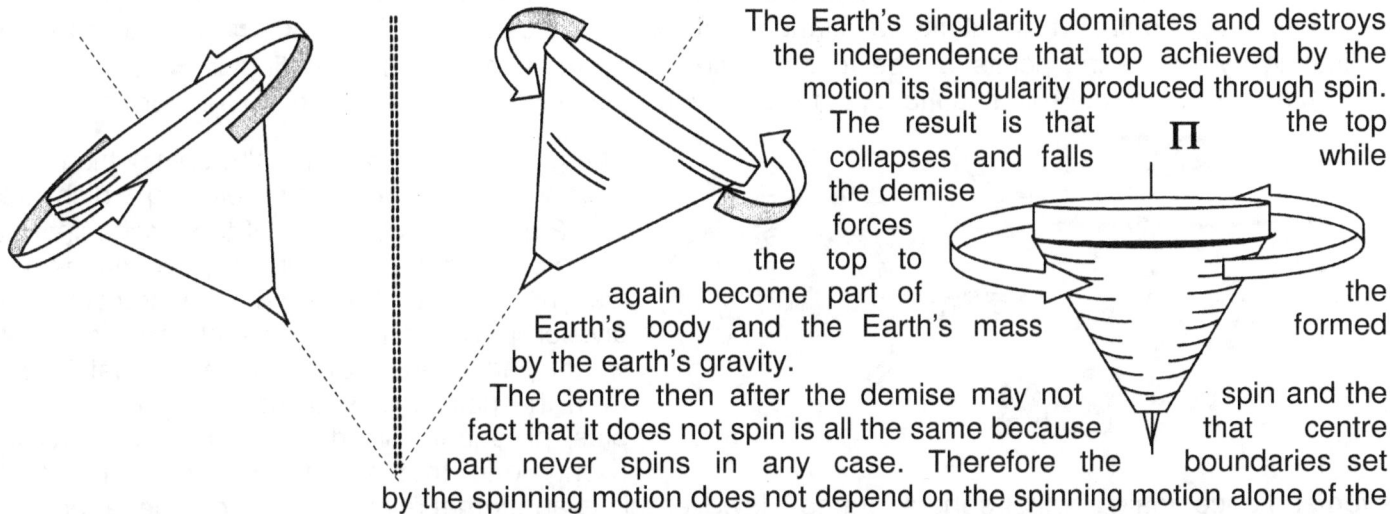

The Earth's singularity dominates and destroys the independence that top achieved by the motion its singularity produced through spin. The result is that Π the top collapses and falls while the demise forces the top to again become part of Earth's body and the Earth's mass by the earth's gravity. The centre then after the demise may not formed spin and the fact that it does not spin is all the same because that centre part never spins in any case. Therefore the boundaries set by the spinning motion does not depend on the spinning motion alone of the object but also has to stand related to another body bringing about a larger spin or gravitational influence.

Granted the fact that the influence the earth has on the top may be that of gravity but if that is the case then surely the sun has also influence on the earth and other rotating objects through gravity. It needs more investigation because it may bring about evidence we are not aware of. This observation places a much bigger question mark on the statement of Newton where he proclaims no influence on two rotating cosmic structures.

We may proceed to the wider picture that the cosmos hold. What is it the Newtonians fail to see? If an electron is orbiting around an atom, the inside of the atom must be a circle. If the atom was not a circle, it then had to be a cube. The electron cannot rotate around a cube; therefore, the inside of the atom is a circle. In a circle, there is a radius that initiates the circle. The calculation of such a circle is $\Pi \times r^2$.

The radius r runs from the very centre of the circle outwards, from a circle centre point towards Π, which is the value of the form of the circle. In the centre of the circle, there is a point where the radius starts. It runs outwards from that point in all directions towards the circle Π. Technically, there then has to be a point where r is invalid, an absolute point of no influence. However, the circle therefore remains Π. The circle does not disappear; it remains there for all to see. It is only the radius that removes.

$\dfrac{\Pi r^2}{r^2} = \Pi$. If one removes the radius from the circle, the circle remains, only holding the value of Π. By removing the value of r, Π becomes singularity with a place but no space within that place to be. Singularity is the place where there is no space to be in place. However, Π remains because once r receives the slightest of space Π will find form. Then the circle will grow to Πr^2 and r would determine the space. Without space, there is no r but even before r is validated and so there is a circle with the value of Π. Singularity is in every single rotating object, be it the proton or the Universe.

This explains the Coanda effect and the application it presents as gravity. Because of the principal in which the Coanda principle works the motion will centralise the sphere and extend the influence of singularity extending the parameter of the radius. This extends the point there is a limit between liquid and solid. By appointing six position around the centre three points will not move while four will move about the three points forming the centre line.

Force on glass →

Force on water ←

The result is that the four points by duplication will reserve the point moving as the next point in singularity because of $k = a^3 / T^2$ singularity will be a natural result of the motion. Then that point will secure a position $k^{-1} = T^2 / a^3$ which will secure six points about such a centre. The centre will bring about four points spinning around three points holding a line singularity. The line in singularity will stand in relevance the contacting factor $k^{-1} = T^2 / a^3$ and the duplicating by expanding points will be four and serve the relevancy by contributing $k = a^3 / T^2$ as space-time only in form. From this the rest of the Universe burst into the next phase of Creation.

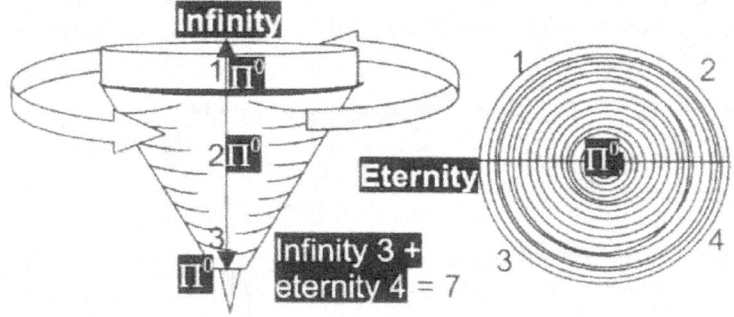

The gravity is in relation to the spin, which is in relation to the four points spinning which are $\Pi^2 / 2$ and that is one part of the Roche limit. It is the dividing of singularity sharing space-time just as we on Earth share singularity by division between the Earth and us others that is not part of the Earth. The total that forms from the point that spawns is seven plus five plus pi square in division of four totalling twenty one that stands related to the first seven and once again another sphere formed. However this is an eternal relevancy that can never break.

Any object in rotation will have a middle point, a very specific centre point that does not spin. That point once again hypothetical yet is a reality as nothing else in the Universe, but none the less must be standing still because every line running from that pint in opposing directions are also in opposing directional spin to each other. Crossing the non-existing centre divide also crosses the Universe in directional spin. Although the points had the same characteristics only seconds before, the motion took them to positions where they oppose the characteristics it had just before and just after the very second in which they are and to which they relate by similar points also in rotation. Due to the spinning nature of such a point with all surroundings changing constantly, with the point varying per instant, the value to such a point can only be Π because of its constant changing while remaining attached to a space less centre.

Using r would specifically oppose another r from every angle. From such a point every other point will be opposing any other point not pointing in the direction to which the first point is pointing, whereby it extends the direction it holds. No matter what the point is or where the point leads, such a point holding a specific direction will be unique in the direction it is rotating because at that or any other specific point wherever, it will be directing not in the direction it spins but in the direction flowing from the center point outwards. Any point will be it opposing itself within the rotating of $180°$ changing every aspect of its previous flowing characteristics it previously had or will once again have in $180°$ from there. While in rotation from the point of an outside observer all may seem static and never changing but to the object in spin every next instant will be a diverting from every aspect it was in very instant passing, and the direction it held in relation to the direction it held the previous mille, mille instant will totally be incompatible with the direction it holds the very next mille, mille instant of rotation. As this happens to all material forming the Universe that proves that no point can be static or constant, all though it may seem that way to outsiders. Although matter is matter, matter can also be anti-matter at the same time.

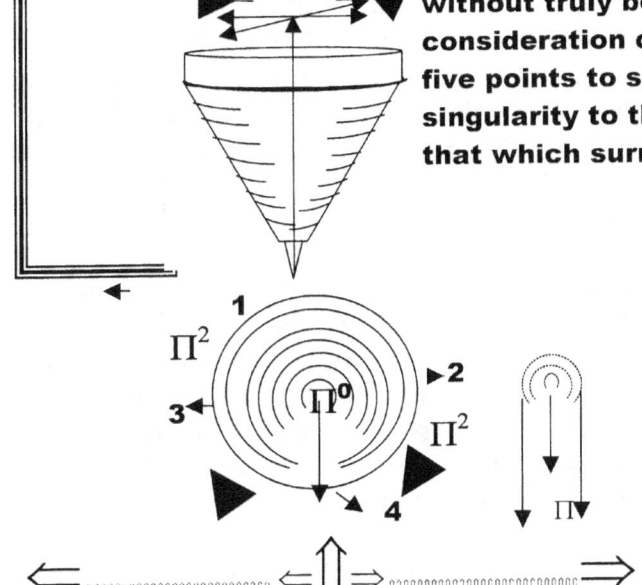

The sphere has seven points. The cube without truly being a cube but is just in consideration of having a cube in form holds five points to singularity. In the centre runs singularity to the value of Π^0, which means that which surround Π^0, holds a position of Π^2

The spinning sphere activates the seven points, which places gravity in relation to a centre. Outside the centre there are five sides by dimension. The sphere has seven points of which four is spinning. The four spinning stands related to the gravity of spin, which are Π^2.

At this stage time is still eternity being interrupted by infinity. Time is not space and science confuses time and space. Time is within singularity that changes allocated positions in order to indicate a new beginning and end an old set-up. Space is that which time or gravity or singularity constantly changes as it redirects the flow of material to new allocated positions in relation to every other position in the entire Universe. To say the Universe is or was 13.5×10^9 years old is shear Newtonian thinking. Was it 13.5×10^9 years and how many days in the year of our Lord and what about all the years that passed since this date was revised? Time was flowing according to interruptions in eternity changing from what was to what is to what will be. Time

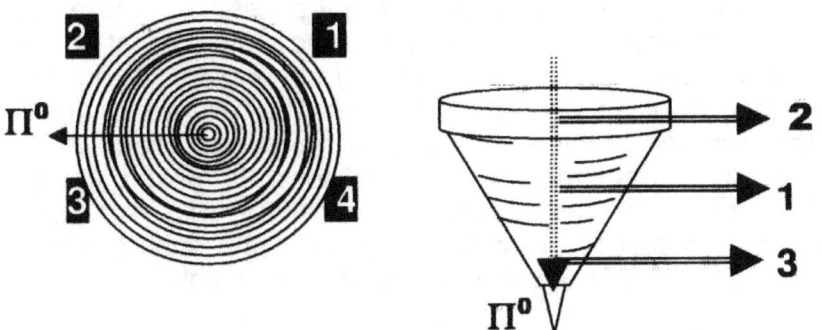

is a norm that comes as things in the Universe change about things that are places around and scattered throughout the Universe. We may presume time at this point somewhere became a factor since sphere sprouted from points on sphere edges and differentiation in development came in place. Time in directing flow of space is as part of gravity as singularity is and movement is. Considering the role that the Roche limit played one can see how points in singularity grew from contraction and secured ever-stronger centres by divulging near points within the realm of singularity control. When a point in form developed at a position that was close than the original Π^0 to Π, the singularity in control took control. Consider what happens to a star that developed closed than the Roche limit of Π to $\Pi^2 / 4$ would allow, it is easy to see how the singularity centred grew by concentrating the heat the points in singularity brought about.

It is most important to recognise the fact that in the sphere there forms a line without space having a singularity factor of 3.

Then a circle forms a value of four positions and this has total relevance with singularity. This circle conducts the movement and without the movement there is no axis. Therefore the spinning four takes charge of the singular three

This is the backbone of how gravity forms where gravity forms as a result of Π coming about.

However time at the value of 3 also forms part of the cycle of space development.

That is why the value of seven repeating three times forms 21 where 21 plus singularity (.991) is the value of Π. As the book develops I will explain how the value of Π forms as 10 + 10 + 1.991 in relation to

7

While every line is circling and brings about time in space to the value of Π repeating Π to form Π^2 at the

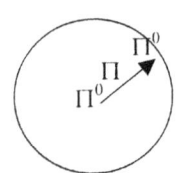

same time Π is extending in one specific centre to the value of Π^0 and only the spin value keeps Π not becoming r. In cosmology there is no r but only $\Pi^0\Pi\Pi^2$. The spin keeps the immovability from becoming Π and maintaining Π^2 by performing duplication, but with any slightest reduction in spin influencing the value of Π^2 in the range of $\Pi^2 / 4$, then Π will start extending and as one can see from the behaviour shown in the Roche limit, the heat will be concentrated at the centre and the singularity in the centre will grow in concentration.

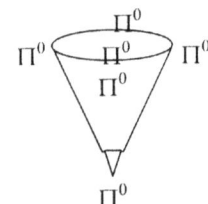

Only at points exceeding Π in diameter was time as Π^2 be able to retain form and also grow. From that space slowly develops because at Π could Π^2 bring about a form which provided motion. In the centre there develops Π^2 and Π^2 keeps all form at a safe distance of Π to bring about the needed solid immovable centre with the form $\Pi^2\Pi$ about the double Π could $\Pi^2 + \Pi^2$. That secures the makings of the atom by applying the Coanda principle of enticing gravity in the centre of motion, which then provides space-time by measure of $(\Pi^2 + \Pi^2)(\Pi^2\Pi)3$.

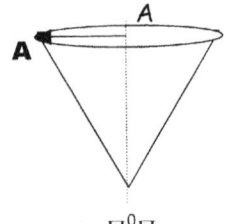

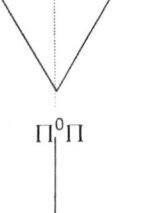

This totalled seven in dots and with three of those seven circling singularity at 1.9991 the atom came about. I again wish to repeat that the centre in the sphere forms singularity as the centre of the sphere and this forms the centre of the Universe. This is where the realisation comes from how the Cosmos started. If there were gravity at the very first instant then there was a sphere at the very first instant because gravity can only be in the sphere because of what it represents.

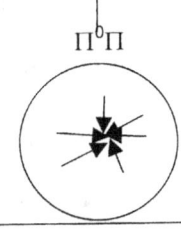

The truth about gravity in the cosmos and that gravity will contract the cosmos one day is the fact that the sphere has singularity as a natural substance. The entire form that is the sphere is controlled from a centre within the sphere. That centre holds the sphere in form and shape. Therefore the strong form is dictated from that space-less centre where there is no space and no form left. The natural inclining is in the form of the sphere. It is part of the roundness that the overall shape of the sphere represents and this structural strength is carrying down to the very centre. Because the circle is forever reducing that reducing which is inherently part of the form of the sphere becomes a tool in distorting of space in the sphere and is eventually removing all forms of space from within the centre of the sphere.

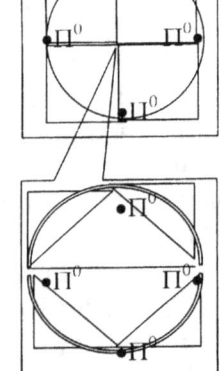

The very centre ends up as having no space because of the reducing that continuous down to become the space less inner centre. The all roundness is the ingredient that forms the backbone of the absolute strength that the sphere has and that is the component that the sphere is so famous for. The form the sphere has, allows the sphere to have a control that is coming from the centre deep inside the sphere where the space vanishes and being without space seems to keep the entire structure rigged. From the centre the sphere shape shows strength that the shape as tough as it is. How does it work in its most basic analyses?

The key to the relevancy is heat and space. When matter heats, it expands therefore it takes more space. When matter is cooled it shrinks, therefore takes less space. That is the relevancy because matter in any form is heat. Heat produces the increase of space and reducing space produces an increase of heat. That is the relevancy.

That is the secret of the Universe. That is the secret of gravity. That is the secret of momentum and every other aspect within the Universe. Time stood still in eternity, then after a command of the Creator time started to move by overheating and eventually formed the relevancy of the proton $(\Pi^2 + \Pi^2)$ the neutron $(\Pi^2\Pi)$ and the electron (3). That is the relevancy. That will be whatever position there is in the Universe. The electron (3) may become obsolete, the neutron $(\Pi^2\Pi)$ may become obsolete in neutron stars and even $(\Pi^2+\Pi^2)$ the proton will become dysfunctional as space reduction completely disappears from the star's space-time occupation. However, those stars may be dark, and beyond our vision, but are not non-existing.

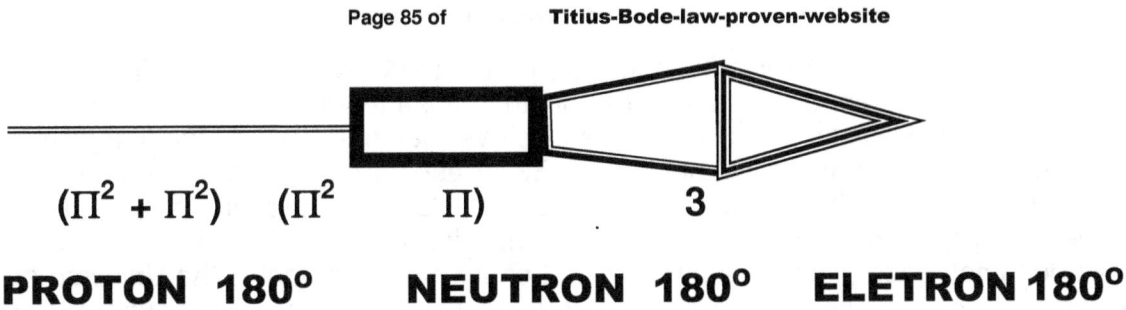

$$(\Pi^2 + \Pi^2) \quad (\Pi^2 \quad \Pi) \quad\quad 3$$

PROTON 180°   NEUTRON 180°   ELETRON 180°

The relevancy of $(\Pi^2 + \Pi^2)$ $(\Pi^2 \Pi)$ (3) = 1836 will remain but the mass of the electron and the mass of the proton will change in every space that time applies. Cosmology thus far was incomprehensible because it was incorrect. When applying natural laws, it becomes so simple that a person as ordinary as I am can understand and explain it. Believe me, if I can understand it, every other non-brainwashed human on earth should understand it.

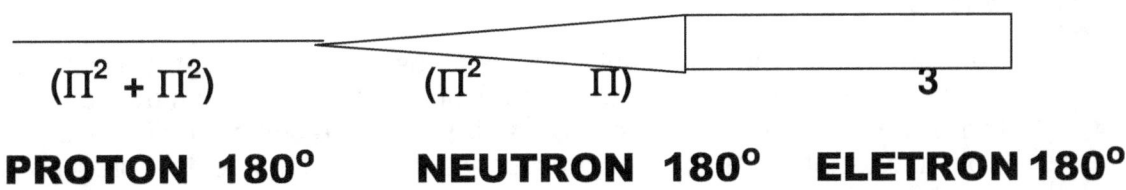

$$(\Pi^2 + \Pi^2) \quad\quad (\Pi^2 \quad \Pi) \quad\quad\quad 3$$

PROTON 180°   NEUTRON 180°   ELETRON 180°

The relevancy of $(\Pi^2 + \Pi^2)$ $(\Pi^2 \Pi)$ and 3, is a dimensional reduction of the flow of heat from space back to time. The flow of heat becomes necessary to prevent solid matter from overheating. By removing heat from the gas of space, through the neutron, to the solid of matter, space reduces as the intensity of heat flow requirements increase.

PROTON   NEUTRON   ELETRON

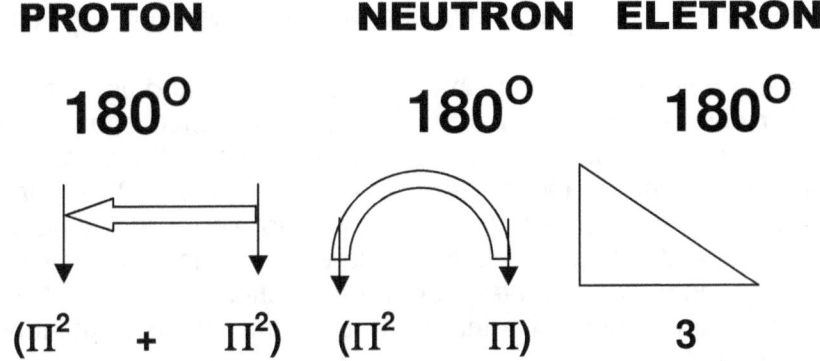

$$180° \quad\quad 180° \quad\quad 180°$$

$$(\Pi^2 \quad + \quad \Pi^2) \quad (\Pi^2 \quad \Pi) \quad\quad 3$$

What was within the Universe at the start will be in the Universe at the end. The Universe holds all; maintains everything and combines the lot. In Afrikaans we call the Universe the "Heelal". It is a combination of two words namely "geheel" and "alles". "Geheel" means everything and "alles means everything. Therefore the "heelal" directly translated from Afrikaans to English will mean the "Everything of everything". Nothing can be added and nothing can be lost. It is all-inclusive. With this fact so commonly known and accepted, how can the Universe grow? How can the Universe expand? Well, it cannot, and that is yet another illusion the Scientists create through misunderstanding. What is in it is in it and it cannot grow, as much as it cannot shrink. It cannot expand and it cannot demise. It is only a consistence of changing relevancies, where the relevancy flows away from one part of eternity or singularity (space) to another part of eternity or singularity (time).

Every dot insignificantly small as it may be, is a part of another Universe as much as it is part of the accumulative Universe and every dot in the infinity holds singularity, which we translate as " nothing" being " darkness". There cannot be "nothing" just as much as there cannot be "darkness". There cannot be something big or small, but it into relevancy of perception, and then the relativity of perception becomes the question. There cannot be hot as much as there cannot be cold. The sun FREEZES hydrogen to a liquid at six and a half thousand degrees Celsius and Universe boils over in the form of the Hubble constant at the temperature (we presume from our vantage point) at minus 273 degrees C. If we Humans cannot or will not abandon our human perception and our manly perspective, we may as well return to astrology for all its worth.

Every point in the infinity we may not observe is not merely part of the Universe in not being "nothing", but is the point where the Universe started representing singularity. It is the very first point where everything began so many eternities ago, because after all, how can we ever determine where the first point was, as they were very much equal and alike at the beginning. Every aspect of the Universe started with the fundamental fact that no point in the Universe can represent "nothing" as a number, because every aspect in the Universe represents singularity in what ever form it may hold in that specific spot forming space-time. If man does not reach a conclusion where that conclusion is matching the Universe and stop to match the Universe with man (and man's incapability), we may all go back to caves and become starving hunter-gatherers again, because we will never find a way to progress to the ultimate understanding of the Universe. $1^{245196783712} = 245196783712^{\circ}$ and that is the first dimension I singularity.

Singularity is as close as any spot can ever come to zero BUT IT CANNOT EVER BE ZERO. From

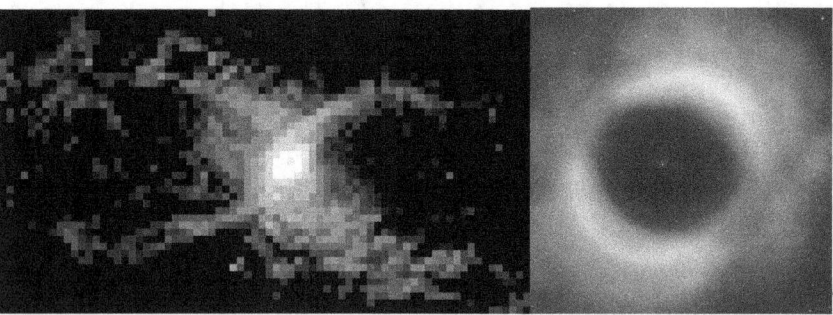

singularity diverts space-time and there cannot be space without time as much as there cannot be time without space, not withstanding the size of space or duration of time.

Through space-time singularity connects as much as relates linking the Universe into a network of influences beyond what ever we can ever conduct. There can be no spot that does not participate in the curvature of space-time. From the point of singularity runs space holding time to the prescription singularity dictates.

With singularity connecting singularity will or cannot relieve or release the connecting other than by a method we humans refer too as an explosion, but space-time separating as much as joining singularity dividing can change by applying time too space in a changeable manner, stretching and shrinking the time aspect by changing the density of the occupying heat, which creates the space allowing the spin of the occupying heat creating space setting the time.

As singularity is eternally cold, it also has to affect the standing between matters, as matter diverts from singularity. The more occupied space there are between singularity and space ends, the more it will be to the advantage of the effect we consider heat, and the less space allowing time the more it will advantage the statement we refer to as cold. But hot and cold are human relevancies not acknowledged by nature. Science does acknowledge a flow from singularity towards space by reducing heat; therefore admitting singularity must hold temperatures indefinite.

Looking at a star overheating it is obvious how singularity breaks into heat. By demolishing singularity it means Π demolishes the very point holding singularity

The question never asked is why would stars overheat? We can blame pressure, but pressure would not bring about a star disintegrating from the centre, as the star depicted here clearly does. A burst from pressure should blow the sides out. If we consider the evidence we see that stars overheat because that which comes out of a star, as its inside is heat that became clouds of heat sprayed into outer space. It clearly was compressed heat in a liquid that blew out to become a cloud of heat that is sealed in time.

Stars we call Super Nova has blowouts. That man knows since before writing began, but since of late this phenomenon becomes more and more seemingly misunderstood. If stars blow as stars should and as we can clearly see from the picture just above, then the explosion happening to the star comes about from other principles, surely. It is very obvious the two occurrences are not a result from the same basic method the Universe uses in destroying stars.

Positioning, finding, allocating and valuing

Singularity

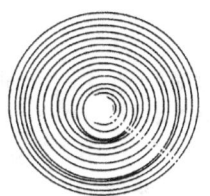

In the **precise middle** of all **objects in rotation** is a precise centre dividing the object in sectors that will **start the spinning initiation** from that centre point. Thus, the spinning object **will have a middle point**, a very specific **centre point that does not spin** and only holds Π as a specific value because no radius can apply. But also the one value such a line **cannot have is zero** because the line **is there and holds contact** to the rest of the material bringing about that **zero does not start any** line and therefore the **value of the line must be infinite**, just as described in **accordance** and by **the definition of singularity**

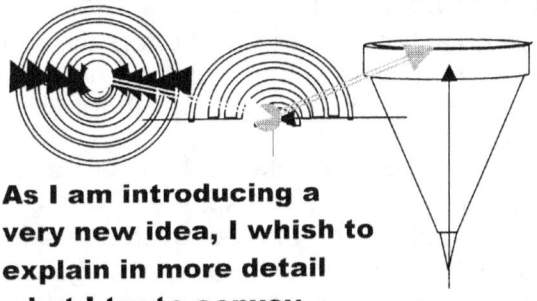

Move the rotating line progressively to the middle by reducing the length the line have from the edge to the middle. At one point all further reducing ends.

As the rotating direction moves inwards, the rings will become smaller and smaller.

As I am introducing a very new idea, I whish to explain in more detail what I try to convey.

That point albeit hypothetical, is also as much a reality none the less and is placed where that point must be standing still because every line **running from that point** in **opposing directions** are also **in opposing directional spin the other or opposing side.**

In considering the spinning motion in the fraction of time in the detailed instant every aspect of rotation will turn in every instant of change in time. Although the points had the same characteristics only one instant before, they oppose the characteristics it had just before and just after the very instant in which they are and to which they relate by similar points also in rotation. The fact of the graph proves my point in quarterly opposing dimensions and values,

There must come a point where the ring is infinitely small, where it can reduce no more, where it reached its ultra limit, but at that point it cannot be zero, because the point is there. Understanding all the following is connected intimately and all conditionally to the fact of accepting that all individual particles in the universe use motion and therefore spin.

The spinning of Π^2 around the centre Π^0 establishes Π and Π is what produces the singularity Π^0.

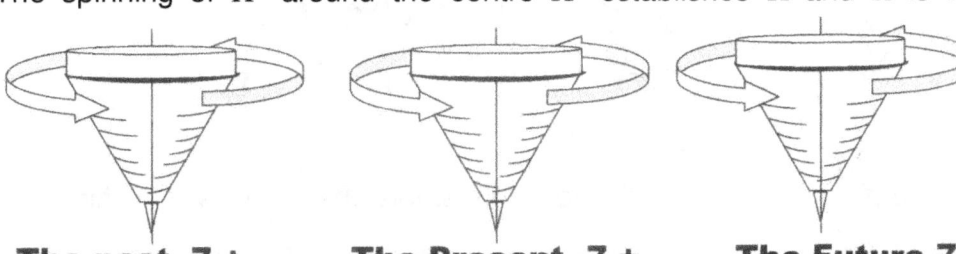

The past 7 +   **The Present 7 +**   **The Future 7**

Singularity is always present in the form the sphere presents however; in the sphere singularity presents no influence except in the case where the sphere starts to spin. In the spinning a line comes about which caries the term of an axis. The axis has never been acknowledged as the most vital piece of any particle in the Universe mainly because the role that the axis plays has been neglected and subdued by the incorrect attributing Newton placed on the motion and the axis.

The moving of Π^2 to by duplicating Π involves relegation and not motion as we consider motion. Forming Π was Π^0 getting a side or form and that is the start of space. There was no true side but only a form that came into place. The dots have no sides have no length or diameter.

At the start of everything was in singularity being without space or form. In the very beginning there was no measurable space or measurable time involved. The time could have been a micro, micro second as much a trillion millennium because time had no relevance. Time started as eternity interrupted by infinity, as it still is the case, however the line that eternity followed was no line because there was no space to hold the line although the was and is in place. The line was momentarily interrupted by infinity, however with no one there, there was no one to notice. The lines were not lines but relations to sides being formed.

Inherent to the form the sphere offers, there is a specific location of singularity where the radius first goes single $r^0 = 1$ and then form goes into the realms of singularity $\Pi^0 r^0$. This form offers the biggest density support compared to every other shape. The cube also may have such a point but having such a point does not connect directly to six points located on the edges of the cube or any other form the is.

In relation to such a centre where $\Pi^0 r^0$ forms singularity there are always four cubes related to such a centre where the centre is part of seven points in total representing the sphere. Every cube has lost one side to a point of the sphere where the sphere takes control of form and removes one side of the cube. In relation to the time factor that is inherently part of singularity by the extending of singularity there are five sides connecting to four points standing related to singularity by the Π^0 factor and that gives 5 X 4 = 20. That is always directly in relation to seven points singularity offers.

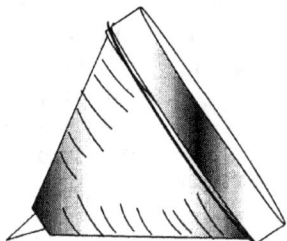

The top lying still holds the same singularity principle that the sphere holds because if the shape the top has. The roundness protects singularity at a seventh position deep inside. However the top is a dot Π or even going down to a spot Π^0 and is only by the form the material has which puts singularity in place.

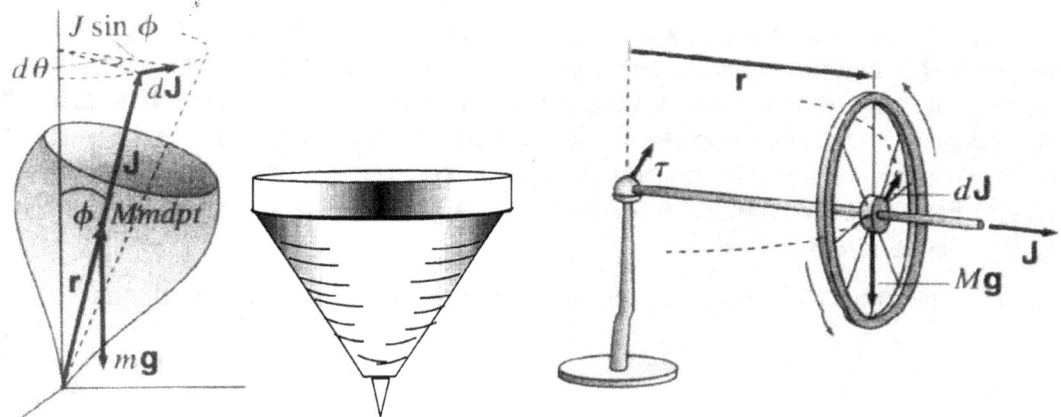

According to Newton it takes no effort $\dfrac{dJ}{dt} = 0$ to get the top from where the top was motionless to where the top is spinning. I say this on the work that Newton suggested comes about from the effort it takes the top to circle.

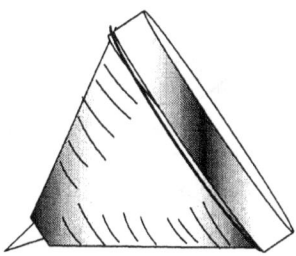 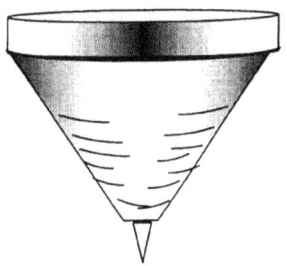

From a position of a lying down top to a position of spinning erect does not take "nothing" as an effort. Coming upright by spinning motion puts the top in a new and independent Universe as the top then finds courage to fight the gravity of the Earth up to where the last "breath" is fought.

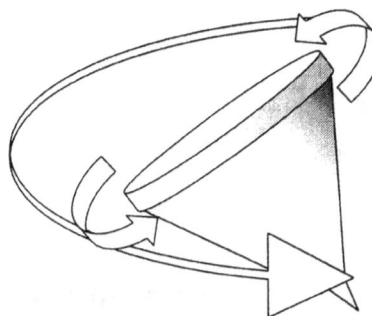 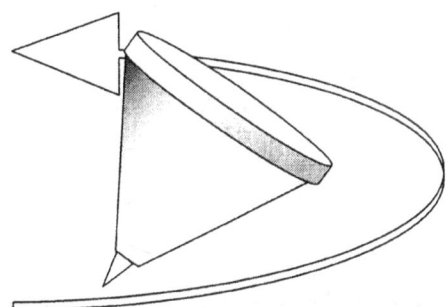

In the motion a line comes to life running in the centre of the top. This line is not just another line but can focus the top to spin upright and erect and defies the earth's gravity. The line was not there when the top was motionless on its side. By supplying motion the line can concentrate an effort that will unleash such dependence to the top that the top will come into a position where the top have the tenacity to take the Earth gravity on in a dual. This is a very important trademark of the four cosmic laws.

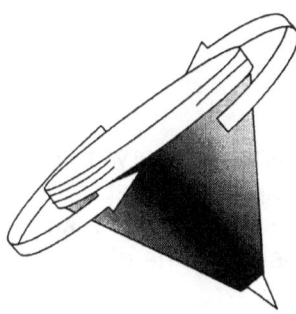 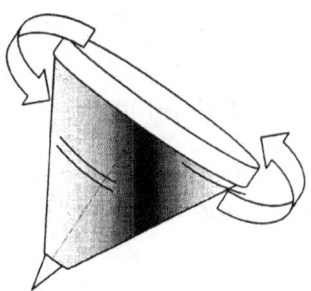

From every which angle one look at the top the top seems energised or "possessed" and I can even be slightly forgiving towards Newton for calling the drive forcing movement as a force because although not a force the stance the top takes when spinning upright leaves one with an impression of forcefulness being part of the situation.

In this one must not see a force but one should see the manipulating qualities of life extending to the top and by life's ability to manipulate space-time and control motion in space-time with space-time, the throwing of the top is as little a cosmic event as the apple Newton saw falling from the tree.

Both instances were life that controls events and as far as there are proof then life duplicates the laws with which nature controls gravity. If you start to imagine about life in the Universe you may just as well start believing in ghosts fairies and all other fantasy creatures. Life is completely alien to the cosmos.

There is zero proof of life being anywhere but on earth and that is a cosmic fact. Science must decide whether they wish to speculate about the fancy fill, but in such an event distance their fantasies from science and reality, or stick to science in reality and believe only in facts as science present facts.

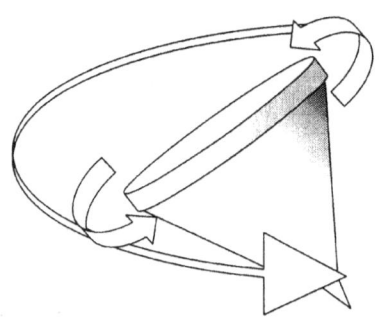 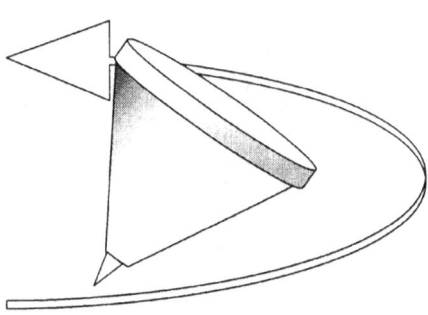

Let's consider what are facts with the top spinning as the top does. This is no fantasy or life coming from some imaginary source but it is a cosmic reality which life found a way and means to manipulate.

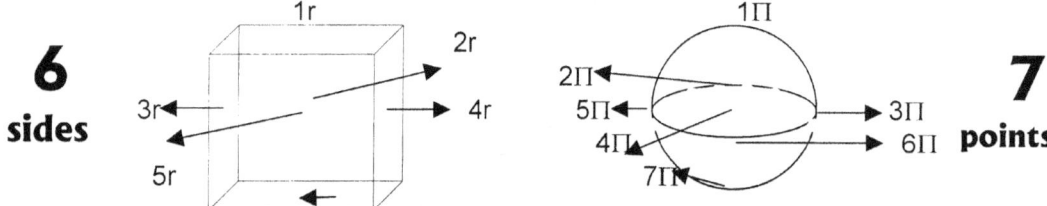

The effort it takes the top to spin gives the top a distinction of extreme significance. The top is promotes to a star in motion because it charge singularity into existence where singularity then control space-time.

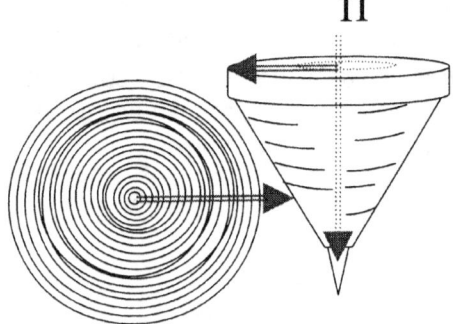

Singularity is a mathematical point hidden in the very centre of every sphere. Singularity is very much inactive in every sphere forming space but is there to keep a centre of structural bonding in every sphere. Something happens to the top in having a singularity just like any other sphere has to a point that takes charge with all the cosmic dynamics the sphere may show. The movement of 4 establishes a line of 3.

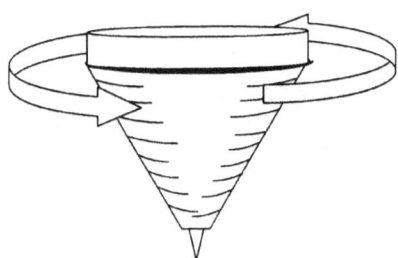

The top is charged with an energy, which not only takes charge of the top as well as the body of the top, but also the immediate space surrounding the body of the top. When a person with skill manages to put a high degree of spin into the motion of the top the top then spins in the surrounding space so vigorously the top stars to whistle vigorously therefore vibrating the space surrounding the top.

As the top comes to motion the top find the characteristics that the top shows very indicative of the characteristics that all moving objects in the cosmos shows. The top spins in a straight line that bends by a 7^0 inclination. Then with all this new information in view, how does this effect gravity? We have to start where Albert Einstein started and then proceed from there. Let us start where we associate with our Universe and see where does light place us in the Universe

Have you ever stood outside and thought how is it possible to get that wide area of space that you see into something as small as your eye? If you have never thought of this you have never thought about the cosmos because this is physics and the start of physics. Contemplating that is how one starts to understand the cosmos and on this thought I base everything I introduce as physics. You are about to experience ideas you have never even thought about and I am introducing to you concept that you are unaware of that it exists. It starts by understanding how an eye fits all the light outer space offers to see.

How does the vastness of space fit into the eye of a human? I have been doing science research about the Universe for many decades and I uncovered how the Universe forms space as it grows. The process is well known to science and has many names but up to now no one in science ever understood how the principles apply. I do not wish to venture much into this detail since it is lengthy to explain and covers

many details but should you wish to know more it is detailed as such in the books I sell on the market.

The lack of understanding about how these principles function can only be contributed to mistakes in science about science at this moment including facts science ignores and these mistakes are what

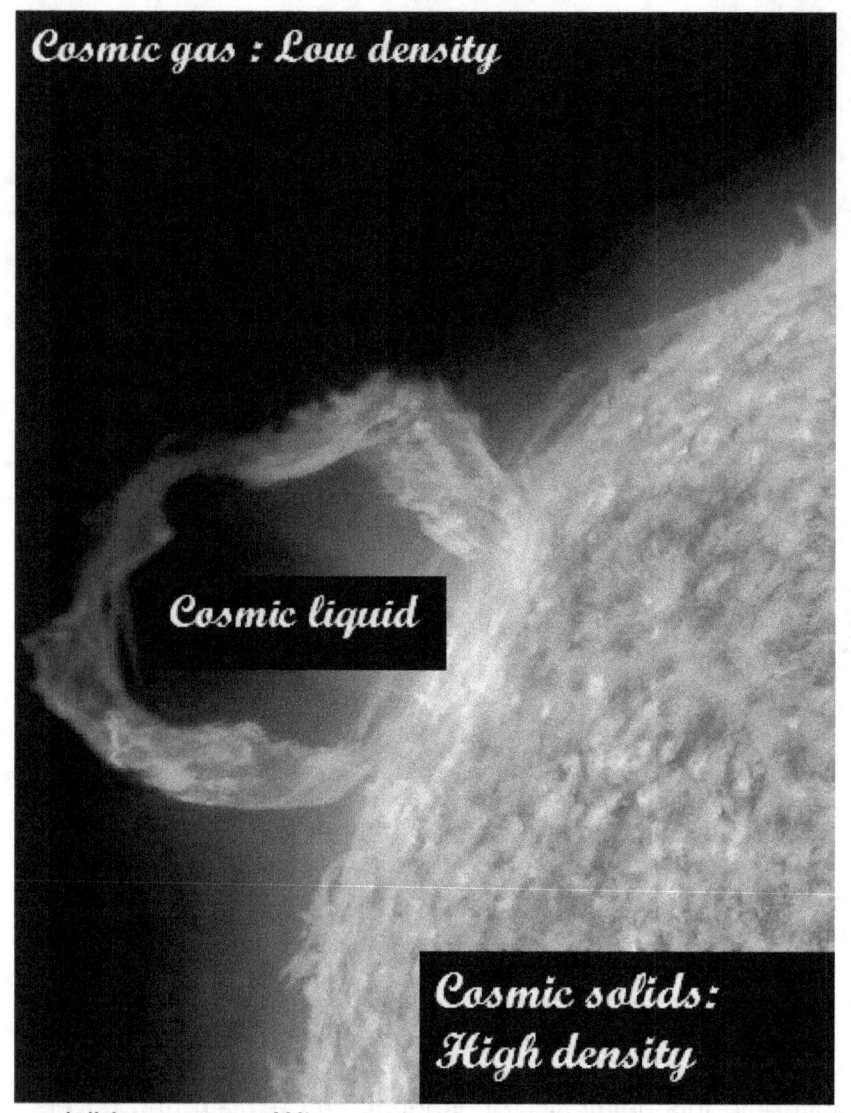

Cosmic gas : Low density

Cosmic liquid

Cosmic solids: High density

science so far are unwilling to admit. These mistakes academics embraces in the theories they apply and accepting my theses details would be equal to their denouncing of the correctness of every thesis written on cosmology. I denounce the credibility of the idea of mass within the cosmos and that is the last concept anyone in physics would embrace. I have found irregularities in science and I reveal it but not only that; I give the correct approach as well and where necessary I supply the proof thereof. My concept changes science that those in science are unwilling to let their ideas become discarded in favour of accepting mine when they embrace my work. Since I find no acceptation or any endorsing about my work I now bring my work to the general public. I wish to reach all persons reading on a level as normal persons do read to understand concepts so that I can introduce the approach to science I take. In this detail come the levels by which I have to do the planned introductions. I explain this somewhat tedious because I wish you to know what theme your firm will promote should we share a future in this venture.

I guess the question is why would there be light in the first place and darkness to form a visible contrast. Why am I able to see light coming from the sun and the darkness of space I see as darkness? I farmed in Africa for most of my life and I saw animals see by the light that we humans think of as darkness. When attacked in a pitch-black night animals run as if they see in daylight, just as we are able to see when we are in the light of day and that darkness is light they use to live by. They are unable to even see the light of a spotlight shining on them or shining in their eyes.

They hunt and eat and graze and mate during the dark more often than in our visual light. Most animals only come out at night because they are unable to see the light by which we see. I realised it is our arrogance and stupidity that makes us believe darkness is not light whereas darkness is more light than the light that enables us to see. I grew into the habit of answering any conclusion I draw with a question: why? Why would they see better in our darkness than we can see in our light? The question that follows this conclusion is why there are two forms of light within the Universe.

The Universe is there for all to see and the Universe makes sense when everything that is senseless becomes fully understood. If one question remains unanswered, the rest of all understanding will be engulfed by the one point not understood because it will bring to us darkness since the light of understanding has vanished. I learned to never be clever but instead be scientifically inquisitive.

In the picture above on this page we can see there is a gas in outer space and there is a denser liquid atmosphere and an absolute dense earth and moon. It is obvious that there are differentiations and only Newtonians that live in a dark-age Neanderthal science world holding onto mass as if it was all they have will not be able to observe this. The darkness produces more light since the darkness covers more space and if we don't understand the darkness then that shows how little we understand about cosmic light. The "light" we see is heat approaching us and the " darkness we see is heat losing density by the Hubble constant reducing density and therefore moving "away" from our view. As the light holding 3^3 is light filling the space we think of as outer space and light we think of as light that we see is $3\Pi^2$. I guess I will have to explain this briefly to be able to reach some small level of understanding.

The Universe can't expand because whereto will it expand since the Universe is everything there is. It can't go anywhere since it is everywhere. The Universe is heat that form density at different levels. Heat in outer space moves to form liquid that we see as an atmosphere and liquids become flames as light coming from the sun. The sun draws cosmic heat as gas from outer space and condenses the heat into a liquid by which the sun absorbs the liquid as it applies the Coanda effect forming gravity.

The heat forming outer space shifts by gravity to form a solid and that is to cool singularity from overheating. But as the concentration of heat shifts from outer space to solids the region we think of as outer space loses density and material acquires the heat to further the density of solids. It is heat shifting from one form to another form in order to regulate the levels of equilibrium in heart. As heat density is lost space grows because space and heat are sides of the same coin. The loss of density forms an increase in heat and the gain of density becomes a loss in heat. With that I must also state that outer space is the hottest area in the Universe and the Black Hole is the coldest area in the cosmos. This I base on the principle that the hotter the area is the more it will expand and the colder an area is the more it will shrink. I explain this much better further on. However I will indulge the reader by explaining this in a very brief manner.

Planet	Mass per Earth unit	$k^{-1} = T^2 \div a^3$ Movement	a^3 of space volume	T^2 During time units
Mercury	0.06	$T^2 \div a^3 = 0.983$	$(a^3)=$ 0.059	$(T^2)=$ 0.058
Venus	0.82	$T^2 \div a^3 = 0.992$	$(a^3)=$ 0.381	$(T^2)=$ 0.378
Earth	1.000	$T^2 \div a^3 = 1.000$	$(a^3)=$ 1.000	$(T^2)=$ 1.000
Mars	0.11	$T^2 \div a^3 = 1.000$	$(a^3)=$ 3.54	$(T^2)=$ 3.54
Jupiter	317.89	$T^2 \div a^3 = 1.000$	$(a^3)=$ 140.6	$(T^2)=$ 140.66
Saturn	95.17	$T^2 \div a^3 = 0.999$	$(a^3)=$ 868.25	$(T^2)=$ 67.9
Uranus	14.53	$T^2 \div a^3 = 1.000$	$(a^3)=$ 7067	$(T^2)=$ 7069
Neptune	17.14	$T^2 \div a^3 = 0.999$	$(a^3)=$ 27189	$(T^2)=$ 27159
Pluto	0.0025	$T^2 \div a^3 = 1.004$	$(a^3)=$ 61443	$(T^2)=$ 61703

The table above is the conclusion Kepler drew after 80 years of study. Because Kepler proved that space a^3 is distance making T^2 time and k the directional movement because $a^3 = T^2 k$ or movement $T^2 k$

and therefore singularity forms $k^0 = \dfrac{a^3}{kT^2}$ and then the spin determines the space in movement

$$T^2 = \frac{a^3}{k} \text{, the movement determines the space } k = \frac{a^3}{T^2} \text{, the movement reduces the space } k^{-1} = \frac{T^2}{a^3}$$

From this it is clear that k holds two values being $k^{-1} = 0.983 = \Pi^2$ and $\Pi^0 = 1$ in the next one $k^{-1} = 2.98$

At present Mainstream science only acknowledges that the table hold the value of $T^2 \div a^3$, which helps hiding Newton's misconception of $T^2 = a^3$ and this is blatantly cheating with mathematics. Nowhere in mathematics can the second dimension value be equal to the third dimension value. This is as if saying that $2^2 = 2^3$ or as if the image I see in the mirror being second dimension is equal to me standing in front of the mirror that forms the third dimension. It never can be even if it is Newton trying to sell this hoax.

From the table we see that space divided into time where the planet moving in the circle represents the time factor and the space it moves through represents the space in which time moves. The table clearly shows a negative directional flow of space going towards and being within the sun. Since we can see from the Titius Bode law that the shift in space does not bring about the decline in distance of the orbiting planets, it therefore has to be the negative directional of space that moves towards the sun.

By space dividing into time it then shows how space goes singular in relation to time.

PLANET	PERIOD (Years) (T)	MOVEMENT (T^2)	DISTANCE	SPACE (a^3)	RATIO k
Mercury	0.241	0.058	0.39	0.059	0.983
Venus	0.615	0.378	0.728	0.381	0.992
Earth	1.000	1.000	1.000	1.000	1.000
Mars	1.881	3.54	1.524	3.54	1.000
Jupiter	11.86	140.66	5.20	140.6	1.000
Saturn	29.46	867.9	9.54	868.25	0.999
Uranus	84.008	7069	19.19	7067	1.000
Neptune	164.8	27159	30.07	27189	0.999
Pluto	248.4	61703	39.46	61443	1.004

PLANET	SEMIMAJOR AXIS $a\,(10^{10}\,m)$	PERIOD T (y)	T^2/a^3 $(10^{-34}\,y^2/m^3)$
Mercury	5.79	0.241	$k^{-1} = 2.99$
Venus	10.8	0.615	$k^{-1} = 3.00$
Earth	15.0	1.00	$k^{-1} = 2.96$
Mars	22.8	1.88	$k^{-1} = 2.98$
Jupiter	77.8	11.9	$k^{-1} = 3.01$
Saturn	143	29.5	$k^{-1} = 2.98$
Uranus	287	84.0	$k^{-1} = 2.98$
Neptune	450	165	$k^{-1} = 2.99$
Pluto	590	248	$k^{-1} = 2.99$

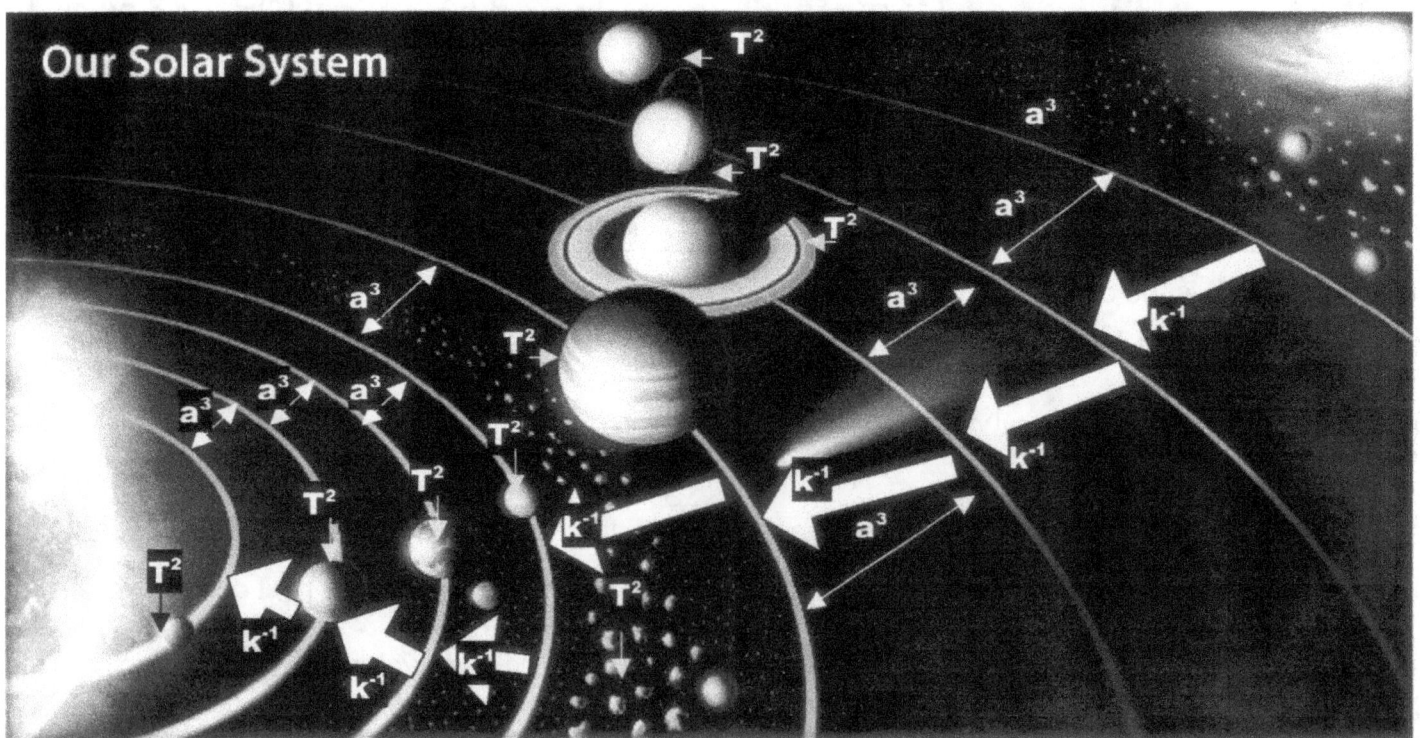

The value of T^2 represents the planet spinning around the sun and in this it forms by rotational movement the time factor T^2. The space in which the planet revolves around the sun then is a^3. With that being the case we find space a^3 compresses as it moves towards the sun by the value of k^{-1}. As the tables show there is a shift of space moving at a rate of k^{-1} all the way to the sun and this is how gravity moves space by condensing heat. When something is hot it expands and when something cools it contracts; that nature. This was the finding of Johannes Kepler years before Newton ever came into any picture and this is how nature forms the solar system and the study took eighty years to complete by Tycho Brahe and afterwards Johannes Kepler. This is the scientific information that the cosmos no less mathematically provided Kepler and this cannot be inaccurate since this is nature with no bias or a shift in blame.

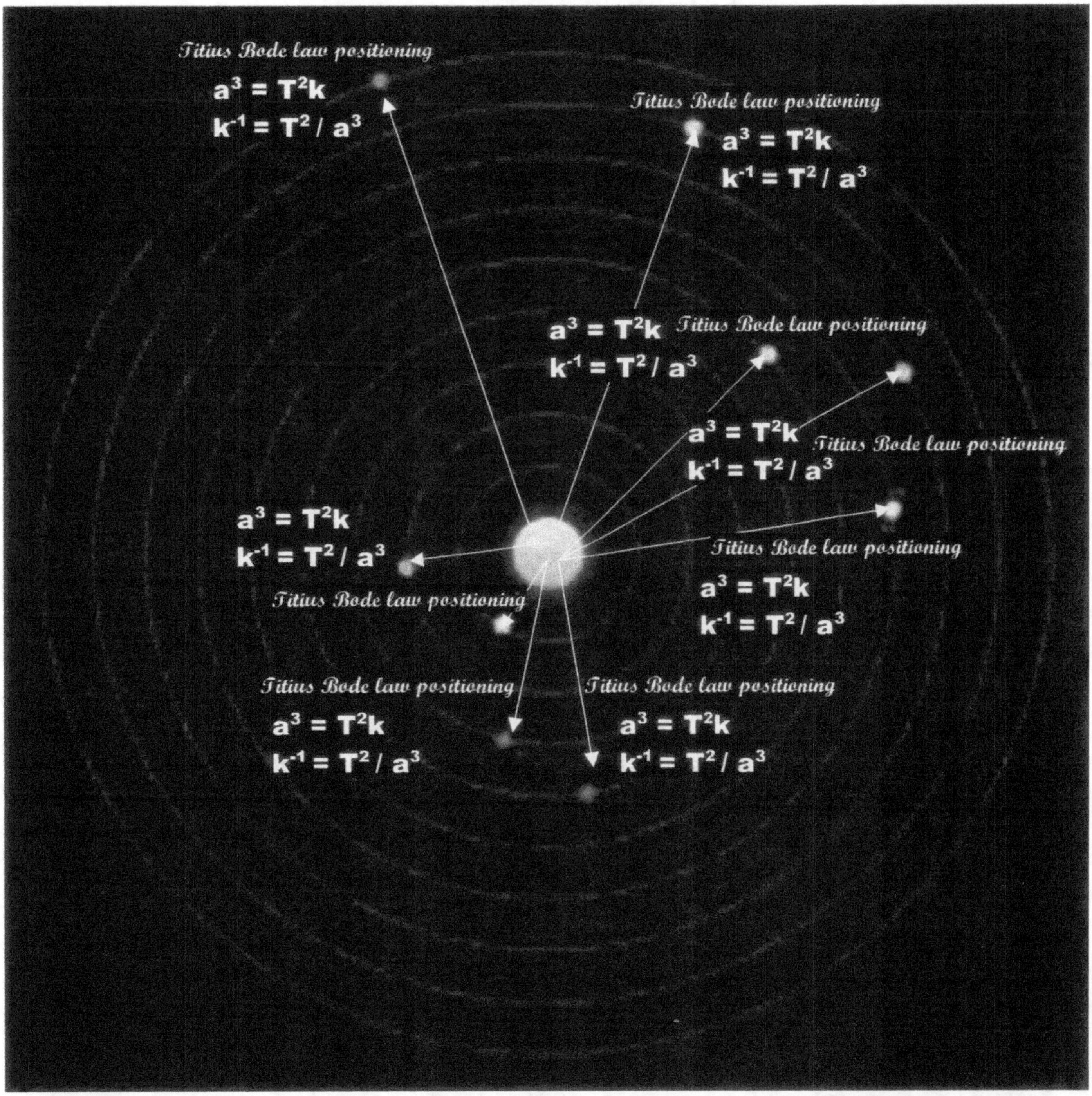

Titius Bode law positioning

$$a^3 = T^2k$$
$$k^{-1} = T^2 / a^3$$

Titius Bode law positioning

$$a^3 = T^2k$$
$$k^{-1} = T^2 / a^3$$

$$a^3 = T^2k$$
$$k^{-1} = T^2 / a^3$$
Titius Bode law positioning

$$a^3 = T^2k$$
$$k^{-1} = T^2 / a^3$$
Titius Bode law positioning

$$a^3 = T^2k$$
$$k^{-1} = T^2 / a^3$$

$$a^3 = T^2k$$
$$k^{-1} = T^2 / a^3$$
Titius Bode law positioning

$$a^3 = T^2k$$
$$k^{-1} = T^2 / a^3$$

Titius Bode law positioning

$$a^3 = T^2k$$
$$k^{-1} = T^2 / a^3$$

Titius Bode law positioning

$$a^3 = T^2k$$
$$k^{-1} = T^2 / a^3$$

By the rotation of the sun, the movement of the sun freezes space from throughout the solar system and far beyond the solar system up and to the boundaries of the solar system to move the heat by compression into the sun and finally cooling the point holding singularity within the sun.

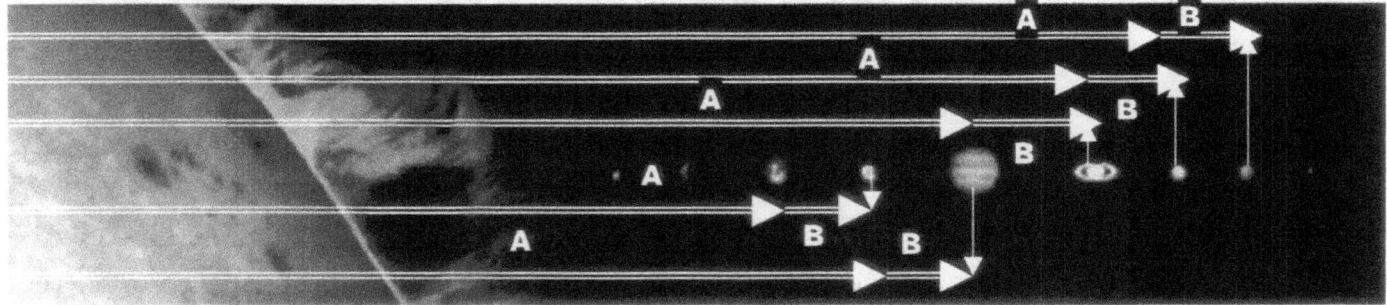

In reality in nature A *and* B *is the same distance. If* A *is seven and equal to* B *then* B *must also be 7*

As I will explain further on, by moving double 7 going square in relation with Pythagoras the space becomes 10 that forms between two planets and the sun. This placed all movement of 7 relative to 10.

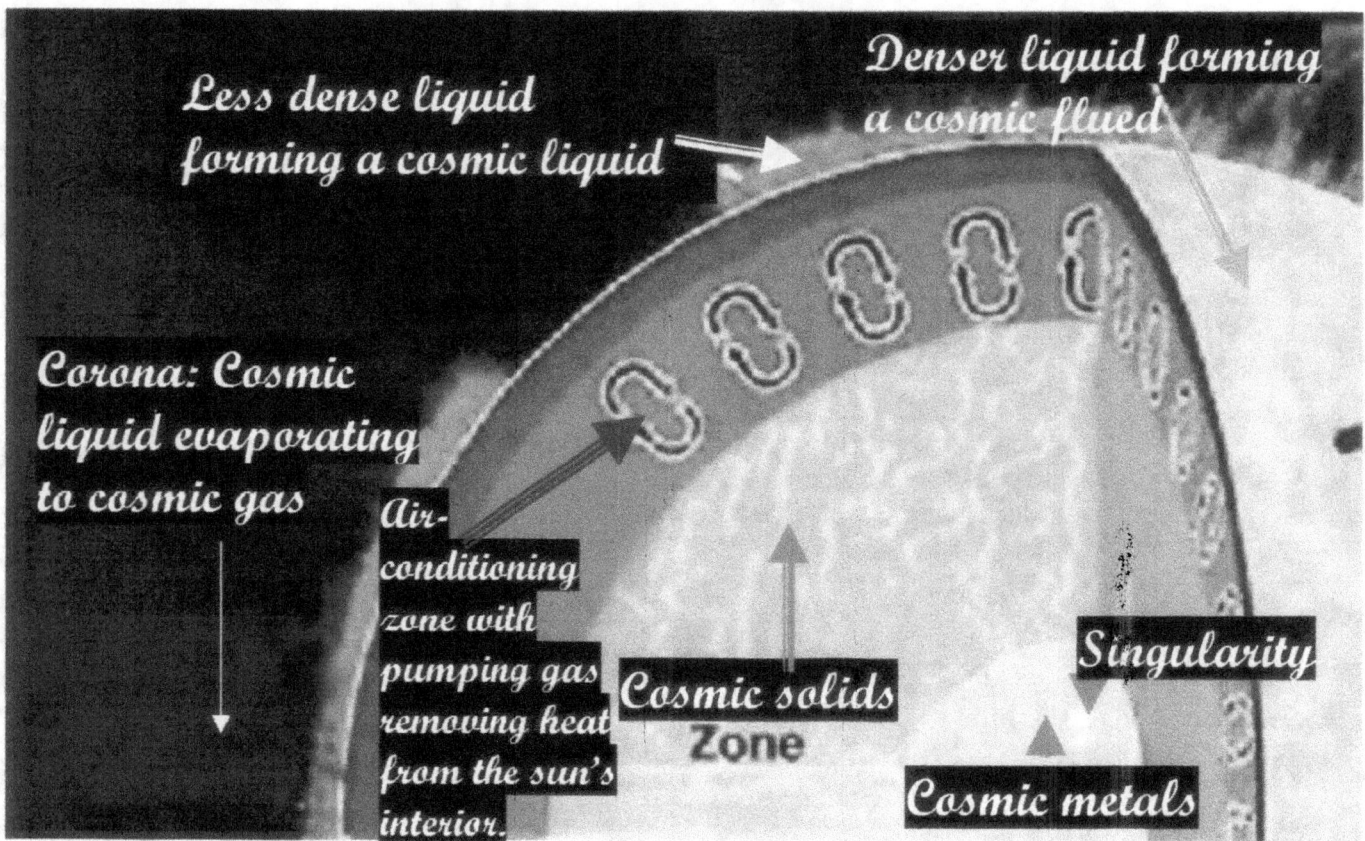

Less dense liquid forming a cosmic liquid

Denser liquid forming a cosmic flued

Corona: Cosmic liquid evaporating to cosmic gas

Air-conditioning zone with pumping gas removing heat from the sun's interior.

Cosmic solids Zone

Singularity

Cosmic metals

As the sun spins it cools the overheating heat of outer space. The heat called outer space moves from all over to fill the centre and by the spinning freezes to form a concentrated liquid around the sun. The sun freezes the gas forming outer space into a liquid by pumping gas in the form of Hydrogen and the hydrogen cools the star while forming a semi solid semi frozen cosmic fluid on the surface of the sun.

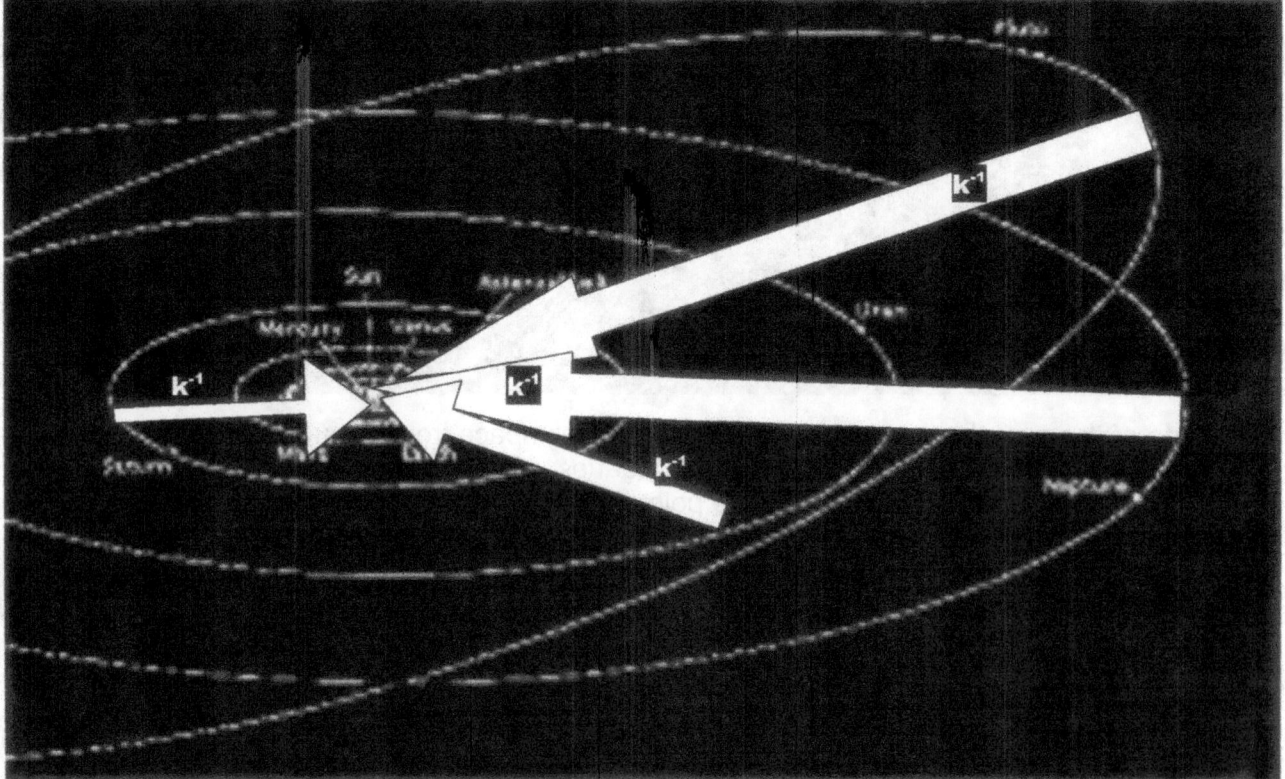

From as far away as Oord's Clouds the space forming the solar system loses density while the sun absorbs the liquid heat in outer space to help the material grow and to prevent the sun overheating. By maintaining the sun's heat within the sun this maintains the sun's structural integrity and prevent it going Supernova. Going supernova is where a star overheats inside and explodes blows the solids as liquids into cosmic gas. If the gravity cannot maintain the heat levels the star simply explodes back to gas.

With an explosion, the release of heat causes an enormous expansion of space where heat converts into space. This converting of heat into is the only means of energy drive and science do not even recognise this principle at all. Heat converts to space and space converts to heat and when condensed the heat is cold notwithstanding what we as humans think and when the heat expands to space the heat is hot because heat expands and cold condenses. When an atom goes nuclear it release heat because the spinning of the atom is so rapid that it freezes the heat inside the atom into solid material The points

holding singularity forming space is frozen solid and in that space it spins far beyond the sped of light. When they "split" the atom they slow the movement down allowing the atom to overheat. The spin of the atom becomes too slow and the atom as a structure overheats, just as a nebula does and it expands the space it holds to increase the space as it "unfreezes" the heat inside the atom.

The release of heat inside the atom is the most fearsome manmade event we on earth can witness. In the case of the explosion, where the expansion of heat to space is approximately at or close to the speed of light, the violence is totally destructive and fears enough to cause brain damage.

After 1945 the world realised that as atom is condensed heat and the release of such heat is the most destructive warring tool man has and could ever devise. Yet inside the atom is compacted heat that by massive spin going faster than the speed of light, holds so much heat it can destroy massive cities. Very obviously one will recognize the similarity between the nuclear explosion and what we witness to the outside of the sun.

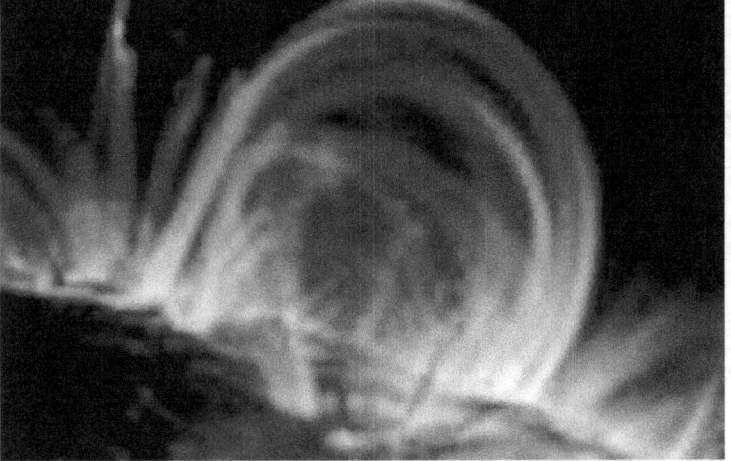

The sun, just like all other stars are one big sphere of compressed heat. It I a container filled with heat to a point of overflow. See the heat in liquid form squirting from the semi frozen outer layer of the sun. The gravitational density or gravitational movement freezes the vapour of outer space into an icy cover of heat where when the liquid overheats underneath the surface it squirts into space as it flows and return to the surface again. Some of the liquid that was condensed heat turns to "steam" as it forms a vapour that becomes the cosmic gas the outer space consists of.

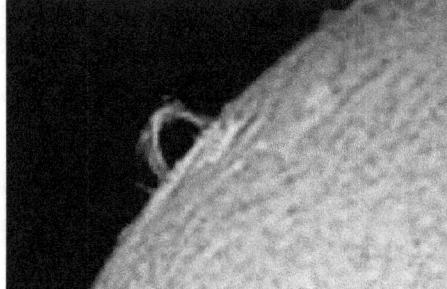 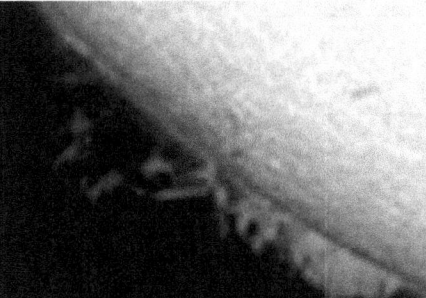 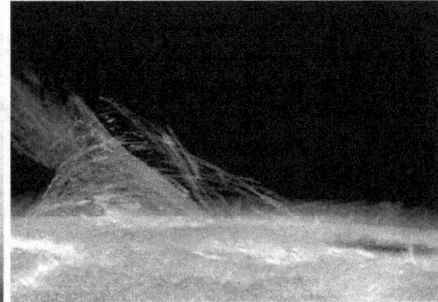

Every picture we have of the sun shows a bowl of frozen and liquid heat and is condensed to the limit. It is clear that the sun collects heat from outer space as it condenses the heat into what we see the sun is. I wish to remind the reader that the sun is a gas refrigeration pump the collects heat from outer space and the freezing is so dynamic that it becomes the biggest deep freeze in the solar system. We have to apply conditions befitting gas freezing and not a coal boiler and forms steam as it vaporizes water. That image is as far in the past tense as much as Newton's "mass pull mass" idea is and the sun or no star can ever run out of fuel or "die" or "starve" due to the lack of fuel. There are two forms of cosmic material that forms the Universe. The one is "earth" or solids and the other is "heaven" or cosmic liquid. There is no other.

We use breathing to manipulate cosmic movement by exercising muscles. By burning heat the body creates more muscle. The sun provides raw heat as a condensed energy and sends the heat in the form as daylight, which life converts in sustainable energy. The plants refurnish the oxygen with heat it gathers from sunlight and then releases the revitalised oxygen into the atmosphere. The body then uses the heat in the oxygen to burn as fuel. It does not use oxygen because if fire burned oxygen the earth would be out of oxygen billions of years ago. The muscle uses the heat that oxygen carries and that heat is the fuel the muscle uses to produce movement. If there at the moment is not enough sells the supply oxygen to the body to deposit heat as a fuel for movement the body has to produce more in storage to hold when the body requires a sudden supply of heat. We use the heat in the oxygen to burn the fuels the body supply as replenishment but that too is a matter of understanding nature, which is what Newtonians in science are not good at. The food that is carbon cannot be burned because while you might think it can, carbon does not burn. Even something as explosive as what nitrogen is, does not burn but the outburst of heat stand related to the way it holds and release heat. There is relevance between atoms move to form a liquid or matter This relevance is the basis of all gravity and gravity is the basis of everything time is and time is heat in some form. Carbon is as much a solid as iron is a gas and tungsten is a fluid.

It's a SOLID	It's a LIQUID	It's a GAS	
Hydrogen 1	melts at -259^0 C,	boils at -252^0 C,	The solid material remains the same but that which is in between the solid material increases to form a gas or changes density to form a liquid or gas. The liquid can become so dense it can almost only the solid material remains. It is not the material that becomes a gas but the way they are parted. This totally contradicts Newtonian thinking.
Helium 2	melts at -269^0 C	boils at -268,9^0 C	
LITHIUM 3	melts 180^0 C	boils at 1300^0	
BERYLLIUM 4	melts at 1287^0 C	boils at 2770^{0}C	
BORON 5	melts at 2030^0 C	boils 2550^0 C	
Carbon 6	melts at 804 ^{0}C	boils at 3470^0 C	
Nitrogen 7	melts at -210^{0}C	boils at -195.8^0 C	
Oxygen 8	melts at -218.8^0C	boils at -183^0 C	
Fluorine 9	melts at -219.6^0 C	boils at -188.2^0 C	
Neon 10	melts at -248.59^0 C	boils at -246^0 C	
Sodium 11	melts at 97.85^0 C	boils at 892^0 C	
Magnesium 12	melts at 650^0 C	boils at 1107^0	
Aluminium 13	melts at 660^0 C	boils at 2450^0	

There is a ratio between every element and the heat in which it is that will make it classify as a solid or a liquid or a gas. There are solids that spin inside liquids and liquids when less dense will be a gas.

Hydrogen 1is a liquid at when it **melts at -259^0 C,** and becomes a gas when it **boils at -252^0 C,** Hydrogen 1 is not a gas! Hydrogen is not a liquid. Hydrogen 1is what it will be when it holds any particular relevancy with the second for of substance in the Universe, which is cosmic fluid that also can be gas depending on the density at the point of movement.

Helium 2 is a liquid at when it **melts at -269^0 C** and becomes a gas when it **boils at -268,9^0 C.** **Helium 2** is not a gas! Helium is not a liquid. It is whatever it will be under a specific temperature and the temperature it is depends on the movement that it holds.

LITHIUM 3 1is a liquid at when it **melts 180^0 C and** becomes a gas when it **boils at 1300^0**

BERYLLIUM 4 1is a liquid at when it **melts at 1287^0 C** and becomes a gas when it **boils at 2770^{0}C**

BORON 5 1is a liquid at when it **melts at 2030^0 C** and becomes a gas when it **boils 2550^0 C**

Carbon 6   1is a liquid at when it **melts at 804 ^{0}C** and becomes a gas when it **boils at 3470^0 C**

Nitrogen 7 1is a liquid at when it **melts at -210^{0}C** and becomes a gas when it **boils at -195.8^0 C**

Oxygen 8 is a liquid at when it **melts at -218.8^0C** and becomes a gas when it **boils at -183^0 C**

Fluorine 9 is a liquid at when it **melts at -219.6^0 C** and becomes a gas when it **boils at -188.2^0 C**

Neon 10 is a liquid at when it **melts at -248.59^0 C** and becomes a gas when it **boils at -246^0 C**

In spite of Newtonian ignorance about nature there is no _natural_ gas or _natural_ liquid or _natural_ solid. It all depends on a ratio between heat and material movement. Carbon 6 holds a specific relevance to heat and that is why it melts at 804 ^{0}C and becomes a liquid and at 3470^0 C becomes a gas when it boils. These values apply only on earth and not even everywhere on earth! It depends on the density of the atmosphere applying at that point. The density again depends on the movement causing heat. However this density depends on a ratio between solids and non-solids. Carbon holds a relation with heat when it is a gas and when it becomes a liquid it has another relation with heat. <u>Oxygen 8</u> melts at $-218.8\ ^{0}$C and only then is a liquid at when it and <u>boils at -183^{0} C</u> when it becomes a gas. When the body inhale oxygen it takes a certain amount of heat with it and that heat the body uses as a fuel to move. The oxygen holds relevancy with heat at a specific value because at that value there is specific movement applying to the oxygen in relation to the heat it holds. When oxygen loses so much heat that it then has to use two particles to unite with one particle carbon the ratio of heat used is very much different. The loss of heat the oxygen experience has gone to feed the fibre of the human. Then the loss of one particle carbon begins to be very significant. By breathing one particle carbon is lost and by eating carbon is gained depending on the volume of intake and the particles lost in the process. In that there is balance and by exercising the body is forced to produce sells in which to store excess heat so that it can be readily available when time brings need and requirement of such heat as to produce movement. The idea of taking in so many calories in relation to that much food is a jargon of Newtonian bullshit in the same sense as mass that produces gravity. Any person exercising with me can eat as much as they like so often as they like because if they eat too much they will vomit for hours on end and with the practise I give them they will replace all the fat with pure muscle. In fact I force my athletes to eat seven meal a day because they need the carbon to replace the carbon they lodes by exhaling. Aging is in the tempo there is between food intake (carbon intake) and breathing which is getting rid of carbon fibres.

I found out why the four phenomena are in place instead of Newton ideas. Science never mentions these phenomena because science can't use Newton and explain these phenomena or use these phenomena to prove Newton. These four phenomena that the cosmos uses as we speak have been in place ever since the Universe formed. Since science can't explain the phenomena and the phenomena destroy the credibility of Newton science avoids these phenomena as if it brought the plague. When I came along and was not impressed by others telling me and did not take fore granted what others told me what to believe I became part of what was the intellectually deprived because ***I did not understand Newton...and I am proud of the fact! I might not understand Newton but I do understand nature and nature also does not understand Newton because in nature you don't find Newton! Have you ever thought about the following as you understood Newton while not giving science a second thought or test what you understand about Newton or understand about science?*** What I am about to explain is not how the cosmos started because that is immeasurably more complex than the depth I am going into with this conversation. My intensions are to show why singularity is so important, why seven is connected to the birth of the Universe and how ten became the liquid of space. Looking at the night sky you see many flickering dots and spots and light coming from afar. The one star you see seems to be a near visible dot in the picture while the dot might be hundreds or might even be many thousands of times the size of the sun...and we think of the sun as big. The dot is then that much bigger than the sun because the star we think we see could be a galactica hundreds of times the size of our Milky Way galactica but that shows in the sky as one little dot and yet that entire structure as big as it is many times our Milky Way, fit into our eye socket. But that is not all...there are trillions of such light images and they all fit into one eye socket. What we see is immeasurable and yet we see it effortlessly in the space our eye holds...how can that be? It is because we see dots and spots that form a Universe. The fact that I can see the entirety through a gap in my eye proves that I am not cosmic but I am part of singularity. Again I wish to paint the picture because if anything is important then realising what this entails is really important.

I have asked this many times before but I ask again: Have you, the person reading this, ever thought how it is possible to see that much information that you see at night when looking at the sky with only your eyes? Ever thought about how you are able to see when you see everything in the night sky and how that much light information can fit into such a small space as your eye? Have you ever sat back and think what vastness in information it is that you see when you see the entirety of the Universe when looking at the Universe at night and what the size is of everything of that which you are able to see? But that is not all...there are trillions of such light images and they all fit into one eye socket. What we see is immeasurable and yet we see it effortlessly in the space our eye holds...how can that be? How is it possible to fit what we see into the space of our eyes we have? Think how much is the entire information that is visible at night and think about how all of that fit into the space your eye holds? This is what the Universe is and that is what the Universe represents and when I put that into a mathematical equation I wish to see those persons designing space whirls to calculate the volumetric space reduction with physics. Consider how big is what is visible and put that space into the size of what your eye can hold and ask your mathematically educated Professor in physics to find some ratio between what you observe and the size of your eye.

The ratio is astonishing, but more-over what is truly astonishing is the arrogance of man to think of his position, as being important while the space man holds is beyond any comparison in ratio to everything we see in the Universe we see. Think how small we are when we are able to see the entirety out there! Even if there was other life out there, what is the worth of it in comparison to what there is that we see? Tell your physics professor not to reinvent the Universe will his brilliant mathematics, but just too mathematically formulate how all the light forming a Universe, the visible Universe can fit into my eye. No less telling is the next thought I present you with. Why would all the light coming from all over and everywhere meet you in the precise location you hold? The light left its place of origin and travelled for in some cases 12 billion years or more at the speed of light no less to meet you in the position you are at the location you are and all the light coming from everywhere possible comes directly to you. Makes you feel important does it not? What a thought this is for the further boosting of the ego that man cultivates!

It is not what science declares that is important but it is always what scientists don't declare that holds prominence and more so the reason why science keeps a silence about the information they do not disclose. It is never about what they say but it is why they don't say other things they keep quiet about. You will read how they never disclose the entire truth because science is about promoting one-sided and selectively opinionated information forming fraud no less. I have been per suiting a new cosmic theory that I partly present in an eight part theses, of which the investigating research began in 1977. In 1999 I compiled my theory and searched for a publisher, however as I don't kiss arse it was all for nothing. I use Kepler's formula but not as Newton did when he raped the living daylights out of it because he had no idea what it was about. By using Kepler's tables correctly I show where to locate the very first spot that forms the point where the Universe started. The question about the Universe is how can whatever is in view, come stored as a parcel in an electron, and tell the entire story about the entirety out there locking all that data into the space of a photon I use to see things. That is physics and however you may try, not one person searched for the ability to calculate that part, although the cosmos gave Kepler the formula.

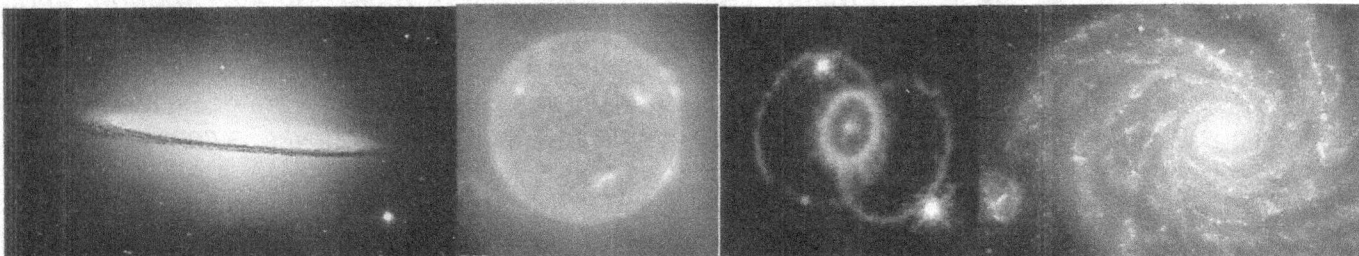

There is a common denomination in every one of these pictures and this runs throughout the entirety we think of as the Universe. There is blackness and then there is colour and what is colour is also round while we know the blackness must therefore also be round. However in the blackness the roundness is without limits making it round while the colour shows limits because of all the roundness it displays.

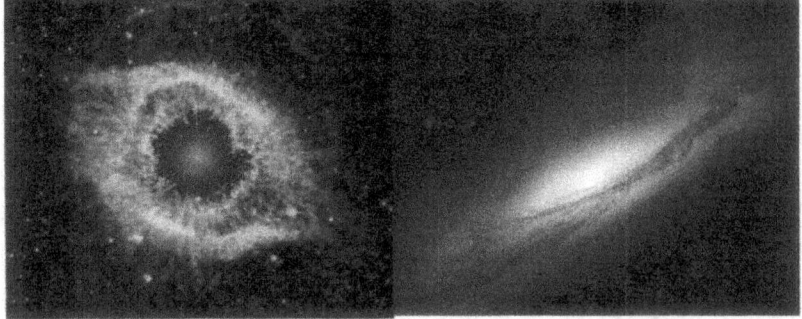

Everything that these pictures tell has one theme and that is heat being displayed in many forms and colours. It is a density variation and that results from objects holding space that moves at different rates in relation to space surrounding objects. Every picture of a Supernova tells a story of the loss of density in the liquid part of the star because materials can only lose density when it bursts by a nuclear explosion. In every cosmic picture we see heat flowing from a less dense and bigger space to a more dense and reduced space and that is what gravity is. It is cooling the Universe by applying movement.

By moving material removes heat from liquid to solids by leaving behind a cosmic gas. Look at what the inside of stars held before the blowout. Stars are filled with heat on the inside and what the star released with the blowout is heat. It is heat that comes out and that is what a star is. It is an unsealed container of heat and gravity is freezing space into a condensed heat. The reason why we think space is hotter within the earth is because we experience the heat the space releases form the confined space within the earth.

But by releasing the heat we experience the space is colder because it is finding relief from the heat it cannot carry. It cannot carry the heat because it is cold.

Outer space can absorb as much heat what is cast into it because it is so hot it can absorb whatever overflow comes its way. The cosmos is a variety of density caused by differences in movement within any specific confinement of space in relation to any and all other space formed by heat. I show what forms the Universe is small spots forming dots and these dots when compressed as it was during the Big Bang form heat in abundance but when it expands it relieves heat by expanding into abundant space. That is what the Big Bang was about. Space overheated and formed a means too cool and that became space.
The Universe is formed by heat that we call light and light is dark when it moves away from us (expanding or becoming bigger thus moving further apart and by that it is drawing visible light inwards) and light is bright when it moves towards us (concentrating by gravitational contraction because contraction makes the light denser) but everything in the Universe is a form of heat. I am about to show that in the Universe there are two substances filling the Universe.

One is material that forms clusters in atoms or stars and the other is non-material and there is no such a thing as vacuum or nothing. The Entirety of the Universe is about the relevance of density variation. The faster any object moves the slower will the solid move becoming intensely more dense until it stands still in relevancy to the liquid space that increases in movement where the density will rise to appoint the liquid is as dense as any solid is and it then reaches the speed of light. However at that point the solid will go into singularity and singularity in immovable. That then becomes a Black Hole where the solid is immovable and the density removes all forms of what space is in the cosmos. It is the ratio between occupied space and unoccupied space that increases.

This places me in life and in singularity. The fact that I can see the entirety proves mathematical I am in singularity. I am 1^0 and the entirety is $9257107472863087369421977849386523186956385 76^0 = 1$ and that makes me equal to all points forming the Universe holding singularity Π^0. If I was part of the cosmos in material I could see no further than that which I can touch with a few of my atoms holding atomic material and that is very little. This argument gets as complicated as anything and I take it much further in books I wrote. The dynamics of this not only changes a view on the cosmos but how we see life.

This past sixteen years plus saw me go without an income as I tried to get my theorem recognised as well as get my warning about the incorrectness of science noticed. During that entire time if I want something as little as a cold drink I have to ask my wife like a little child has to ask his or her parent and that broke me spiritually almost completely. I am left without self worth because those Academic Mafioso physics controlling corrupt gangsters that hide corrupt secrets about falsified facts and their corruption makes me go completely without any form of income for thirteen years. In 1998 I had my entire theory developed and since that time I added nothing substantial except get the information commercially understandable. However I made no progress in any aspect. I started a fight with academics getting them to just bloody read what I had to say and that drained all my effort. If I stopped fighting that would be just what they wanted because they wanted me to sit in a small corner go unnoticed and vanish from all records. However as I am dying I am still kicking… and fighting I will go down while exposing their corruption. The true comprehensive work is filed away and I don't think I will ever be able to publish that because that is as revealing as it is a jawbreaker to comprehend. What you are about to read is commercial information and nothing is very difficult to follow. Going all the time without a steady income left me almost destitute, penniless and morally worthless. I had no money to start a publishing effort to get my work out and my concept across to find the attention of influential readers. Now I found a way by incorporating investors to publish six books electronically as I try to get around the stranglehold of Newton dominance and Newtonian bias that controls science in every aspect there is. This total control is in place worldwide and with the help of investment I now aim for the attention of the public to have the general public decide to choose between Nature and Newton because between the two in choice there is one Universe of difference.

I compiled **a new cosmic theory** by which I eliminated all the incorrectness that Newton has burdened science with but with this being my opinion I did not find a garage full of academics supporters waiting to applaud me and to uphold my views on the matter. Gravity rests on movement of material in relation to other material also moving. The movement of the sun provides the earth with movement but not only that al movement going straight becomes circular movement and circular movement takes place within the circle in which it moves going forward as a straight line and in that idea of a circle becoming a straight line and a straight line becoming a circle the entire concept of cosmic gravity is vested. According to the Big Bang theory the Universe expands and there is no evidence of pulling bringing about a Universe contracting or becoming smaller. The Hubble constant is sole evidence of this proof of expanding Therefore I challenge the concept they build on the fact that mass attracts mass and everything is pulling everything else. Yet still I was not going to be ambushed by their relentless stonewalling my efforts and blocking my efforts in introducing both the incorrectness and the new cosmic theorem I concluded. My views are founded on what is there and what I can calculate as it is applying.

In the end all space are equal but the movement it has makes the difference it presents in relevancy. All space structures hold in the centre singularity concentrating heat and from that centre that all material holds comes all the drive. I can go on and on but heat in the centre couples gravity to space-time, just like Kepler said before he was spoken for on his behalf and without his permission or his agreeing to it. It would have been much more palatable if the Newtonian views were based on some form of a possibility but it works on total fabrication of facts that has one purpose and that is to mislead and to mesmerise by concocting untruths in the use of untrustworthy mathematics and meaningless formulas that no one can legitimise or prove! Newtonian physicist know all their mathematics are senseless because it proves only that they understand nothing except it makes them feel equal to God Almighty because of a feeling of total superiority the mathematics allow them too experience. This we can see from how they tell the Universe what they want it to be. The flowing is the result of using much superior mathematics to prove they know nothing about what they know.

Science forever hides the truth of any subject behind this veneer that the public is too stupid to understand and therefore they disclose their opinions about whatever field they wish to promote with explaining how they came to the conclusion they have. Not once ever have they lifted the cover of what they hide and about all the incorrectness of what they say. Every person on earth takes what science says in declarations as truth but because it is culture to accept their word as the final truth on all matters they never admit or divulge any incorrectness on their part. It is the information about the unbelievable oversight of Newtonian mistakes I disclose that proves how they keep the oversight of Newtonian mistakes silent and why they don't divulge that which they keep silent about. It is about them never committing to the entire story saying what all sides of the subject may offer by giving an all-round presentation of everything anyone would require to know as then to be in a position to evaluate with insight. This idea inspired my effort to find a partner that will fund the promotion of there books or a single book started off as I realised it was about time to inform the public about a science misconception.

I furthermore challenge any person to prove the solar system or indeed stars form in the manner of mass as the Newtonians teaches. I challenge every person to prove that the Titius Bode law is not the way the solar system forms. Whatever they dispute or however they ignore me I am the only one since time began to explain as much as prove how the solar system forms as I explain and prove how the Titius Bode forms. But the Titius Bode can only form by implicating the Roche limit, the Coanda effect and the Lagrangian points and notwithstanding the importance of my uncovering of science principles that were never yet understood before science and Scientists ignore me and give me the cold shoulder. I can explain gravity for the first time ever and in science no one shows interest because if they give me any credit for being correct then they have to admit that all the information they approved so sincerely was falsified facts implemented since the time of Newton and think of all the rotten egg they then will have dripping from their faces.

My work changes everything anybody ever understood about God's Creation and when going deep enough I don't only prove HOW God created the Universe but I prove mathematically that there is a God and only because of this God being the Creator could the Universe have been Created by His thoughts, but that is going deep into my concept. In another book I show that the Universe started exactly and precisely as the Bible says it did and if you change one word the mathematics I use does not add up. This I accomplished because I mathematically found the manner in which the Universe formed before material formed. I can explain how the four pillars hold the Universe in place and how the four pillars form in the process they form a Universe.

The Big Bang only came about when light formed the Universe and there became a partition between everything that spins faster then the speed of light in atoms holding neutrons and protons and we have that which spins as fast as the speed of light which is electrons and photons and then all other space forms gas as it moves way slower that the speed of light. This is why the Bible sates the Creator ordered, "Let there be light" but there already was a Universe in place by that time.

If I made a statement that Newton is wrong about gravity, which person would believe me? If I said all those in science know very well that Newton is wrong about gravity but is hiding this fact for personal benefits in order to ensure their work remains to be accepted, who would believe me. If I said that everyone in science are aware that the formula on which all science are based $F = G \dfrac{M_1 M_2}{r^2}$ is as false as a politicians' honour, this fact will then come as an astonishing surprise to everyone and I get blamed for smearing the characters of the most honourable group of persons God ever thought to put on Earth. .

This $F = G \dfrac{M_1 M_2}{r^2}$ is the formula judged to form the basis on which the entirety of physics rests. Yet, nothing can be more inaccurate than this formula that science form its entire basis on. As students learn this formula off by heart it starts off a brainwashing process and this procedure becomes everything that science represents. Should students not accept this formula as the gospel truth and as if it forms the only concept that could represent accuracy found in the entire Universe, that student would be sent off branded as not to be capable of understanding the fundamental basis of science.

That student will go home labelled as stupid without any further possibility of studying physics in the future. I prove not only that this formula is rubbish but also that there is no mass at all. Every one sharing the Newtonian vision of a contracting Universe is dreaming of a Universe where the lot would one day again come together and Creation will end where they say Creation started some time ago. The presumption is that the Universe has mass that is pulling mass towards one another and we are in the centre of an ever shrinking Universe. That is what this formula $F = G \dfrac{M_1 M_2}{r^2}$ represents. That is what the lot of us must think we know… we think we are forming the centre of the ever contracting cosmos where every Newtonian can vividly see with his or her eyes through any telescope that all Newtonian minded scientists are sharing the centre stage of the ever collapsing Universe. Newtonian science holds the view that the Universe is about to end where all mass contracts into one huge lump of material or that is the basis for Newtonian science.

There was one culprit that was responsible for this information getting everyday news and he was E.P. Hubble. Before Hubble got so outspoken about his findings the entire world of physics new they were bullshitting the public blind about Newton's anomalies but was getting away with it for centuries. Just go and question "Kepler's" laws that Newton founded and you can see how far Newton went off the track.

E.P. Hubble saw through his telescope that the Universe was moving apart. Before this everybody agreed with Newton that the Universe was contracting under the load of mass. Newton saw the moon coming closer by the value of mass as the earth was getting closer to the sun by the value of mass and to be equal to God's abilities those intellectuals could just apply $F = G \dfrac{M_1 M_2}{r^2}$ and redesign the entire Universe according to each ones personal individual liking. If you had the mass you could gauge when the Universe began. If you had the mass you could gauge when the Universe will end.

By having the mass one would find the force that drives the moon towards the earth. Before the event of Newton's miraculous discovery of forces driving planets around the sun to the tune of gravity, this was mainly God's prerogative to have such knowledge. No one gave a fart about understanding anything this formula said because if they did they would know the correctness thereof was a joke. So with nobody having the intellect to argue the correctness thereof, it was considered brilliant just to pretend one do understand and then become a member of the Brainy Bunch society ruling physics. You only have to make believe and pretend what nobody can ever understand to get so wise that you can assemble your own Universe by applying mass.

Then along came a man that had a good look at the Universe. The discoverer that researched the Universe night after night had a name and a position of seniority, which prevented others from pushing his opinion aside. The man was E.P. Hubble. Hubble saw the Universe expanding while Newton said the Universe was contracting. Then along comes this man Hubble. All Hubble had to do is to tell those not believing him to look through his enormous telescope so keeping Hubble quiet was not easy and to try to discredit Hubble would be most stupid.

Through his telescope any one could see that the Universe was expanding and the expansion was most rapid. He looked at the sky and came to a conclusion the lot was not shrinking but it was expanding. Any one that would look through his eyepiece could clearly see the lot was not shrinking. The lot was growing apart. In some cases he said the lot was racing apart.

The Universe was growing by miles and not shrinking into nothing. This man Hubble with his large telescope wrecked everything science was based on and never even flinched about it. Everything was going well and everybody was enjoying being clever as long as you were stupid enough to be a Newtonian. Then this man unveiled a fact that declassified Newton as a flawed worthless form of stupidity. The Universe expands and does not contract. Now once again only God was flawless with Newton flawed throughout. This the Brainy Bunch was not going to appreciate and this did not go down well with the most esteemed the Newtonians. They were not going to give up their ability to redesign the Universe that easily.

In order to cover up for Newton's misperception an entire variety of reasons are established, each accepted as a possible truth. The fact that Newton's principle goes begging never gets mentioned, although the only reason why it would never get mentioned is because it is the only valid conclusion and that they don't want. All other reasons they mention is overruled by Newton's principal of mass pulling. Mass pulling is the founding law that all other factors rest on. The earth slinging the Moon away can't be a factor because the mass of the earth is too great. The mass that pulls reduces the radius by the square.

This unleashed a problem the world had no name for. I blew physics apart. It cracked what was most solid before. Hubble proved everything is expanding in contrast of general opinion about a contracting Universe. Everything known to science became at that point devastatingly unknown to science. The Universe was expanding and not contracting which made the Universe quite wrong. The blame had to go to the Universe. It is impossible to have any vision about Newton being wrong. Newton could never be wrong because Newton was never wrong ...so if the Universe is out of step with Newtonian science, then Newtonian science will correct such an abnormality by finding a way to defraud science and postpone the correcting that the Universe had to comply with since the Universe owed the Master Newton some apology. The Universe had to carry the blame for the audacity to have insufficient quantities of mass.

Did the Universe not know that he whom never can be wrong has the name Isaac Newton! Decisive action was needed. At this point I cannot believe that the most brilliant minds were so naïve and therefore I must suspect deliberate deception. Hubble was far too prominent to blow away and Newton was found wanting. At that point they put the onus of proof not on Newton but turned the focus away from Newton to what they presented as the guilty party. They had to determine when would the Universe confirm its incorrectness by affirming Newton's obvious correctness and start to contract like Newton said and not expand as the rebellious cosmos is doing! If they had to admit that Newton was wrong, the most intellectual science then had to admit they had nothing to show for all their minds brilliant work. Then they would be uncovered by their stupidity being revealed and not be praised as the only wise in the world.

Science that was defying the likeliness of a living God stood bare and naked in disrepute for all to see. They put the onus of proof and converting onto the cosmos. The question was when will the cosmos come clean and prove Newton correct by having enough mass to contract as Newton insisted it must do. When will the cosmos admit to a mistake and set its crooked ways straight. When will the cosmos meet its diverting from Newton and reach a point where the Universe will finally come to comply with the standards that Newton demands. It is the cosmos that is wrong by going about expanding therefore it is time to find out when the cosmos will correct its manner and adhere to the superiority of Newton. To deal with such a task they needed a man with a bigger ego than he had an IQ.

They needed a person that thought more of his abilities than his ability to grasp any complex situation. They needed a man that was presented as a genius without ever proving his genius. They had a man that filled the centre of the Universe, which then placed the man in a location so high in status from where the man could see the entire Universe. They had just such a man. He went by the name of Albert Einstein. For all the genius Einstein had, Einstein failed to see the most simplistic and tiniest mathematical rule. Einstein failed to realise that if there was insufficient mass at the beginning of the expanding Universe, the growth of the Universe will reduce the influence of such mass as a factor further because as the radius grows, such growth will restrict the gravity by rendering the mass progressively more incompetent. However. It was more important to acknowledge that Einstein filled the centre of the Universe because that is the only place Einstein could be to calculate all the mass he saw that filled a Universe.

If the Universe is expanding as Hubble indicated, the growth of the radius will reduce the influence value of the mass as every second passes. The mass will become more and more wanting for such a task. Yet with this obvious shortsightedness about the most fundamental mathematical principle the mathematical it was this that the genius Einstein failed to acknowledge, Genius Einstein saw him fit enough to calculate and measure something as overwhelming as the Universe. As in the case of Newton, Einstein was an ego driven maniac that saw his abilities fit to measure and master the Universe while his mind was too simple to recognise the most basic principle of mathematics, the principle of relevancies or ratios. If you put something in an equation in division of the top it is the bottom part of the equation that is most important and not the top part. In $F = G \dfrac{M_1 M_2}{r^2}$ the size of the radius by the square will determine the outcome of the equation's probability and not the top part holding mass. What a mathematical genius that one turns out to be. While the radius enlarges, at the same proportion does the influence of the mass factor reduce and the mere fact that the radius increase shows that at no stage further into the future can the mass stem the growth of the radius because the radius overpowered the mass factor already. Unless there is new material entering the Universe at a point, which is impossible, the entire concept is fraud.

Science goes even much further. They sequestered Albert Einstein to measure all the mass in the entire Universe to find out when will the Universe start contracting and come to the end of its life cycle in accordance to Newton's gravitational pulling principles. This action shows that there is no limits or ends to which they will not go to find the end or bring finality to whatever they try to establish. To measure the end of the Universe is going much further than to establish when the earth and the moon will meet their gravitational destiny, with Newton and his theory of mass pulling mass so fundamentally proven. But there is no end to their resolution for they never stop with their inquest. When the Critical Density Investigation did not deliver the results that would bring satisfaction, they went in search of Dark Matter. I am going to go much deeper into the Critical Density and the search for Dark Matter later on in the book.

No outcome of sorts was necessary because they only had to shift the blame of wrongdoing from Newton onto the cosmos and the bluff was on. The idea was never to admit wrongdoing on the part of Newton and Newtonian science but to post pone, delay and divert attention away from the truth. If there was not enough mass to start with, no dark matter can kick in later on and start secondary mass frenzy that at that stage will then be enough to bring about the required mass potential that will turn the Universe around from expanding to contracting. To establish a scenario that would hide all deception they got the man that has a bigger ego than an IQ, they tell the world this man is a genius while the fool does not know the least of mathematical principles because his Master Newton did not know the least of mathematical principles and so they got him to measure the Universe. While they did not even have any device (and will never have such a device) through which anyone would be able to see what the entire Universe holds, they set of a scandalous misconception that Einstein could calculate all the mass in the Universe.

Off course as can be expected, there was not enough mass and there will never be enough mass because there is no such a thing as mass in the entire Universe. When the deceit played out to the full, the fraudsters being the paternity of physics elaborated on the delusion by trying to find dark matter that is hidden. If the dark matter did not develop enough contraction by this time, there is no chance in the future to develop enough gravity because the factor of what mass supposedly should have is tarnishing as the Universe expands. The bigger the radius becomes the less would the mass effect be notwithstanding.

With everyone in science saluting Newton's gravitational contracting there was an extended effort by Albert Einstein to find the critical density of the Universe. That is the backbreaking effort that science took with painstaking accuracy to find the density that the Universe must have to start to contract as Newton said it happens. The critical density idea did not pan out and that left science high and dry for answers about science.

They did not stop there, no Sir, they conspired more by exploring onwards in darkness to find the answer. The answer had to be in place so that Newton must come out in the end as the only idea that could be correct about mass pulling mass closer. They left no stone untouched to come up with a conclusion…and yet not one person in all that time started to think that the contraction will be much better monitored by researching the moon reducing of the radius it has between it and the earth. Would it not have been much easier to study when the moon will splash into the earth and from there work out when Newton's attraction will have the worst collision we can think of happening in our backyard? No one was after the truth because it was all about vindicating the correctness of Newton. So someone got very smart and even more conniving! Someone invented dark matter. If you can't see dark matter it means you can't not see dark matter either. So if you can't prove dark matter exists no one can prove there is no dark matter in existence. This was brilliant, befitting the thinking power only physicists could achieve.

Then science goes further and tries to detect the untraceable that is invisible. They try to find dark matter hiding in unseen places in a shady Universe. They spend billions on detecting non-existing dark material or dark energy research but not a single dime goes the way of finding out when the moon and the earth will have a gravitational self-destruction event. They tried to locate dark matter that was waiting to pounce on an unsuspecting Universe to start to pull it! Do you not find that very odd...or is it just I wondering about what? Someone in science thought it was worthwhile to study what would happen in the very solar end when the sun comes to a final exploding conclusion. Someone in science took time to bother with the end of the solar system and what will be the applying conditions during the finale era of our sun having a solar system. They even measured how big and how red the sun would be in the end but never do they bother about what is more obvious, when will the earth and moon get together. That seems very odd and suspicious to the less informed...is it because they knew already mass wasn't pulling anything in particular to anywhere?

Think of what effort it took and still takes to find the reason why the earth is growing in size as it does and why the moon is expanding by growing more apart from the earth? ...And what human effort will go into understanding all the reasons why that is. They are relentless in their quest to find answers but only when it is to cover misconduct or to instate the correctness of what can never be correct. Yet, ask them when will it come to the time when the moon and the earth will finally collide because of Newton's claims on mass pulling to bring contraction and we find only silence. It is the sort of silence I describe when I point out what to look for when one goes in search of a conspiracy. I find not even a whisper...in astronomy and cosmology, dark matter is a hypothetical form of matter that is undetectable by its emitted electromagnetic radiation, but whose presence can be inferred from gravitational effects on visible matter. This is totally fiction and is as fabricated as modern science could be. According to present observations of structures larger than galaxies, as well as Big Bang cosmology, dark matter and dark energy could account for the vast majority of the mass in the observable Universe. This means if they can't see it they can't show it or disprove it and that is brilliant to fool all the sceptics.

One thing it does not answer is if the dark matter does have mass it must have gravity as pulling power. If it is and it has mass it must pull this very instant and that will bring contraction because that is what mass does. Then what is the dark matter waiting for to unleash the gravity by mass to pull the Universe back into forming contraction. Why is it not pulling now if it is going to pull at all? Either it pulls by mass or it does not pull by mass but it can't have some retarding switch that will kick in at a time when it pleases the Newton's. What makes the dark matter slumber mysteriously while waiting to jump on the poor defenceless little Universe and force it to comply with Newton once more. No one can prove the dark matter is not there since no one can prove it is there. This is how one go about to devise a conspiracy. You keep it quiet and while everyone smells a dead rat, no one even thinks of looking in the right direction. The conspiracy is a success if everyone accuses anything but detect the true conspiracy.

There is no mass in the Universe but it all comes down to specific density. Because the faster moving stars associate with more space per time unit the overall size of the star decreases because the space the star runs through increases. However, this allows the density of the star to increase because the material then becomes more solid as it moves through more space. It becomes more massive because it becomes denser but because it becomes denser it becomes smaller in relevance. As the object moves faster it also moves through more space and moving through more space makes the space that material hold become less in comparison to the space that the object goes through while moving that fast. As the space the object goes through increases the temperature of the object decreases and the object cools down because of the movement that increases the cooling of material.

It is the same as putting a fan on that blows over the radiator of a car whereby the blowing of air increases the space the water circulates through and the water cools more because the increase in the volume of space moving in relation to the water cools the water more. When any object moves it duplicate the allocated position more as the movement increases. It fills more space in duration of time but during the instant in time it holds less space per single time unit. In this we have space-time because time going singular produces space. The shorter the period of time is that an object fills space the less space it would fill in the instant but the more space it will fill in the total duration. In other words it shrinks the space it is in by extending the space it holds relevance too. The space material moves through increases as the space material holds shrinks. This will go on until only singularity is left. This means the size of the atom inside a star reduces as the star movement in spin increases and this is why stars that move faster or is gravitationally stronger will reduce in size but increase in density.

When we go in search of what principles applies to form the building material in the Universe we better look and see what is it that the Universe shows us most graphic and we better stop telling the Universe what it is that we want to see and what the Universe should offer us that we wish to see. We better stop telling the Universe it must get mass and start to see what the Universe tells us what it has to offer us to see. If stars burst by releasing heat then stars are constructions that confine heat or cram heat into a small space. If this is true then gravity must be the process of freezing heat by turning movement and displacing space into compacted heat making gravity a process whereby space freezes as it condenses.

Looking at the Universe in singularity from this angle as Newtonian science portray the singularity picture the idea seems most beautiful and by applying the magic of mathematics it seems to be so real except for one problem we encounter and that comes when we use a dash of logic to ask Newtonian science for some straight answers. When using equated mathematical formulas one include a certain part of the Universe by excluding the rest of the Universe. There is no formula that has the ability to contain the Universe in its inclusive entirety because the Universe is eternal no matter how one would look to appreciate what one sees. Looking at any part of the Universe the distance we don't see removes the quantification of the picture that we see. What ever forms a picture of the Universe excludes untold many, many times more than what the picture we see reveals because it is shear stupidity to think what we see in any picture represents everything there can be.

This is the notion Einstein and his group of wise men had about his mathematical formulating of singularity. Lets see how they see what singularity is and don't get me wrong for it is not singularity I attack but it is how they and their mathematics come to conclude singularity. Singularity is a mathematical dimension of 1 being 1^0 or 1^1 or 4578^0 or Π^0 but the final value can't exceed the total value of 1. Having any number above 1 goes into space and singularity represents space less ness. If we look at this grid that Newtonians form their opinion about what is the position on singularity we see a form based on what looks like some Scottish kilt of some sorts and the kilt we see has waves. If there are waves there is a depth and if there is a depth there has to be space within this grand picture of singularity without space. Space has to be three-dimensional because space has to have six sides.

If I can see the topside there has to be a bottom side because the topside begins and the bottom side ends whatever fills in between top and bottom. With a top and a bottom the sides must end to have the bottom separate from the top and then it must have a left and a right-hand side. That proves they put substance in between whatever holds the topside away from the bottom side. That then is material and not singularity. If the grid criss-crosses then there is space because to validate criss-crossing grid pattern requires space and in singularity "where the Universe goes (flat?)" there can be no space so how can they with all the brains between those oh so wise find a dimensional grid with waves and all. Then when I explain singularity by involving mathematical principle I am scowled at in the phrase. The Newtonian mathematical stupidity starts with having the idea that a line or a graph starts with zero. If it starts with zero it never can end because it can never accumulate that with which it started. Whatever a line starts with the same value has to repeat in order to get any mathematical sequence. Start a line with a dot to the value of zero and this happens: $0 + 0\bullet = 0 + 0\bullet = 0 + 0\bullet = 0 + 0\bullet = 0 + 0\bullet = 0 + 0\bullet = 0 + 0\bullet = 0 + 0\bullet = 0 + 0\bullet = 0$ and then start the line with singularity continuing to form a Universe this is the result.

I would love to see what is more basic in mathematics than the mathematical line by compiling the line as follows:
$1^0\bullet + 1^0\bullet = 2 + 1^0\bullet = 3 + 1^0\bullet = 4 + 1^0\bullet = 5 + 1^0\bullet = 6 + 1^0\bullet = 7 + 1^0\bullet = 8 + 1^0\bullet = 9 + 1^0\bullet = 10\bullet\bullet\bullet\bullet\bullet\bullet\bullet\bullet\bullet\bullet$
Mathematically singularity is anything and everything to the power of zero excluding the number zero and please informs the wise about this fact. On this principle I found the way to prove how the Universe started before all material that was already present started the Big Bang process. This is mathematical reality. Whatever the reader brings to mind must not be of cultural bias because whatever mathematics brought to mind and whatever physics brings to mind the dimensional quantifying does not exist in the realms of singularity. That which connect the Universe are innumerable lines all connecting by criss-crossing as circles representing singularity. This is what made me realise how the Universe started.

There was one $\bullet$ that became one dot $\bullet$ that that duplicated in size $\bullet1\bullet2$ and the movement of duplicating what is singularity became two $\bullet1\bullet2$ dots that became three $\bullet1\bullet2\bullet3$ dots that became four $\bullet1\bullet2\bullet3\bullet4$ dots that connected as five $\bullet1\bullet2\bullet3\bullet4\bullet5$ dots after which six $\bullet1\bullet2\bullet3\bullet4\bullet5\bullet6$ dots came about and then seven $\bullet1\bullet2\bullet3\bullet4\bullet5\bullet6\bullet7$ dots formed a line$\bullet\bullet\bullet\bullet\bullet\bullet\bullet\bullet$…but it was a line and the line had dimensional references as it still has to day without having dimensional qualities as it has today by forming space. This is the essence we find when finding a triangle has the same value as a straight line and the half circle has equal value to the two factors mentioned.

The line always refers to two$\bullet1\bullet2$ (the half circle coming from movement) showing relevancies not only from where it came which is the triangle $\bullet1\bullet2\bullet3$ but also putting references applying to where the line is going because while being two the being part represents the triangle that represents the three of two going to two 2^2 bringing about the movement taking 1 going to 1 (1+1 =2) but also going in the opposing direction of going forward. This produced the double half circle $\bullet1\bullet2\bullet3\bullet4$. This is the basis by which the Universe started as it remains in movement.

In the Universe there is only one fuel and that is cosmic liquid. Anything that moves is because heat is traded for movement. When coal and fossil fuel is burnt the process depends on the release of solar energy stored in the fossil fuel for millions of years. The sun compressed light and the compressing was more than what the sun in its atoms could absorb. By not using all the heat condensed some of the heat escapes into outer space because the space in which the heat is can't manage the heat since the space got cold. A star is not a coal stove but by pumping hydrogen and other gasses on the outside is much more an air-conditioned than a boiler burning coal. When a fire burns oxygen it is not the element of oxygen that burns but it is the association it has with heat making the density into a gas that burns when oxygen releases the heat it carries as a gas.

If fires did burn oxygen there would be no more oxygen left to burn. The sun does not "burn" hydrogen but pumps hydrogen and by pumping a gas such as hydrogen this action removes heat from the interior to the outside. However it is sunlight that forms the fuel part in fossil fuel as the heat of the sun was trapped in the fossilised remains of carbon life that goes back billions of years. There are three forms of adaptation forming a relation between cosmic substances where there are two substances, one being material and the other being non-materials.

When it has no space in-between atoms we think of it as a solid and to be a solid the material must be cold, which means it, lacks heat. When the mixture of heat or non-materials and materials rise we think of the substance forming a liquid. The non-material heat is then more represented in the mixture than was the heat present when it was a solid. When it forms a gas there are much more heat present in the space than there are solid atoms and the density is very low because the non-material substance is overwhelming more. .

Everything in the Universe moves and to be within the Universe forming part of any idea within the Universe there has to be movement going straight and movement going in a circle. That is where we find the essence of Creation as the cosmos informed Kepler mathematically. To understand the cosmos we have to understand why 1 + 1 is 2 and the cosmos started with 1 growing to 2 and therefore becoming more than what was before. We therefore must know that the following value must be 1 + 1 + 1 is 3, but why would it be 3. We have to know why 2 + 2 is 4 and moreover why is 2 x 2 = 4. We have to know why is 2 + 2 = 4 + 1 = 5. Understanding this that 2 x 2 = 4 and 2 + 2 = 4 and 4 + 1 is 5 is the most basic but also the most important aspect of creation. Not knowing why 2 x 2 x 2 = 8 while 2 + 2 + 2 = 6 indicates a total lack of understanding the dynamics why the cosmos is what it is in all the dimensions it holds.

As any one can see a star is a container of pure heat but it is not a container that holds pressure. Pressure is associated with space and not with heat. When the star or heat container gets pressure it explodes as we all see in Supernova outbursts. The moment the star gets pressure the space it holds increases and that indicates a sudden increase in temperature. If the star explodes by increasing the temperature then gravity is all about reducing temperature because as temperature reduces the space will reduce accordingly. Temperature is not some scale on some meter some scientist devised in human terms but heat expands when it gets hotter and heat contracts when it gets colder. The space the earth holds might seem hotter to us but that is because it is more contracted. To be contracted (not pressured because it is not a metal container such as a boiler is) the heat has to reduce and increase shows heat levels rising. If space expands the heat increases and if space reduces the heat decreases and that is cosmic science. The levels of heat drops because the heat transmits away from the space in which it are. We can't transform heat values we have to the cosmos. We must adapt to science laws and not the other way around. True science rules say when things expand it is hot and when things contract it is cold.

A star is a cosmic atom and a star is filled with little pumps called atoms. Every atom in the star pumps heat by spinning faster than the speed of light pumping heat from the outside we call outer space to the inside of stars. This process of contracting and condensing outer space, which is expanded heat, we named gravity. The star spins slower than the speed of light in order to harbour atoms that spin faster than light. Every atom is a pump and the protons condenses space or heat or singularity in conjunction with the neutrons and how this works is it compresses expanded heat to smaller space that condenses heat by confining the heat into a smaller area which we see as a star. The star has the condensing ability or the pumping capacity equal to the combined effort of all the protons and all the neutrons and all the electrons within the container we see as a star and in that capacity we see how gravity works.

If there is movement it can only come from an exchange of cosmic liquid turning to gas or taking cosmic gas and converting that to liquid. But driving anything or moving whatever comes about as a result of cosmic fluid we call plasma and a many other names but it is condensed heat. Heat differentiation translates to movement and gravity is movement. Gravity applies when an object drops by movement or an objects keep on tending to move from its position on the surface of the earth to the centre of the earth. The frustrating of movement by stopping further descending brings about mass and that is measured by weight notwithstanding any other definition to try to correct or hide the misconception science attaché to mass. Mass is a measure and not a factor while gravity is a movement and forms a factor that results in mass. Galileo proved that all things fall equal and by all things falling equal this idea eliminates mass completely notwithstanding the corruption of a feather falling with a hammer in a vacuum. Yes the feather will fall equal in vacuum because the feather holds a different density to air because it is larger and that changes the relevant density it has to the density the hammer has. But a car falls at the same rate as a person and just as fast as the person carry bag. To stop the falling one has to open a parachute and the parachute alters the density of material in relation to the space the parachute confines and it is the density that changes because whatever is tied to the parachute did not lose mass in the process.

By going into the air as a hot air balloon is heated is pure evidence that gravity is because of density and density is the changing of mixture between solids and non-solids. By increasing the air ratio in a form of heat which is what air is and confining this increase within the parameters of a balloon acting as a container the density of the solid material changes in favour of the air and the increase in air brings about a change in density and not in mass and with an increase brought about in the density the solids and the bag becomes a gas where the lot rises into the air as if it is a gas. That shows clearly that gravity is the cooling of space because when lifting the balloon becomes ant-gravity and lifts up. If gravity is "pulling down" then anti-gravity must be "lifting up".

Let us humans first detach culture from facts. Take the argument to iron, which we know well. Iron cannot boil, iron cannot flow or bend and iron cannot brake. Iron is an element like all the other elements we know, not one element can do any of the above, in sharp contrast to human belief. As indicated in this book the limits we should find to guide us we ignore for the reason that we cannot see it. We may not be able to ever see singularity, but with intelligence guiding mankind, we do not have to see everything to believe everything. It is because we could not see religion, but still practised religion that set us apart from the other animals. At the start one would find iron and iron in a "natural state" as we find iron on earth being a human produce on the surface of the earth it will be a solid, suitable for man to handle with bare hands. When such a piece of iron is left in a desert in the midday heat, the human hand cannot handle the iron any longer without aid of covering the skin of the hand. Our perception is that the iron became hot, but that is not the case and our view is a culture contribution and not scientific fact. By heating the iron artificially with combined gasses (acetylene and oxygen or what ever) we now can over heat the iron to a state of flowing like a fluid.

To unlock scientific truth we first have to dispose of scientific misconception

In the two pictures we are seeing disposing or releasing heat creates space. We may call it plasma or shock waves or what ever, but in the final analyses it is heat turning to space. Whatever you wish to call that which lies between the particles comes from being a solid, then with adding heat, the solid *"whatever"* becomes liquid and that is the white and orange plasma that we find. That white and orange is heat in a liquid form, just as all flames and smoke is heat in a liquid form. But that liquid does not remain liquid because the governing singularity cannot enforce a commitment ensuring the liquid heat remains liquid. The liquid *"whatever"* you call the heat in fluid form then further overheats turning the heat to space. The space created must be equal to the heat reformed. That is a law of energy where energy equals equality everywhere it is. The only "energy" is to transform heat by expanding or by contraction to movement.

Our human culture tells us the iron now is melting. That is a misconception! Lines forms singularity and it is these lines that holds singularity that forms concentrated heat or expanded space as condense or expand heat and singularity expands heat or concentrates heat. After introducing artificially even more heat with more heat releasing gasses we may artificially form a condition where the iron would become a gas. Again it is not the iron that becomes a gas, it is the space the iron finds itself in that became hot enough to become a gas.

The iron particles remain the same; it is the condition surrounding the particles that changes form with overheating. Important to note is the fact that iron in a solid state will surround itself with solid matter in space applying a solid space. By introducing conditions producing *more overheating* the space or connecting between the particles become concentrated heat forming a liquid substance! It is not the iron that turned liquid but the wrapper containing the iron that concentrated so much it formed liquid fluid by the introducing of more heat to a point where the overheating created a fluid. It is considered that the oxygen burn and by that the iron heats up.

NOT TRUE!

If oxygen burns no oxygen would be left on earth by the time man arrived on earth to use it to the benefit of intelligent life. The oxygen remains oxygen while the oxygen merely does a task in nature where oxygen carries heat to a specific space. On the other hand it is the task of nitrogen removing heat from the point of overheating by means of flames whereby it creates space. One can feel the "wind blowing" as the flames generate created space. In the extreme the creation of such space we call an explosion. In the process where the space between the iron particles still further overheats, it becomes a gas. It cannot be iron that becomes gas, because depending on mixing heat with iron it will be as much a gas as iron will be a liquid or a solid.

It is the space covering the iron particle separating the different iron particles, which will convert and sustain form. The gas is as invisible as space because the gas is the form space holds. It is the relation that materials form not by heat or cold but by linear or circular motion that forms density. There are two forms holding substance earth (solids) created and heaven (heat or gaseous/liquids) created. These are the only two forms of substance that is the Universe solids and non-solids, where non-solids is what increases to form liquids and gas. It is not the solids going liquid but it is more of the liquid in ratio with the solids in between the solids that make a structure go solid or gas. There are cosmic solids and comic non-solids. It is movement going at or slower that the speed of light or solid atoms going faster than light.

Iron is a solid. Introducing more heat the iron becomes more a mixture between liquid heat that reduces the density to the point of concentration where it became a fluid. The iron remained what it is, neither a solid, nor a fluid nor a gas. By introducing more heat it becomes a gas. The gas we cannot see because the gas is space. But so was the fluid space. The introducing of heat brought about the turning of a solid to a liquid to space and every time more space becomes part of the picture. Iron is in its normal form a solid. That means the space, which the iron particles are in, is solid and that disallow the iron to alter the form in which it is. By introducing considerable heat the iron melts changing the form of the iron from solid to liquid. One will find that whatever group one chooses there are gasses and there are solids. If mass was attracting mass then the strongest mass must be attracted to the strongest mass and the least mass must float in the air. $F = G (M \times m) r^2$ hardly can even begin to explain the fact that there is a gas that is more massive than iron but floats in the breeze just as hydrogen which is the least massive element.

Nitrogen 7	**melts at -210^{0}C**	**boils at –195.8^0 C**
Oxygen 8	**melts at –218.8 ^{0}C**	**boils at -183^0 C**
Fluorine 9	**melts at –219.6^0 C**	**boils at –188.2^0 C**
Neon 10	**melts at –248.59^0 C**	**boils at –246^0 C**
Sodium 11	**melts at 97.85^0 C**	**boils at 892^0 C**
Magnesium12	**melts at 650^0 C**	**boils at 1107^0**
Aluminum13	**melts at 660^0 C**	**boils at 2450^0**
Silicon14	**melts at 1412^0 C**	**boils at 2680^0 C**
Phosphorus 15	**melts at 44.25^0 C**	**boils at 280^0 C**
Sulphur 16	**melts 119^0 C**	**boils at 444.6C**
Chlorine17	**melts at –101**	**boils at –34.7 C**
Argon 18	**melts at –189.4^0 C**	**boils at –185.8^0 C**
Potassium 19	**melts at 63.2^0 C**	**boils at 760^0 C**
Calcium 20	**melts at 838^0 C**	**boils at 1440^0 C**

Every element clearly has a different density point It is not mass that plays any role in the measured formation of elements. Elements can be on earth prone to form a solid better than it holds a liquid form but that is related to gravitational conditions we have on earth. If man would land on Jupiter there would be no man landing on Jupiter because the conditions on Jupiter could never sustain any form of life. All the above elements would have total different for limitation on Jupiter than they would have on Earth.

If one looks at mass there is no connecting point putting a realistic ratio between what is heavy and what is light? If gas floats in the air we have to presume it is lighter than air because if something floats on water the density the object has to float makes it less dense than water is and therefore it is lighter than water is. In this same manner we have to look at gas and irrespective of the "mass" value Newtonians grant the element to have, if it floats and it is airborne like a gas is then it is less dense than air which makes the element lighter than air. The hot air balloon fills with more hot air and that reduces the density that allows the balloon to float in the air. The balloon is lighter than air notwithstanding it never lost mass.

Looking at stars the way Newtonians do we they relive the coal burning boilers. They see coal furnaces being stoked to burn and heat boilers. In the days of Newton coal stoves were the nuclear science of the day and while all other departments in science moved on and away by developing away from 17th century values and from coal stove principles Astrophysics and cosmology remained true to Newton by reinventing the coal stove in so many ways not even the coal stove could think of the facets it can go through. Newtonians see stars being fuelled like coal boilers and such steam boilers can run out of fuel. This is so much Newtonian backwardness as mass forming gravity and the moon coming closer and the cosmos shrinking and we falling into the sun because of non-existing dark matter making up what is required to make Newton not to seem the idiot that Newtonians are because they make him and his contraction theory to be less foolish that what it apparently is and they overbearingly are. What is of vital scientific importance is that there are three fundamental dimensions controlling the universe. The three are beyond intermingling and one confirms a status in relation to the others but not intermingling in status. From singularity comes matter and forming space-time in own accord. By matter not controlling time, space grew uncontrolled and the third dimension came about. That dimension birth we now recognize as the Big Bang, but the Big Bang is the last of a three prong cosmic growth. Science has to recognise the dimensions of densified (singularity), occupied (matter behind the electron) and unoccupied (space-time outside the orbiting electron boundaries) forming three points of cosmic recognising space-time.

Every atom holds (I am guessing), as many dots as the sun has subatomic particles per atoms and that would still be a very conservative guess. Every dot is a controlling centre selecting a regional centre where every regional centre selects a centre. This goes on as long as there are spots forming groups as individuals unable to survive independent. The others that was unable to group formed heat that became space, which became the broken dots. The dots form groups to survive and as a group, the survival depends on doing what the group has to do to remain cool. In another book, I reserve one chapter to explain the phenomenon what I called the Lagrangian atom. These dots arrange in a manner that they could favour either the space duplicating aspect or the space dismissing aspect.

This can only be the result of the fact that even in the case of the sun, the inner space is almost entirely liquid heat and the liquid heat produces sufficient space to dismiss as the centre that holds the heavy metal particles, where all the dismissing is done. The liquidity provides motion while the solidity removes motion in the centre of the star. The dismissing going on is in the space factor where the space leads to a denser heat within that space because there are insufficient material to accommodate all the heat by the dismissing factor T^2. In that case motion far outweighs dismissing $k>T^2$ but a time comes in every star that the dismissing takes absolute charge. $k<T^2$ That is when the star goes dark. The Earth is mainly about duplication of space much more than dismissing of space and so is every structure in the solar system.

I would suggest we think of stars in the following terms. A star that generates and transmits a lot of light is weak on gravity because their progress started recently. They command a lot of space-time but the demand they have to keep their cooling acceptable is very low. In that they can generate a lot of light but with the demand on cooling low and the gravity in the centre not very developed, those stars cast a lot of light back into outer space. It is just because of the size the stars hold that tell the that the stars are still young and have a weak developed governing singularity. The stars will have very prominent hydrogen and helium layers, with the inner core not very prominent. The control of the star is still very much in the individual atoms and in that the motion the atoms have to produce in order to maintain their individual singularity will only come about through motion. The atom has to make contact with as much space-time through motion as possible since it has a very poor ability in contracting space –time in support of the cooling system.

There cannot be something big or small except in the relevancies of perceptions and then the relativity of such perceptions becomes questionable. There cannot be hot as much as there cannot be cold The sun freezes hydrogen to a liquid at 6500 ^{0}C and outer space boils over at 0 K. If we humans cannot or will not abandon our human culture driven perceptions and our mankind's pre-programmed perspective we may as well return to astrology for what the future hols.

There are so many boundaries out there ready to destroy us because of our lack of insight, as did the challenger disaster. Creation birth started off with one dot so small eternity met infinity within. Then came one more, and another and they continued coming until there were a countless number of dots. The accumulative size of the dots were the same size as one dot because in the true Universe big and small plays no part. The dots were infinitely small and eternally big at the same time because size is a relevancy and without one the other has no size. So in the true perception, there is no difference in size.

 The idea than humans make aircraft that can fly has no meaning in the cosmos. Life as an entity is totally alien to the cosmos and life can only be on earth and nowhere else unless the atheists can prove otherwise. Everything that moves in the cosmos moves because of gravity and gravity is a difference between densities in space. A gas is a lesser density because it spins at a higher velocity and not because it holds more or less mass. A gas floats in space because like the hot air balloon it has a higher relevancy to heat or it captured more heat in-between the atoms forming the compound of the element. When anything moves fast it is colder and because it is colder it captures more heat to maintain the speed of spinning it has. On the other hand when objects spin slower the elements tend to be more solid and the density is much higher.

Every element is made up in form by innumerable many dots that form singularity. Every dot was by itself as well as the accumulation as it currently is the present universe. The earth in itself is a Universe standing apart from other universes such as the moon as well as the space between the moon and the earth.

The moon is a universe.

Rules applying on earth do not apply on the moon and visa versa. When considering conditions with in the oceans and applying space-time another set of rules apply therefore the sea places a body in another universe. It takes the same engendering technology going underwater in deep sea diving that going into outer space. Every dot was a Universe in its own and the accumulation was a Universe.

The earth in itself is a Universe as the moon is a universe, because rules applying on earth do not apply on the moon and visa versa. When considering the conditions with in the ocean and applying space-time another set of rules apply, therefore being in the sea places a body in another universe. The number of universal entities is still countless, as much as it was in the beginning, before dots formed atoms. Every dot insignificantly small as it may be, is a part of another Universe as much as it is part of the accumulative Universe and every dot in the infinity holds singularity, which we translate as " nothing" being " darkness".

The very first instant when the cosmos started the perfect became imperfect. When what was perfect became imperfect the Universe moved as time in eternity split from time in infinity and I show where time in infinity is and where time in eternity is. When the spot differentiated and became differently allocated from the dot the Universe started. When infinity moved away from eternity the Universe started. When the perfect overheated hot and cold formed relevancies that put space in between time in infinity and eternity. Even today this is the fuel that drives the Universe as liquid that parted from solid reunites with solid to form a density difference. Every dot insignificantly as it may be is a part of another Universe as much as it is part of the accumulative Universe and every dot in infinity holds singularity, which we translate as "nothing" but it cannot be nothing. Singularity is what forms the Universe and is the smallest that can be. The light specks we see scattered throughout space at night are stars and stars are allocations where solid materials spinning and with that are by gravity or movement contracting gas to form a cosmic liquid.

Look at the image of this Nebula and see how it glows with heat. It is clearly show the star was a heat container that opened up and had the liquid inside burst into gas. My eyes tell me this picture is a showpiece of heat contained and the container overheated where the liquids turned to gas. In order to cool gas the gas has to flow and our refrigeration and air-conditioning work on this principle. Even the engines of the space shuttle apply this very principle to cool the heat. The star is hydrogen gas frozen to liquid by the spin of the star and in that the star is filled with liquid because the pumping of the star froze the gas into liquid. By contracting the heat surrounding the star the heat condenses and it releases heat because it became cold but due to the spinning slowing down it couldn't maintain the cold and the supernova overheated.

Every star is a Black Hole in development. As the Universe grows the density of material increases by the same margin as outer space loses density and in the end the core of the star unites all atoms within the star where the atoms by movement becomes singular in one unit. Inside every star spinning there has to be an iron core to produce spin because I prove mathematically that electricity is the very same as gravity but it is dimensionally charged on different levels.

At first when I started a study to find out more about "gravity" and the ever, elusive graviton, I came across lazy gravitons, and eager gravitons. It seemed that the lazy gravitons produced little gravity and the heavy elements such as Xenon and Radon are the heaviest inert gasses and therefore must be the laziest gravitons around while Lithium as a solid being the lightest solid must have gravitons so potent you can use them as Superheroes that replaces Tarzan on weekends and Public Holidays. This remark is pure trash only because the graviton is pure trash.

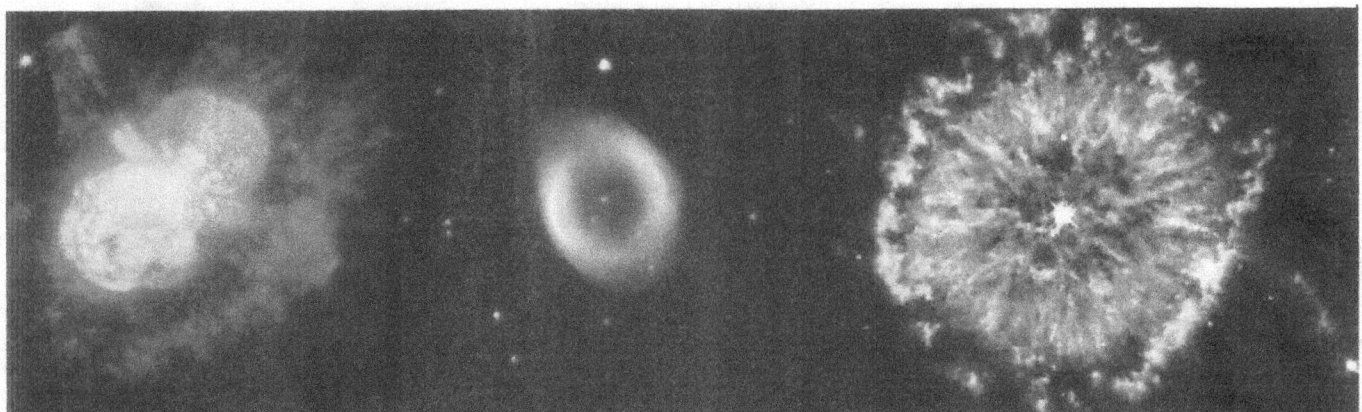

Gravity is a reality because atoms pump heat into material as to avoid overheating that results in an explosion as we can see when looking at Nebula. There are so many facts that accepted science know about, but do not fit into the perspective of accepted science, after which science blatantly ignored these facts. I have indicated but a few examples and the examples are the most basic science offers.

All information I have disproved thus far is science taught to children at school. Therefore, it is not hard to imagine how much nonsense they propagated in the more complex issues. Can you imagine for one second, a star that "collapse under its own "gravity!" …And this rubbish resulted in that two Nobel Prizes were awarded to two Nobel winners because they proved it!

It is clear from all the images that the liquid that was inside the star before the explosion froze in liquid form when the explosion thrust out what was inside the star and clearly it was heat forming space in the form of liquid that came about because what was remained on the rim of the circles. The liquid that was

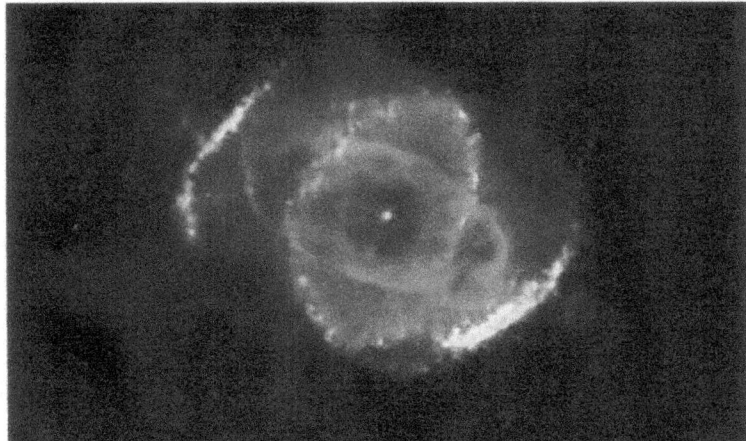

inside clearly remained liquids that froze in space resulting in liquid forming the outer edges of the layers and as it turned to gas the gas formed more outer space. Clearly the density variation is visible as what was inside formed either a denser liquid or a darker gas. It is also clear that one layer after another layer overheated and expanded by causing an explosion as we can see from what remained. This explosions and what comes about from it as "shock waves" carry many other Newtonian names but in the end it is heat going into gas as the cosmic liquid forms cosmic gas and cosmic gas becomes more cosmic space as it expands. In the picture of the sun the sun shows liquid and so does the galaxies and it is all too obvious that space in the form of gas freezes into liquid as gravity reduces the space into frozen liquid. Moreover still, we can see the point of the remaining core holding the incredibly dense material that forms the singularity within the star.

Only in Newtonians' Science as primitive and backwards the outlook is will a coal-stove fire need a thunder- storm to start the fire. Please allow me to explain, and believe me, I am not making this up as I go along, just to ad humour to this otherwise very dry letter. I did not invent this fiction in order to create a bit of comedy; this is truly Xepted Science. There apparently is a point at which point the star performing like a coal stove starts because gravity lit a fire and the fire is burning the coal (core) to get the star going. Take into consideration that every idea science clings onto today was an idea put forward when ships were driven by winds blowing and in Africa there were believed to be dragons.

This is pure hot science as was practised in the 17th century but very current. At the time everyone saw steam changing as advanced as nuclear science is viewed today. They new the boiler and steam was going to change life and the essence of it but that was then projected this massive new knowledge of the day into astronomy because only the brightest minds could understand the working of steam. If it formed the edge of science astronomy had to use it. So steam was the fuel of the Universe because Britain was the centre of the Universe. As in the case of all fires, the star (or stove) ignites with a small spark. It had to be that the fire grows as the hydrogen heats up, just as it would do with the coal that has to heat before it can burn. Depending on the amount of fuel available, and the heat to ravage the fire, this star can burn from anything between a few million years and eternity. What a lot of Neanderthal bollocks fit for fairy tales and other fantasies. The fuel the Universe holds can never run out because it is time forming space. They thought it is the mass that determines the life cycle of the stars and the bigger the star the more mass it had so the shorter would the lifespan of the star be. Today we know big stars are very soft in terms of mass while the smaller the star is the more potent it would be in terms of mass. This is because the relative density of the star increases as the overall size of the star reduces. This is all determined by the "weight" of the matter in the star. In the case of the sun, the star will "die" leaving behind the ashes of a helium core, with the density of 10^{17} g/ cc. How helium can remain as helium with such density, is yet another unsolved riddle. Every idea Newtonians believe in is centuries old. Whenever they discover something new such as the Titus Bode law, the Roche limit, the Lagrangian points, the Coanda effect, the Hubble expanding or nuclear physics it is burdened by either covering the importance, hiding it by preventing it from becoming common knowledge or turn it into a prank. Hot air lifts balloons and when this clashed with Newtonian mass pulling things down it became a joke by referring anything that overshadow its purpose to be hot air. Can you believe that they can believe a star can die? Only life and what comes from life can die because life holds time at a limit but not the Universe. To their view the Universe is limited but the cosmos shows it is eternal. Black Holes prove that stars can't run out of fuel!

So ignoring hot air they stuck to the story of mass pulling things down instead of thinking that if hot air makes things rise it must be cold air that makes things drop and that would have brought my theory to the forefront two centuries ago! To put this into context, they had the Newtonian stove burning coal and the stove starts with a small match, then it simmers, while the heat produces more weight and pressure and the coal burns until much coal burns bright in all splendour and glory. With the fire raging, the flames will light up the night sky, and the light is visible for miles around. The more the coal the stove has the more the glow is of this red, hot fire. After a while the stove runs out of coal, and with no one to stoke it, it starts to simmer, after witch the glow disappear altogether and the stove dies out. Believe it or not but this in all the out datedness it represents, this is still modern cosmology, preached by people that has so many degrease they can put it to use as wallpaper to cover walls!. It is clear that the twenty first century has not reached cosmology, because if ever I heard a Medieval, old wives tale, then this must be it.

However, this is not where the stove comparison ends. I have books in my possession, where one of the worlds most accomplished and renowned cosmologists, is of the opinion that the hydrogen falls on the red, hot core of the star and then ignites. This will happen to cooking oil that drops on the red, hot stove, but not on the inside of a star. Science only chases money! Somehow, somewhere someone found traces of minerals on the red planet, and this unleashed the feverish interest that is taking place currently. For many decades and years, there was a lack of interest in manned space flight albeit to the moon or Mars. The only interest was getting into space and manufacture semi conductors in "micro gravity" again to profit the Hoggenheimers and the John Dows of the world fit the bill with Tax money. I might sound cynical, which I am not. On the contrary, I applaud every move NASA makes all the way. All I am asking is for honesty. A small amount of earnest will go a long way. In fact, to my thinking, they can take 100 percent of all the money spent on arms each year and dedicate it to NASA. That should not be that hard to do. Tax all sales of arms by one hundred percent, no matter who is baying be it the American, Russian or Chinese government, or the African governments living on charity, yet having enough money to conduct the most horrifying wars. Less people will die and mankind will profit much greater from such an action.

The density by gravity shifts from where the substance is nonbonding to where the heat as a substance is bonded by movement forming materials. The density shifts from non-materials to materials and this flow of heat forms one part of gravity. As the liquid or cosmic gas loses substance to materials contraction the density in the non-material division loses compactness by losing compound and therefore it expands into space while materials collect heat and therefore grows in space. That is what the Hubble shift is about. As materials form density the non-materials loses density and also therefore value. A star does not grow into a Black Hole but outer space envelopes the star into a Black Hole because as the star seemingly shrinks so it is outer space than reduces the effective space that it holds. The more outer space loses compound and gain space by losing density the denser does material get but as it gains in density it loses in relative space. The cosmos is a shift in density running from non-materials to materials and this evidence is proven by the fact that the earth grows in size while the moon drifts further away and this is correlated.

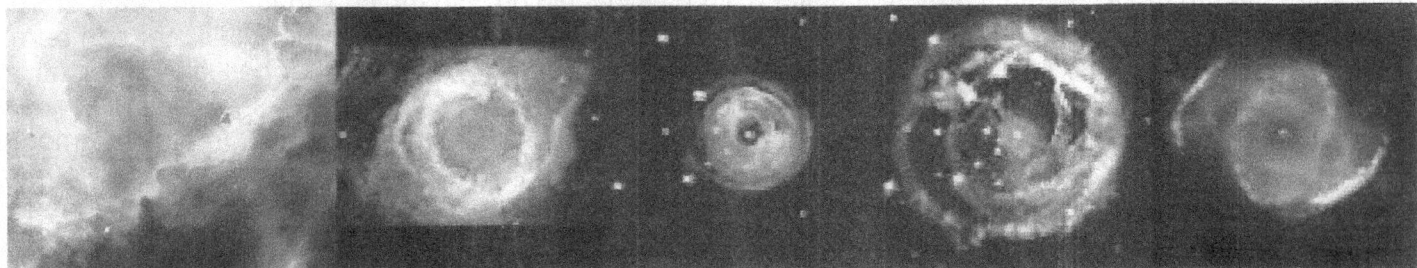

In the star heat serves to keep the atoms cool. Without sufficient heat flow the atoms will get hot and expand. When the atom expands it not only holds more space depreciating the relevancy of space to material within the star but it spins less and the heat within the tar rises even further. This forms more space as heat that expands by overheating produces space and when the space reaches an ultimate critical level the Roche limit is reached and following the Roche limit come the Roche lobe. I explain this process in much better detail elsewhere in this book. Therefore the fuel that a star runs on is not coal producing fire that can finish when the stoker retires from his job with old age and it leaves the star to *"die"*. No star can ever *"die"* because a star is not a coal stove burning as it cooks oil or burning hydrogen. The atoms are pumps pumping liquid into the star and the liquid allows the material to grow while the liquid produces the rise of relevant density where the relevant density matches the overall density applying in the Universe at that moment. A current Black Hole was a star a very long time ago and our sun is a future Black Hole a long while from now but it is in the process of developing.

In every case each one asks for a higher level of concentration or a much higher in depth understanding level than the one before or the one following because to inform and to give backing by proof requires a lot less or more concentration of factual information. The facts presented is the same but the degree of being presented with comprehensive information does alter the purchase field that the book will enjoy. However for the sake of explaining the contract I have in mind I will stick to the use of one title but all titles will be offered in a deal on equal terms. It is the prerogative choice of the investor to choose which market would the investor feel he would receive the highest return on an investment.

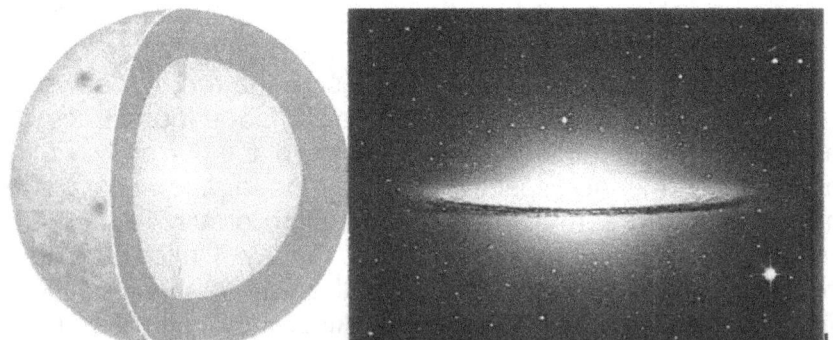

Looking at the stars and galactica through a Newtonian perspective Creation rumbles on without perspective, purpose or destination. It is a tragedy that people will be so obstinate in their programmed mentality to click on one thing and miss the entire picture. Only a definite relation between two balancing values forms the complete Universal relevancy of SPACE-TIME $a^3 / T^2 = k$ as Kepler received it from the Cosmos. However when I started my private studies I got a picture of how science saw cosmology!

Controversy and scandal makes things sell but in my case those in charge of astrophysics smother me and I am silenced because they are so powerful they could silence me up to now. To be published they must underwrite my work and that they are not prepared to do because I bring information that puts everything they say in dispute as much as disrepute. I am surprised about what flurry the moon landing controversy holds because what I have to show overshadows the moon landing controversy by light years. Considering what hoax I present makes the fact that the moon-landing event ever took place or did not take place much insignificant and quite pale in comparison.

This is the original *Rasko Jovanovic`s* formulation of the " Titius-Bode Law " This formulation is that the mean distance R(k) of the planet from the Sun is :
where **k** = 1-Mercury, 2- Venus, 3- Earth, 4- Mars, 5- Planet V, 6- Jupiter, 7- Saturn, 8- Uranus, and 9 - Pluto.
M is **1** (Mercury, Venus and Earth), **2** (Mars, Planet V and Jupiter) and **3** (Saturn, Uranus and Pluto).
N is the number of the "Titius - Bode Law " version :
there are nine planets in Solar System and there are nine versions of the " Titius - Bode Law".

Version	N	R(N)	AU$_N$*10^6 km
Mercury	1	1.2973	57.91
Venus	2	4.5708	108.208
Earth	3	9.694	149.597
Mars	4	21.7635	227.94
Planet V	5	46.8076	417.8796
Jupiter	6	103.8381	778.33
Saturn	7	233.8603	1429.4
Uranus	8	522.8772	2870.99
Neptune	-	-	-
Pluto	9	1163.8905	5913.52

There is a connection between Titius - Bode Law and the Pascal Triangle:

Planet	k	Pascal Triangle	bin(k)
Mercury	1	1	0
Venus	2	1 + 1	1
Earth	3	1 + 2 + 1	2
Mars	4	1 + 3 + 3 + 1	4
Planet V	5	1 + 4 + 6 + 4 + 1	8
Jupiter	6	1+ 5+10+ 10 + 5 + 1	16
Saturn	7	1+6 +15+20+ 15 + 6 + 1	32
Uranus	8	1+7+21+35+35+ 21 + 7 + 1	64
Neptune	9	bin(7) + bin(8)	96
Pluto	9	1+8+28+56+70+56+ 28 + 8 + 1	128

We assume the next value for the Neptune orbit:
bin(Neptune)= 32 + 64 = 96 = 2^5 + 2^6
I do not agree with the value that the value of **k** in this holds. The value that Kepler awarded **k** shows movement and not a numerical placing in accordance with a placing of the number of the planets. The plants' location shows clearly that only three positions connect directly and the others hold no influence as a unit. On the next page is the actual Kepler table showing with numbers in a column that

$a^3 = T^2 k$. The numbers in the columns form values and not positions where $T^2 = \dfrac{a^3}{k}$, $k = \dfrac{a^3}{T^2}$,

$k^0 = \dfrac{a^3}{kT^2}$ and moreover the value is $k^{-1} = \dfrac{T^2}{a^3}$. By showing that **k** goes negative $^{-1}$ it shows a direction indicating a flow and not a number indicating a planet location within a sequence.

PLANET	SEMIMAJOR AXIS $a \ (10^{10} m)$	PERIOD T (y)	T^2 / a^3 $(10^{-34} \, y^2 / m^3)$
Mercury	5.79	0.241	$k^{-1} = 2.99$
Venus	10.8	0.615	$k^{-1} = 3.00$
Earth	15.0	1.00	$k^{-1} = 2.96$
Mars	22.8	1.88	$k^{-1} = 2.98$
Jupiter	77.8	11.9	$k^{-1} = 3.01$
Saturn	143	29.5	$k^{-1} = 2.98$
Uranus	287	84.0	$k^{-1} = 2.98$
Neptune	450	165	$k^{-1} = 2.99$
Pluto	590	248	$k^{-1} = 2.99$

Reading the value correctly as a mathematical statement in accordance with the ratio Kepler placed values as $a^3 = T^2 k$ the change in ratio that then reads T^2 / a^3 has to be $T^2 / a^3 = k^{-1}$. This is just another way where science cheats with mathematics and cosmology becomes senseless. There is a column dedicated to indicate the value of **k** and notwithstanding that, the value and the column is ignored as if it wasn't there. In the one column the values range between 2.98 and 3.01 and in then other it indicates the value of singularity at between 0.983 and 1.004. These values indicate a precise position that **k** takes on in relevance to a^3 and T^2. There is no way anybody has the authority to ignore **k** or change the connection **k** holds to a^3 and T^2.

PLANET	PERIOD (Years) (T)	MOVEMENT (T^2)	DISTANCE	SPACE (a^3)	RATIO k
Mercury	0.241	0.058	0.39	0.059	0.983
Venus	0.615	0.378	0.728	0.381	0.992
Earth	1.000	1.000	1.000	1.000	1.000
Mars	1.881	3.54	1.524	3.54	1.000
Jupiter	11.86	140.66	5.20	140.6	1.000
Saturn	29.46	867.9	9.54	868.25	0.999
Uranus	84.008	7069	19.19	7067	1.000
Neptune	164.8	27159	30.07	27189	0.999
Pluto	248.4	61703	39.46	61443	1.004

PLANET	SEMIMAJOR AXIS $a \ (10^{10} m)$	PERIOD T (y)	T^2 / a^3 $(10^{-34} \, y^2 / m^3)$
Mercury	5.79	0.241	$k^{-1} = 2.99$
Venus	10.8	0.615	$k^{-1} = 3.00$
Earth	15.0	1.00	$k^{-1} = 2.96$
Mars	22.8	1.88	$k^{-1} = 2.98$
Jupiter	77.8	11.9	$k^{-1} = 3.01$
Saturn	143	29.5	$k^{-1} = 2.98$
Uranus	287	84.0	$k^{-1} = 2.98$
Neptune	450	165	$k^{-1} = 2.99$
Pluto	590	248	$k^{-1} = 2.99$

From this it is clear that **k** holds two values being **k** = 0.983 and **k** = 2.98 and not a numerical position in the planet line-up. The concept formulated by Newton to present Kepler's work as $a^3=T^2$ is mind blowing ridiculous and the figures in Kepler's columns prove that $a^3=T^2k$. How does Newton find the audacity to destroy Kepler's figures? The man became power mad with fame and thought he could change the Universe and so he did. That is what science upholds to this day. We have to see the sun drawing space closer and the only way that can be is if the sun was as cold as a freezer. $T^2/a^3=k^{-1}$ shows mathematically space moves towards the sun at a rapid rate. if k^{-1} shows a direction that is negative then the flow is not growing but it is shrinking. We better then have a closer look at the sun and see what there is to find.

We stand on the outside 150 X 10^6 km from the spectacle and from such distance we judge the sun. We don't even judge the sun from what we can see but we judge the sun from what we feel. We feel heat coming from the sun and from that we argue that the sun is hot. We see the sun has heat rising from the surface as a liquid soup. That puts the hydrogen layer as the outer layer in a liquid. Hydrogen freezes on Earth at a temperature, which is the coldest amongst all other elements. Yes, the sun is 6500^0 and that is on the outside. To a human that is hot but a human has no mind judging the sun. If the sun squirts pure heat turned to liquid from the surface and the heat falls back into the surface, the sun is a lot colder than the Earth is. The earth requires an enormous effort to cool hydrogen down to a liquid state. We must mind the way we think of the hydrogen in liquid. The hydrogen remains a solid. The element is untouched by temperature differences. It is the heat environment surrounding the hydrogen that changes from a gas to a liquid to a solid. One remove or one amplifies the heat in which the hydrogen is and that turns to liquid or solid or gas. The hydrogen is untouched in the elements worth.

Yet we see the heat flow amongst the hydrogen as a liquid. Nevertheless we remain adamant that the liquid we see as the sun is a gas and the hydrogen is in a gas and the sun is a gas bowl filled with hydrogen because in our mind hydrogen must be a gas. After all, our element table classifies hydrogen as a gas and that is the way we think of hydrogen. We do not consider hydrogen to be in a liquid state when we see the heat is flowing just like a liquid and shows all indications that it is a liquid. No the sun is hot because the sun feels hot.

In the Universe there are no hot or cold but a state of differentiation produced by time. The Universe parted by parting heat from cold when eternity parted from infinity, when Π^0 singularity parted from Π singularity, when 1^0 parted from 1^1. There is no hot or cold but there is a relevancy where one factor cools and another factor overheats. By retaining the sun is the coldest space in the solar system and outer pace is the hottest there can be.

From since the time that man discovered intelligence (if he ever did) man has been with the presumption that the sun is the hottest centre in the solar system. Later on in the present time, it came to someone's

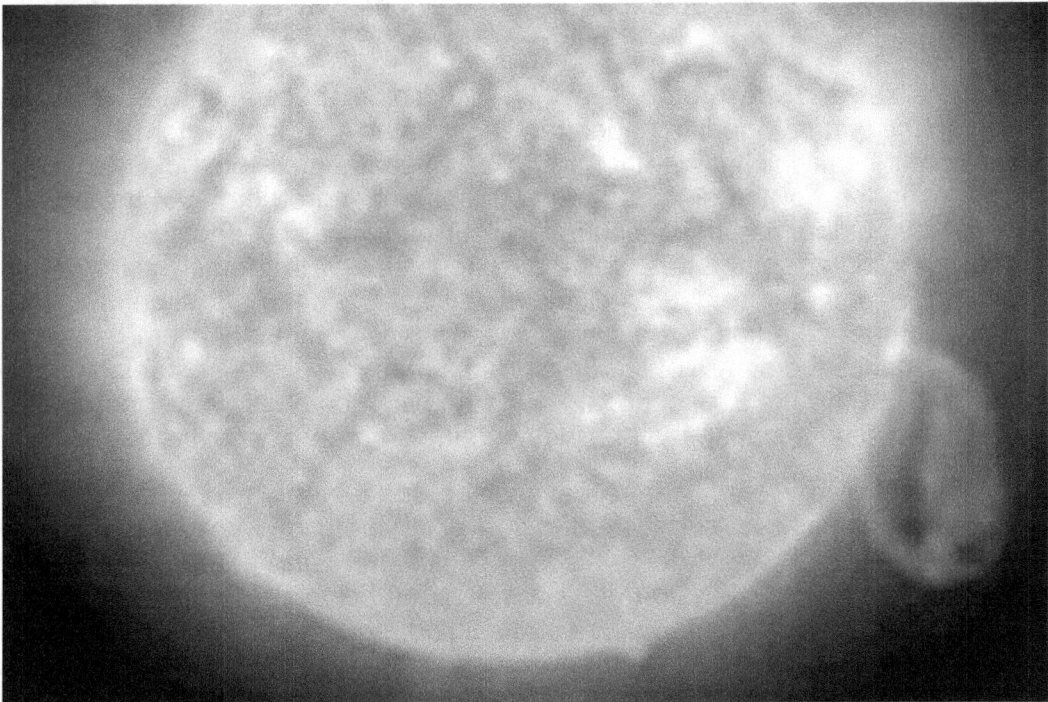

attention that the sun also holds the solar system in gravity. The Earth by its standard and dominating its sphere of which it can control with influence is the hottest centre in the space of its domain and it holds the moon centred to the Earth. The gas planets are the hottest centres in relation with the most heat and they all hold their satellites captured by a hot centre. All space structures hold in every centre there is that is confirming their independence at that point of securing independence the centralizing of the most heat it is able to concentrate and from that centre holds all material captured or controlled in the domain of what that forms the independence of the structure. I can go on and on but heat in the centre couples gravity to space-time, just as if Kepler said before he was spoken for on his behalf and without his permission or his agreeing to it.

$a^3 = (T^2 \ k) = a^{3 \ +2 \ + \ 1} = 6$ with the sphere presuming the position of singularity as part of $k^0 = 1 =$ **singularity**. Einstein proved that at the point where space reduces and such reducing reaches a point where space as a factor in the third dimension disappears into the single dimension (space going flat) gravity is overwhelming. Einstein interpreted this, as the complete Universe going flat but while it may be true that the Universe is going flat, that can only be within singularity since singularity represents the Universe as flat as it can get.

The centre of any sphere has to be at the very point where space completely falls away. It is at the point where all the points of line centres meet by the crossing the centre of their individual connection coming in to contact as a group. In that way one may assume that the lines connecting the controlling points on the other end are crossing on a centre point that all that is participating in the constructing of the sphere is democratically electing such a centre. Please note this conclusion very well because this forms the heart of the Coanda principle. That will put that position where the lines cross which in itself is centralising all space in the sphere at that point, such crossing point will become very distinct and controlling where that point forms in the single dimension and singularity is the single dimension. Kepler also solves another riddle that truly got Newtonians unstuck. This, to which I now refer, is what is referred to when they refer to the Hubble constant.

The growth we see in the Universe is an adding of space in every cycle completed by every cycle, which all the protons complete. The adding is the smallest addition that can come about in the shortest period of repeating by cycle rotation there can ever be. This growth of space-time next to singularity confirms the growth of singularity as singularity recalls the space it uses to grow in the time it grows. The margin of growth will be by the extension of k in the formula $k = a^3 / T^2$. Every cycle completed in the relation to space by the initial value of k. $k = a^3 / T^2$ leaves ultimately a^1 extending as space or as Kepler chose to indicate it as k^1. That too has to be compensated by the duration of time reducing the time aspect by the margin that the space expands. This confirms what is evident in the Hubble Constant. The further one looks at time the more time seems to race because time has the invert properties we give to space.

There is a position that is in motion that is forming the very edge of the outside. To be in motion the position must be in relation to a point from a centre. From the centre, there must be a specific allocated space ending at the object in motion and starting from a centre that has no dimensions. The object in motion determines the one limit and the centre with no sides and no space, which is standing still in singularity, determines the other limit. By that we can see there are only one way of looking at what we can observe and that is from the outside in.

The atom must be the utmost coldest and the proton is even much colder because when that cold escapes it turns to heat forming space that no one can understand. When the spin of the atom allows the cold of the atom to release the heat it has, which is heat it had frozen to space the atom holds but when this heat releases from the containing form of the atom it brings about much more heat than the Human mind can cope with. One may not look at the material and judge the surroundings. The fact that hydrogen remains a gas and so does helium in outer space must serve as enough proof that outer space is hot, regardless of our interpretation of the temperature gauge telling us what we wish to hear. One must look at outer space and judge outer space from the findings only considering outer space. If helium remains a gas, it is hot. The removing of heat makes the centre of the Earth cold although we see it as being terribly hot. The only reason why it can seem to be hot is because it is cold and in such a cold environment, the heat can gather and space can collect heat because the particles find the surroundings extremely cold.

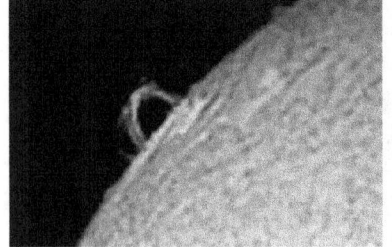

The cold in the earth centre causes the concentration of heat by space reducing, as all cold surfaces tend to do. If it was hot, the space within the Earth would expand and the space within the Earth where we think so much heat is concentrated does not expand therefore it must be cold. To gather and accumulate the space in a liquid means it became much colder being a liquid. Finding the surroundings terribly cold will allow the heat to gather and not expand but when the surroundings are hot, it will not tolerate more concentration of heat and thus will expand to rid the balance of excess heat within space. Look at the sun and see how the sun turned the hydrogen to a freezing cold liquid at 6500 K. Hydrogen is in a fluid state within the sun and is colder than the hydrogen that is in a gas form in outer space. The sun is the coldest place in the solar system. That is when the protons oversupply the removing of space to produce the cold that is so apparent. By the reducing of space, it can concentrate heat to a fluid state by producing the opposing cold that finally freezes the heat to a solid state. The expanding of space is a way of duplicating space without reducing space and by duplicating in the form of expanding it becomes just the opposite to duplicating by motion therefore reducing space by halving space in time. That is what gravity does. By motion, space duplicates and by space, halving it removes heat in space as well as by dismissing space. In all the applying of gravity, space dies. The density of the protons brings about space dense enough to harbour the heat in such quantities and visa versa applies in outer space. Movement places solids into more space thus spreading the heat that forms the space of solids over a larger area. This is why blowing a heated object with a fan cools the object down. The blowing fan distributes the object over a wider area and then this distributes the heat where the heat then flows into a larger area of air. Moving the air or moving the object through air has the same effect because it is in the movement that the distribution of heat gets exchanged.

When an aircraft stands still it holds a specific ratio of body or solids in relation to liquid or air. As the aircraft moves and the faster the aircraft moves the bigger this relation becomes because of the increase in air it makes contact with in the space it holds. What this brings about technically is that the body stretches in relation to the size of the ratio it had when the aircraft was standing still. The faster the body of the aircraft distributes the heat it holds over a wider area the colder the outside becomes in relation to the inside of the body. However the aircraft body can stretch much less than the air can shrink or compress and this balance of space falls much more to space or air compensating for the balance changes that takes place.

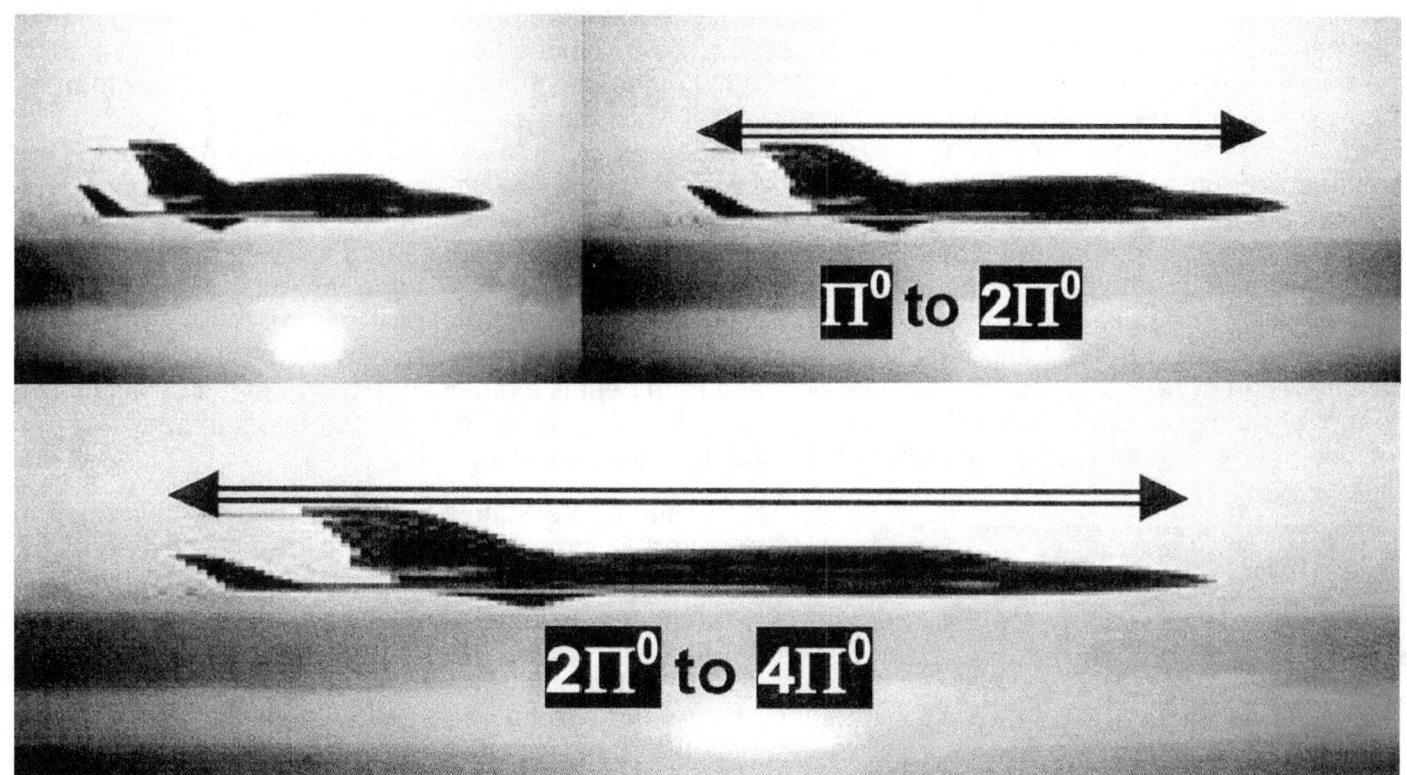

$$\Pi^0 \text{ to } 2\Pi^0$$

$$2\Pi^0 \text{ to } 4\Pi^0$$

For that reason the movement of the aircraft shrinks the surrounding of space it has so much that it forms a cloud around the aircraft. However with the water also being a solid atomic structure the water cannot compensate is the loss of space as much as the concentrated air can and the water becomes a cloudy vapour. However the aircraft has to increase in space and the temperature difference will vaporize the concentrated cloud quickly and so the aircraft discards the liquid.

The application of gravity that condenses space and bringing about heat by the compressing of space we apply in the way we go about tapping into the energy that nature provides. Internal and external engines combustion engines all rely on this application for harvesting motion by driving power. The pistons compress space and with a piston moving up into compression in a cylinder the piston pumps the compressed air into a container and such confining of space will increase the heat by the piston where this effort will reduce the space brought about in the container. The heat coming about inside the cylinder has no relevance to particles colliding because all compressor cylinders cool down. The walls becomes colder because when that cold escapes it turns to heat as the heat releases from space forming a secondary form of material forming space that no one can understand when the spin of the atom allows the cold of the atom to release into uncontrolled space. This release and unification with space that heat does is the heat it had frozen because the motion of spin to space that the atom holds remains in a frozen state under the guard of the spinning electron. When this heat releases from the containing form of the atom frozen by the spin of the electron it brings about much more heat than the Human mind can cope with. One may not look at the material and judge the surroundings.

The fact that hydrogen remains a gas and so does helium in outer space must serve as enough proof that outer space is hot, regardless of our interpretation of the temperature gauge telling us what we wish to hear. One must look at outer space and judge outer space from the findings only considered in the terms which outer space insists upon. If helium remains a gas, it is hot. The removing of heat from the space that contained the heat makes the centre of the Earth cold. In our Universe we see it as being terribly hot because the heat then forms a separate substance but remains a form of material but that is because we see the heat and not the space derived from the separating of the heat. The only reason why the space can seem to be hot is because the space is cold and in such a cold environment the heat can gather in a much concentrated state and space can collect heat because the particles hold concentrated heat in the space separating the particles. There are two forms of space, solids and liquids or as the Bible says, heaven and earth where earth is solids and heaven is liquid sky.

This process mentioned above increase exponentially in the case of the turning sun. By removing such high concentration of heat from the space that used to be expanded heat, the space then must contradict the heat by being extremely cold. We look at the heat in the space, which by that time is another form of material and find the surrounding heat in the space hot while the space is extremely cold. The cold in the Earth centre causes the concentration of heat by space reducing, as all cold surfaces tend to do. The proton contributes to that reducing of space. If it was hot the space within the Earth would expand and explode but the space within the Earth where we think so much heat is concentrated is so much it does not expand therefore it must be cold. To gather and accumulate the space in a liquid means it became much colder when the space parted from what then is being a liquid. Finding the surroundings terribly cold will allow the heat to gather and not expand but when the surroundings are hot, it will not tolerate more concentration of heat and thus it will expand to rid the balance of excess heat within space. The concentration or release of space with heat or space from heat is a direct contribution of the singularity in control of the space-time. The regard of the singularity stipulates the conducing of heat in space or the release of heat to form space by means of bisecting the occupied space.

Look at the sun and see how the sun turned the hydrogen it holds captured in its atmosphere to a freezing cold liquid at 6500 K. Hydrogen is in a fluid state within the sun and yet it is still colder than the hydrogen we find in outer space that is in a gas form in outer space. The sun is without any doubt the coldest place in the solar system…it freezes hydrogen! Look at the picture and see for yourself! That is when the protons oversupply the removing of space to produce the cold that is so apparent in the heat levels that do not join the spell. By the reducing of space, it can concentrate heat to a fluid state. By producing the opposing cold that finally freezes the heat to a solid state, we find that is what matter is. The expanding of space is a way of duplicating space without reducing space and by duplicating in the form of expanding it becomes just the opposite to duplicating by motion therefore reducing space by halving space in time. That is what gravity does. By motion space duplicates and by space duplicating the material must be by dividing or bisecting - halving it removes heat in space as well as by dismissing space and in that concentrating heat. The density of the protons brings about space dense enough to harbour the heat in such quantities and visa versa applies in outer space.

The particles claim more space when heated to preserve the cold. The claim to more space produces more space and reduces more heat. Such expanding brings about cooling. When particles heat or cool motion applies in some form. Motion started at a point when the Universe was extremely hot and there was no space. By introducing motion space formed and the lack thereof produced friction that became heat that became space. It is natural, it is simple, and above all, it makes believable sense.

The application of gravity is that which condenses space by bringing about heat with the compressing of space. We apply the progress we have as a species in the way we go about with our skills to unveil ways we can tap into the energy that nature provides. Internal and external combustion engines all rely on this application for harvesting motion by driving power. Compress space even today with a piston in a cylinder and then pump the compressed air into a container and such confining of space will increase the heat by the piston effort to reduce the space brought about in the container. The heat coming about inside the cylinder has no relevance to particles colliding because all compressor cylinders cool down with time moving and not necessarily with the loss or release of particles. It is not only the discharging of air that will reduce the temperatures inside the container. The time flowing bringing motion about where the motion is not about particles escaping but heat escaping in the replacing of the heat density (not the density of the particles forming the material content within the container) but the space that compressed to heat will also bring about that the heat displaces through the container wall to the outside. This is bringing about equilibrium where heat will always flow from more dense areas to the lesser dense areas. This has no influence on the status of the particles on the inside of the cylinder but only concerns the density levels of the particles inside versus outside. After the pumping of air increased the heat in the cylinder which even can go to dangerous levels, will reduce back to room temperature when further pumping ceases and that stops further air movement into the cylinder and such surging of pumping air is what brings about heat stabilizing.

Mainstream physics ignored the clear connection completely, notwithstanding it being so very obvious. There is this far in their recognising of principles in natural physics not one single reference made to prove their appreciation of this matter. They are bent on particle colliding. When particles collide, such collision forms an atomic thermo release and that action we call an exploding atomic bomb. What principle this argument about particles colliding ignores is that all atoms use negative charged electrons forming the atomic limit on the outside forming a definite border to the boundaries of all atoms and in both electrons from different atoms are being negative charged. In being negatively charged, it means both will come out and totally reject the other. The closer they come the more violent the rejecting will be and such rejecting is the production of heat that will turn to space. The electrons repel other negative charged sub atomic structures, which the electrons are that form the outer borders of all atoms. With all electrons highly negatively charged (being as negatively charged as any possibility will allow to match the utter extreme) such electrons could not touch.

It is about time scientists start looking with their minds and not their eyes at the Universe and see what is truly out there to see. All the difference we find is seated in the human mind. We humans set differences because we look at the cosmos by placing humans and the life we find on Earth in a pivotal centre in the cosmos instead of placing singularity in the centre and life where it belongs; only found on Earth. Einstein proved mathematically that in the presence of a strong gravity such a strong gravity slows time down. Surprisingly with that evidence being around this long nobody in science since Einstein's discovery took those statements and made any further progress from that. It seems to have been left in some drawer to dry. Science still sticks to the opinion that time did not change, not even slightly, since the beginning of the time and holds the same pace ever since the start of the Big Bang notwithstanding the implications this concept carries. Before the Earth took one year to circle around the sun and even before the sun was there a year was still the same duration of one year. How odd... don't you think ... that the only aspect in the entire Universe that is beyond change is the aspect of time? With the entire Universe including all the gravity now present and not excluding one Black Hole or dust speck pressed in such an area that was possibly the size of a lepton even then the gravity extending from that circumstances must have been beyond what words can ever describe.

When everything was that small when the Big Bang took charge, the gravity at the time was beyond light, because even today in the Black Hole the gravity is beyond the speed of light. If the gravity was that high and Einstein already proved that strong gravity slows time down, then there is one logical conclusion and that is that time was in fact at the time of the Big Bang standing still. Mathematically it is incorrect to allow gravity to compress the Universe into a spot smaller that an atom and exclude any other factors and relevancies to change.

Should the moon landing be a hoax, then one department in the U.S.A will be damned by the public for one presentation but the hoax I uncover puts the entirety of physics concerning astrophysics in the open as the biggest hoax invented by man in the history of mankind. The fraud (and I call it fraud for there is no better description) that science hides from the public overshadows the controversy about the moon landing by many miles. I have worked on this subject that I present ever since 1977 and I also bring the solution to remedy the matter. Yet no one in science wants the public at large to read my work and be informed about the many details that I uncover and what I bring to the table to solve the unproven dogmas they present as fact because then they have to admit there is a problem with their science and that it is based on fiction and not on facts.

I compiled **a new cosmic concept** by which I eliminated all the incorrectness that Newton has burdened science with but with this being my opinion I did not find a garage full of academics supporters waiting to applaud me and to uphold my views on the matter. Gravity rests on movement of material in relation to other material also moving. The movement of the sun provides the earth with movement but not only that al movement going straight becomes circular movement and circular movement takes place within the circle in which it moves going forward as a straight line and in that idea of a circle becoming a straight line and a straight line becoming a circle the entire concept of cosmic gravity is vested. According to the Big Bang theory the Universe expands and there is no evidence of pulling bringing about a Universe contracting or becoming smaller.

The Hubble constant is sole evidence of this proof of expanding. Therefore I challenge the concept they build on the fact that mass attracts mass and everything is pulling everything else. Yet still I was not going to be ambushed by their relentless stonewalling my efforts and blocking my efforts in introducing both the incorrectness and the new cosmic theorem I concluded. My cosmic concept is that the Universe is about heat forming densities. It is the density of hydrogen making it a gas as much as it is the density of the massive Krypton, Xenon and Radon that makes these elements gasses although the gasses are the heaviest inert gasses and Lithium as a solid being the lightest solid.

The gas I mention is many times over as massive as the lightest solid is and yet with all that massiveness, it is gas and gas floats in the air. The gasses form what they are because they are a mixture of heat and material putting a factor such as mass completely out of the picture. All materials are solids as they are liquids as they are a gas because they can be frozen into solids and melted into liquids and vaporised into a gas. In each case the density of the material changes from forming a solid to going into a liquid or becoming a gas. It is about density putting a relevancy between the status of materials.

One must realise that time forms space as "space" forms the history of time left as light in "space" and that the "space" between the earth and the moon is not "space" but it is what time left behind as space. To know the age of the earth and the moon one must take the expanding that happens every year putting "space" between the earth and the moon and put that in relevance to by what time leaves as "space" and from that find the true age of the earth. As the earth "grows" by becoming "bigger" so the moon and earth forms distance by "space" according with time moving. This movement "away from" and "going bigger" is the true and only measure of how much time developed the Universe and that we see as "outer space"

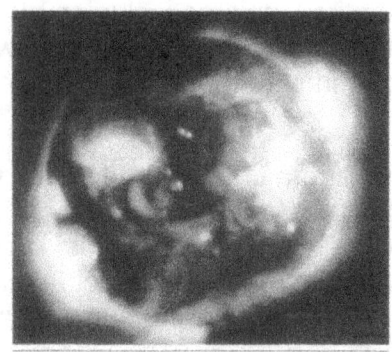

The sun reduces space to a liquid and that the pictures show with liquid and not gas squirting from the star. Instead of looking at what they see they look at the 6500° C and according to that scale declare the sun as hot. Hydrogen is a liquid on the surface of the sun and as it comes into contact with outer space the friction caused by the movement makes the hydrogen or some of it overheat again where the hydrogen turns from liquid to gas. The rest of the material that squirts into space does not overheat and returns to the sun as a liquid and as cold as a liquid. The sun moves extremely fast in comparison to all other planets and by moving so fast it freezes the hydrogen in outer space from forming a gas in outer space to becoming liquid within the sun.

In perspective to singularity relating to the centre of the earth it is the earth that stands still because the connection from the centre to any point on the surface never moves as the line that

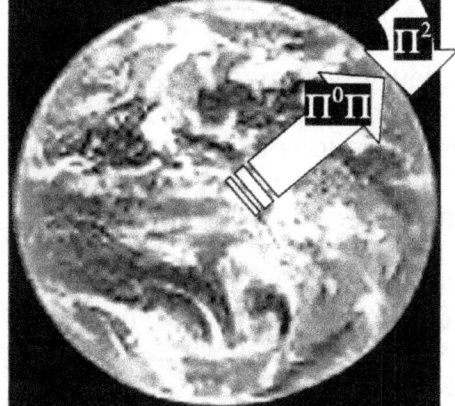

forms a connection between singularity and the surface remains still. That means all movement is in the liquid aspect.

This puts everything that changes as a part of liquid movement although it forms a "mass" connection. Although the line ends at the surface the line running from the centre of the sun is connecting the relevance, which extends to more than a third of the distance going all the way to the next stars, which is Alfa and Beta Century. This line connects every planet to the sun and that is why the Titius Bode law positions all planets. However from the perspective of the sun it is Alfa and Beta Century that moves at a rate the sun can't actually cope with while the sun is standing dead still. The moon by moving in a twenty-four cycle (according to the earth centre the earth is holding still and the moon is moving in a double shift) as well as forming a cyclic connection of $7° \times 4\Pi°$ it also rotates once every (about) twenty eight days but it is the moon that moves on both accounts and the earth is dead still.

Gravity is about movement freezing space and we better forget we feel heat and start to think as the cosmos operates. The cosmos is not human and holds no human concepts. Kepler shows in the tables that space a^3 reduces k^{-1} as the sun spins T^2. When space being three-dimensional divides into movement, which is square, space, declines or reduces indicating the relevance (k) goes negative or the distance becomes smaller k^{-1}.

Mercury	$T^2 \div a^3 =$	0.983
Venus	$T^2 \div a^3 =$	0.992
Earth	$T^2 \div a^3 =$	1.000
Mars	$T^2 \div a^3 =$	1.000
Jupiter	$T^2 \div a^3 =$	1.000
Saturn	$T^2 \div a^3 =$	0.999
Uranus	$T^2 \div a^3 =$	1.000
Neptune	$T^2 \div a^3 =$	0.999
Pluto	$T^2 \div a^3 =$	1.004

This is a mathematical statement showing physics reality that not even Newton can break because this is physics and not that three dimensions a^3 is equal to two dimensions T^2 as Newton stated by declaring $a^3 = T^2$. Anyone stating this as accurate has no mathematical sense or has no inclination about sensible physics even if the idiots name is Isaac Newton! The Table I show is Kepler's finding and that shows who is correct, I or science that upholds Newton's views that $a^3 = T^2$.

Individual movement of material occupying specific space is forming density in relevance to all other material moving at various but specific speeds and the faster any atom or material moves, the denser form the movement will make the material to be. Seeing relevancies apply in the picture above it is not the mass that increases but it is the density of the material within the star that increase by claiming less space to hold more material in an denser environment. As cosmic gas or also known as outer space expands it moves slower while the density decreases.

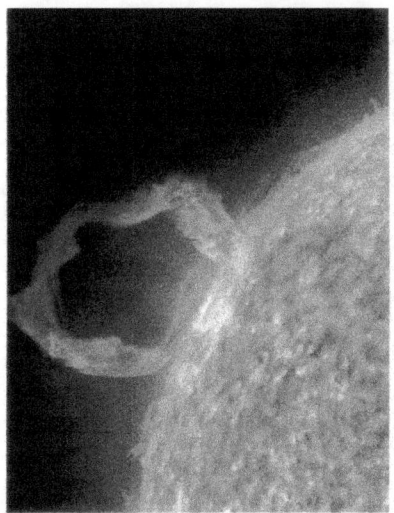

The increase of the density of stars reducing space while becoming denser with more material in less space comes about by more material within less space spinning faster because of reduced space bringing about faster circling of material within a smaller confined space. In contras outer space again is moving slower because the space increases through expanding and more space moves slower. This puts the applying relevance on material to move faster in relation to outer space moving slower and thus material becomes denser as it moves faster while it is in ratio with outer space expanding and thus moving slower. This ratio allows material to move faster and then contract more space in the form of heat from outer space, which is filled with non-material heat. As material compact it absorbs heat from outer space that loses density. That secures material growth and by reducing density secures outer space expanding. The star stays the same as outer space expands and that makes that the space the star claims to occupy remained as it was when the Universe in outer space began to expand in terms of the star contracting.

The prominence squirting from the sun can only be liquid heating up as it touches the much more hot cosmic gas. The fact that it rises can only be because it is heating up or getting hotter. If the sun had "pressure" it will release that "pressure" in a cloud of gaseous steam and the sun would go supernova in an instant.

The sun spins and this movement contracts the space that by spin or gravity becomes a liquid air that came in as a gas then turns to heat being in a liquid state because of the density increase changing the sun's inner space or atmosphere on the inside from a gas to a fluid that surrounds all the atomic solid particles. Our Earth's atmosphere has all the characteristics of a liquid and is then as such also a liquid and that makes the sun's atmosphere so much denser and therefore so much a denser liquid.

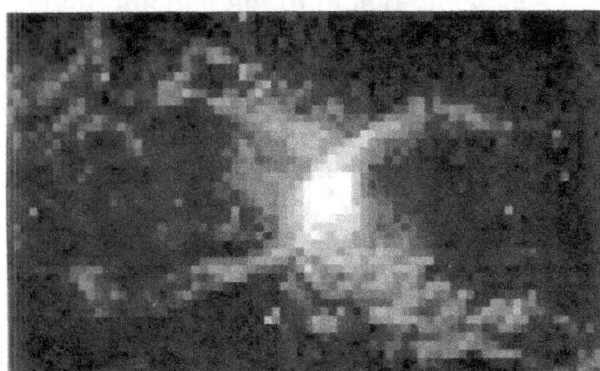

When I see a star burst open I should take note of what the star releases and look for the principles applying that should form such a release of heat. Newtonians are forever copying Newton's style by telling the Universe it holds the planets in formational alignment because of their mass while not size nor invented "mass" plays any part in the process. We must stop playing God and create a non-existing Universe and begin to confirm what there is. The biggest concept of being a Newtonian is to tell the Universe what it is instead of looking what the cosmos says and be told what the cosmos says it is.

At one very specific point gravity compresses outer space from a gas into a liquid and heat in that space then becomes liquid. In the photo's we see the heat returns back to space as it cuts through the sun's curving surface wall bringing about "an explosion". The main issue to realise is that the pumping produces a density increase and the density increase turns the inside from gas to liquid. It is not the oxygen or the hydrogen or whatever fills the container that is a gas or a liquid but it is the amount of space that turns to liquid heat that turns the container from a "planet" into a "star". Even the earth has already some flimsy liquid atmospheres in comparing to outer space. This is the only difference between planets and stars if you insist on having planets and having stars. The sun has no pressure but the excessive movement freezes the gas in the sun into a liquid because the idea of expressing values in terms of temperature is a Newtonian made custom.

When we go in search of what principles applies to form the building material in the Universe we better look and see what is it that the Universe shows us most graphic and we better stop telling the Universe what it is that we want to see and what the Universe should offer us that we wish to see. We better stop telling the Universe it must get mass and start to see what the Universe tells us what it has to offer us to see. If stars burst by releasing heat then stars are constructions that confine heat or cram heat into a small space. If this is true then gravity must be the process of freezing heat by turning movement and displacing space into compacted heat making gravity a process whereby space freezes as it condenses.

When a star burst open it releases massive amounts of heat into outer space. If it exposes heat bursting out then the reason that would apply is it must be because it froze heat into a state of solidity. If the star bursts as it explodes by releasing heat it then clearly overheats. The question never asked is why would stars overheat? We can blame pressure, but pressure would not bring about a star disintegrating from the centre, as the star depicted here clearly does. A burst from pressure should blow the sides out.

What ever the terms used there must be a recognising of the inter relation between heat and space where the reducing of the one will lead to the increase of the other. The star does not apply pressure to bring about fusion it freezes the elements into fusion by applying millions and even billions of degrees Celsius. It is our conception of hot and cold bringing total; confusion about the principles of cosmology.

I do not think that I or any other person is at liberty to try to calculate any on goings with in the star but from what is clear from the outside one may come to some measured idea of the stars position in space – time. Gravity is the cooling of space by duplicating or moving space, albeit filled or not filled. When the star spins too slow it does not cool sufficiently and then it becomes warmer inside. As it reaches a point where it overheats because it moves to slow it burst and by expanding the space it regulates the temperature. At a point when it can no longer contain the confined heat it expands and such expanding we call a Supernova occurrence.

The contraction of space must be equal to what amount of heat the total number of atoms spinning within the star can retain and gravity is that balance. The fact that it can freeze heat to a liquid surrounding hydrogen while holding a temperature of 6500^0 C should be an indication it is not what we seem to acknowledge as normal. The sun is freezing hydrogen to a dense liquid at 6500^0 while space is boiling (expanding through overheating according to the Hubble Constant) at $- 273^0$

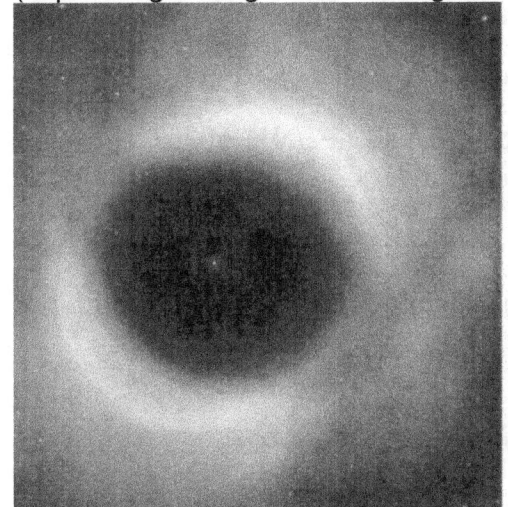

When a Super nova goes bang Newtonians say it is because "_**gravity has gone mad**_" and then they still see their position as being intellectual. Since when has gravity got emotions that can go array or "mad". Still more off the point is how can that be to their ability the best answer they can get up to while remaining satisfied with the effort! Stars we call Super Nova has blowouts. It can't be a pressure burst because there is no material wall enveloping the heat and thereby Stars we call Super Nova has blowouts. Pressure release comes from when material containing compressed space bursts its limit. That we humans know since before writing began, but since of late this phenomenon becomes more and more seemingly misunderstood. If stars blow as stars should and as we can clearly see from the picture just above, then the explosion happening to the star we know as a Supernova comes about from other principles, surely.

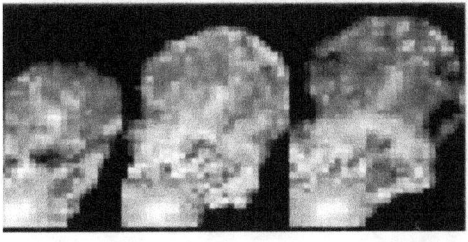

It is very obvious the two occurrences are not a result from the same basic method the Universe uses in destroying stars. When heat surges and becomes too high, it turns into space. That process we call an explosion. It is frequently seen, yet never acknowledged by science. When heat reduces, it relinquishes space in the producing of more concentrated heat, this process we see as cooling.

Science academics have to review there thoughts on relevancies because what seems to be hot is cold under certain circumstances and what seems to be cold to a point of freezing is boiling hot. There are no standard issue and fit all through out the universe. Every singularity attaché different criteria to borders controlling the space-time with in it rule. What fits humans on earth does not even suit conditions on the moon, yet science cannot appreciate that the moon applies very different standards to that of every structure and every structure is a cosmos on its own turf, supplying its own turf.

We must look at nature to find what is hot and what is cold. Something hot is that which can
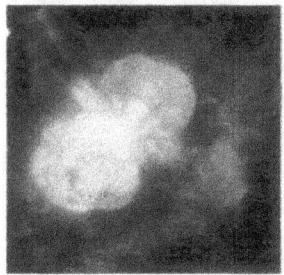
expand no more because that which is hot expands. When something is cold it can contract no more because it reduced space to the utmost limit. Outer space is the hottest because it eternally expands while the Black hole is infinitely cold as it contracted what it could contract and keeps on contracting. It is not the specifics that are of importance because the specifics change considerably when taking into account that hydrogen remains in a frozen state at 6500^0 C therefore it is obvious we have to look at other clues to give some indication of what is in process. On earth in the time we have as a duration we find hydrogen freezing at 269 ^{0}C as where it freezes on the sun at 6500^0 C, which implicate the reduction of space to an enormous increase in time duration.

In conditions on earth the rotating velocity of the electron is 3×10^5 km / sec. With conditions being that different it can not nearly be the same in the sun. As space reduces time increases. By having the space reduced to such an extent that it matches near Big Bang relevancies (a period where heat flowed like water and which is the very same conditions we find within the sun) the space would apply accordingly. We also know that relevancies is all about conditions showing similarities under variables and therefore the space and heat component may seem altogether incompatible but is almost the same given the singularity presence within the sun and comparing that to the earth.

What is applying to stars inside the galactica centre is applying to particles inside the sun. Science sees the nuclear reaction but do not recognise and therefore do not admit that the nuclear reaction is three different phases. At the beginning of the process all the heat is solid, placed in a container by nature and the container has a human name called the atom nucleus. In the atomic explosion there are three ingredients that are distinctly apart. When the solid melts down, it becomes a fluid.

The fluid we gave the name of light. There is not enough space to explain the detail of the argument, but light is not a gas, it is a fluid. The first step of the nuclear explosion is converting the solid to liquid. In the liquid state the star does not overheat. The overheating becomes part of the second phase. That phase involves the turning of the heat-fluid to a heat-gas we call space. Space is heat overheated creating space, as heat is space concentrated creating a fluid or liquid not yet correctly named.

Every one knows that a gas is one dimension HOTTER than Liquid as liquid again is one dimension HOTTER than being solid. If the star is liquid on the inside, and the liquid evaporates when coming into contact with outer space, then outer space is the hottest, notwithstanding what ever boundaries and values we humans attaché to the dimension. Our human standards have to change to accommodate the rules layer down by the cosmos and not apply the cosmos to suit our rules of hot and cold, big and small, near and far. In the case of the Super Nova, smoothing prevented the liquid turning into gas, therefore overheating. The liquid froze as a liquid becoming a cosmic lollypop.

From the offset of the first dot dividing as it became the first two dots, it was bringing about the second dot and the eternal number of dots growing from that means that the splitting of the dots assumed as the dots were growing from infinity in size, which is in fact only part of the relevancy because at the same time the infinity presented eternity, where both locked the same value to the dot. This long sentence structure is an effort to explain that everything is linking to another either directly or through other particles and everything came about precisely simultaneously being eternities apart.

The stars are in relevancy part of the growing cosmos, where the growing cosmos presents a liquid covering all solid strictures. The structures are no more or less particles irrelevant of size, since time places the value and space is dependent on time. But by the continuing process of the eternal overheating, the geodesic cosmos overheat gradually which presents as the Hubble Constant and this process changes time in space. Since every star holds an individual singularity, separating its singularity from the galactica singularity it is within, it remains as a relative liquid while the cosmos changes its side of the relevancy becoming more a gas.

That which prevents the overheating turned the layers into frozen identities not overheating therefore it became a liquid outside the star. This star was turned into a miniature galactica, sustaining billions of individual singularity, because the governing singularity did not destroy, but the singularity of every nature is still in support of one another. From this picture (and others of Super Nova) one can learn a lot, if one is truly interested in applying cosmic law to the picture and not some human response to what we think would apply to an earth-like star that holds gas as an ingredient.

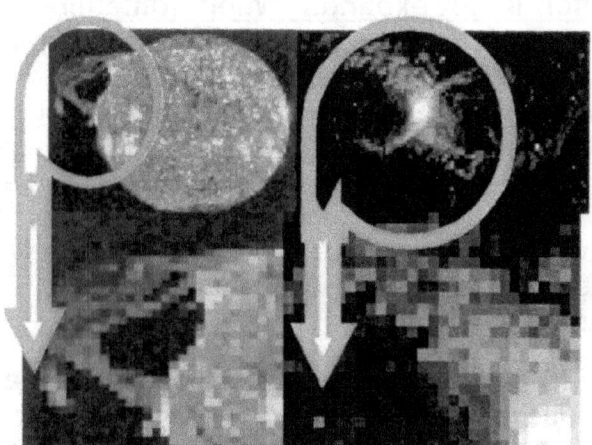

The pictures show clearly the difference of a star NOT overheating being "normal with liquid pouring from it and then becomes a gas as it evaporates. We see the sun exerts heat but still we think of the sun as hot. The sun gets rid of the heat, which means the sun is cold and that is why it removes heat from its surroundings. But because we feel that which the sun rejects we then contribute this heat to the sun by attaching that which the sun removes. It is all about relevancies forming as singularity applying matter in relation to the overheating it started to combat.

Again I have to press the thought that it is singularity determining space-time form through conditions that bring about the state between matter. Matter can be solid liquid or gas, but it is the condition of the space-time derived from singularity that places the form and conditions valuing the form of the elements. Hydrogen can be as much a solid as gold can be a gas. In the next scenario the overheating core is hotter than outer space and that brings about that the heat will flow to a colder region. In this case the star is overheating and with that can no longer protect its individual singularity. The part the official verdict and mine is in agreement on of creation is that it all started small, but I go one step further by saying the Universe was at the same time eternally large.

It started with singularity producing matter and the matter changed in relevancy to one another by becoming solid or liquid in relation to each other. Space still was at a premium because the space we know and we see as gas, was not yet part of creation. Since the Big Bang the fluid heat is in a process of converting to space enlarging the role of time as the Universe still systematically overheats. The entire purpose of gravity is to combat overheating to allow time growth.

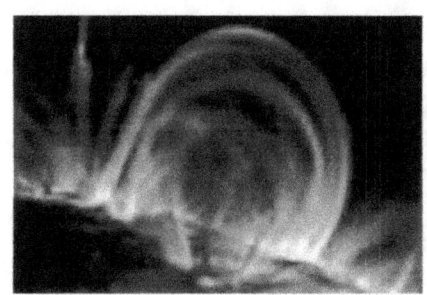

The difference between the star being the dot we can see in the picture on the outer edge and the star being the dot we cannot see on the inside is the time in promoting the individual singularity. First the star in the centre core changes, starting to collect liquid heat, while the outer part remains part of the cosmic structure. As the Hubble Constant grows the star distinguishes its singularity as it protects the singularity from overheating with the cosmic geodesic space-time. The geodesic space-time is also the outer space, but I prefer not to use outer space. At a point the star becomes a separate structure from the liquid cosmos that turned to gas, and then starts using the liquid to promote the generation of matter in a solid state whereby that matter then later turns to space less singularity as space-time completely breaks down forming neutron stars, pulsars and eventually a completely space less point of singularity in the cosmos being an ancient dot once more we think of by using the name a Black Hole.

One has to differentiate between heat and overheating because a star represents the coldest space in the Universe and not the hottest space. Heat and cold are relevant dynamics forming in appreciation of singularity. The sun is the coldest place in the solar system and that is fact. Looking at evidence the sun provides contradict everything science wishes to believe about cold and hot. Science wish to see the cosmos through the eyes of what fits the needs sustaining life on earth and what benefits maintaining surroundings in support of life as one find on earth whereas life has no part in the cosmos except for the speck of dust we call earth.

Looking at the cosmos impartial to life the evidence support another view. Every aspect in the cosmos is the very opposite of what science believe it is. The sun is not a ball of gas but a giant sea of liquid, frozen without any form of gas or air in the interior. Having a liquid interior the sun has no pressure but has the very opposite of pressure to which there is yet no name given. The liquid comes from singularity freezing space-time within the atmosphere of the sun, and such is the case with all stars still in the shining phase. Stars more developed than the sun is frozen solid causing fusion. Isaac Newton was an alchemist and was not a puritan as Newtonians make him out to be. Isaac Newton did believe in magic but this now is conveyed into the 21st century. The Newtonians measure the surface temperature; test the material on the surface of the sun exactly like the Druids did that came just before Newton did during the Dark Ages they decide that by the magic force of gravity this gas "pulls" into a ball. **Space and heat directly relates being the one form of the other**.

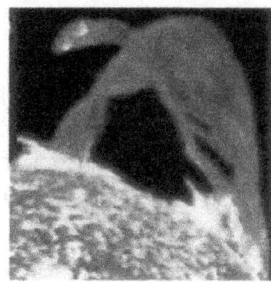

As material contracts the space it spins in this absorbs heat by gravitational condensing which cools the material and the size of the material increase. However as material absorbs cosmic liquid the density of the non-material decreases because of a loss of non-material substance going into material substance and materials grow in size while space expands because of loss in density. That is why material grows (the earth seems to grow bigger) and outer space expands (outer space expands) as Hubble indicated. It is all about relevancy changing cosmic dynamics every instant in time. This is why the earth becomes larger and the moon goes "further away from the earth".

The density of material increases as the volume size of material grows bigger and as the liquid is concentrated into the spinning object in order to keep it cool by controlling the heat. By removing the liquid from outer space the density of outer space reduces as the heat concentration decreases the relevant space increases by expanding. Material grow by removing heat from outer space to the atom and into singularity as outer space expands by losing the quantity of singularity that concentrates heat. In short it is heat that moves by material concentrating the heat from where the density is less to form more density in materials.

In the picture to the left we find not withstanding whatever name we attach to the red liquid substance flowing from the sun into space and back to the sun, that liquid is heat in a very direct form. If outer space was the coldest place in the solar system the heat should immediately escape to outer space and not return to the sun as it clearly does. If outer space were colder the heat would not return to the sun. All elements forming matter in as much as the heat forming an atom is as much a liquid as it is a gas and a solid. There is no hot as there is no cold. It's about storing energy in space or in heat, which is another Cosmic equal being opposing similarities.

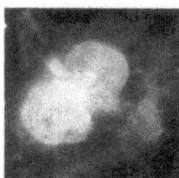

Hot and cold are relevancies brought about by singularity valuating space-time and during the Big Bang the Universe was freezing cold at three billion degrees C. It is the relation matter has with heat that provides the form the particle has at that moment. The increasing or decreasing the heat will alter the form of the element.

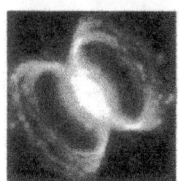

Therefore all elements forming matter is as much a liquid or not than it is a solid or a gas. It is the space surrounding the atom which provide the form the atom find its relativity to the rest of the atoms it share space with. Hydrogen is as much a solid as tungsten is a gas depending on the heat in relation to the space matter is within. Should you reply that it is the gravity pulling the heat back to the sun, then that confirms my theory that gravity is all about collecting heat onto matter with outer space being the hottest place. It is the concentration of heat in space being relevant to form.

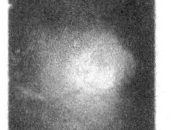

When overheating a star turns its liquid to gas whereby it merely transforms it's interior to a relevancy it has from pre- to post- Big Bang. We humans on earth think that hydrogen is a liquid at -259^0 C but that only apply to the earth. The picture clearly shows the heat in a liquid flowing from the sun and back to the sun. In the sun the hydrogen holds enormous quantities of heat in a liquid at a temperature of 6500^0 C. When a star has its singularity secured the star is bitterly cold because it has heat in a liquid form flowing back to the point of singularity although we may regard the star to be rather on the hot side. The sun (fore instance) freezes hydrogen in a liquid form at 6500^0 C.

If hydrogen remains a liquid at 6500^0 C, just think how cold it must be as the star's interior approaches the point of singularity. Therefore fusing protons comes from cold and not from heat or pressure.   By allowing the singularity to overheat the star overheats and heat within the star flows from singularity to outer space freely. In such an event outer space is then colder than the star because the heat releases to outer space with no intention of returning whereas in the sun it returns as soon as it leaves.

There are two ways to reduce heat; one is to bring about expanding space, as the photographs clearly show. The second one is where heat will reduce when in motion by spin. When withholding or retarding motion matter will overheat. Gravity is the motion of unoccupied space through the dimensional transformation to occupied space. Motion and space therefore is the anti-, the opposite the negative to heat being the positive. With singularity overheating the expansion of the singularity drives heat into space, creating space to compensate overheating

That is a natural phenomenon. The only reason why **heat will** rather **flow back** to the star than **escape**

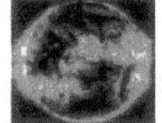

to outer space once the star released it into outer space is **if outer space presents more heat than does the star,** because **heat always flows from what we think is hot to from what we think is cold** no matter what influences may arise. However it is different to the cosmos because it works the other way around. Outer space must hold more heat than does the star but the accumulation of space in relation to heat makes it seem colder bringing expanding of heat to become space. The cosmos is all about **converting space to heat** which we see **as gravity** and **returning heat to space** as a **control mechanism** always **keeping** a very delicate **balance** which we see as **a star shining or being normal.** The purpose of the converting of space to heat is to supply the core where singularity is with heat. **It turns space to heat** sustaining matter but sometimes singularity overheats and then matter converts to heat allowing heat to convert to space. That we call many names amongst others exploding into super nova.

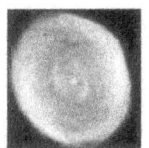

Whatever the names used is less important because the **process rests on space and heat interacting to form energy**. That was what **the Big Bang** was and **the Hubble Constant** is all about where **matter converts heat to space**. I show that **space and heat is the very same thing** and there **is no such a thing as pressure** but releasing **heat produces space** and **concentrating heat reduces space** with the two interacting on singularity demand setting time to space with time being the spin or motion of heat in space. **Heat and space form the second singularity** caused by the **fragmenting of singularity to compensate overheating during the pre-** Big Bang matter forming era. That is what we see as **light and space**, which again is the **same thing and is fragmented singularity forming radiation and heat, where the star re-transfers heat back to space due to an overload.**

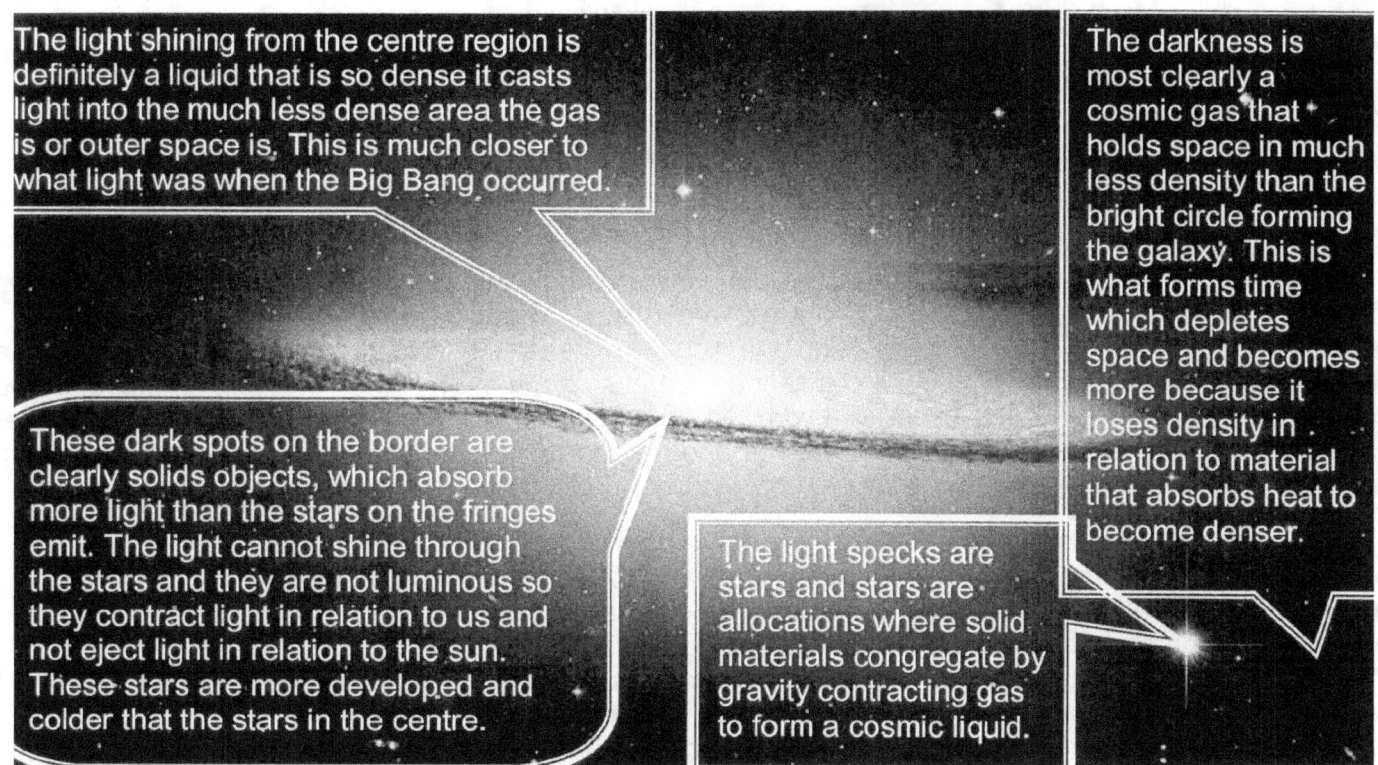

The light shining from the centre region is definitely a liquid that is so dense it casts light into the much less dense area the gas is or outer space is. This is much closer to what light was when the Big Bang occurred.

The darkness is most clearly a cosmic gas that holds space in much less density than the bright circle forming the galaxy. This is what forms time which depletes space and becomes more because it loses density in relation to material that absorbs heat to become denser.

These dark spots on the border are clearly solids objects, which absorb more light than the stars on the fringes emit. The light cannot shine through the stars and they are not luminous so they contract light in relation to us and not eject light in relation to the sun. These stars are more developed and colder that the stars in the centre.

The light specks are stars and stars are allocations where solid materials congregate by gravity contracting gas to form a cosmic liquid.

The reason why there is something such as gravity is there is a transfer of heat by material movement. Studying the tables Kepler left us it is very clear that space and not material is moving towards the sun and therefore the sun is contracting space where space is in a process of reducing in volume while the sun is compressing space by a similar margin. Material can't move towards the sun because the Titius Bode law prevents material to come closer to the sun. It is this and three other laws I explain for the first time since the science started. Look very good at every picture of the prominence flowing from inside the sun. What you see is not vapour forming a cloud of mist but a liquid squirting from an even cooler inside.

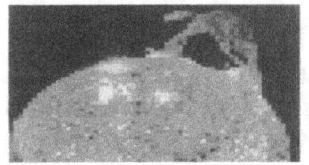

On the inside the movement of the sun freezes the gas forming cosmic gas, which is what is between the earth and the moon to a liquid filling the sun. Again it comes down to enforcing standards our Newtonians apply to life onto the cosmos and the result is stupidity only a Newtonian wise man could be capable of. See the fluid push out of a bowl of liquid, spilling both sides as it falls into liquid. The inside of the sun is not gas but it is fluid. In all of nature there is no NATURAL GAS as much as there is no NATURAL SOLID. Hydrogen is as much a liquid as iron is a gas and neon is a solid. It depends on the element relating to the space/heat in the circumstances surrounding the substance at that very precise instant in time. We have to stop telling the cosmos to show us what we wish to find and start accepting what the cosmos is telling us to find. This shows science must place much less emphasis on life and much more on reality. Life as a comic reality is non-existing and life not withstanding corruption plays no part in the Universe.

Only science, Newton and God in that order is never wrong and that is religion. Lately someone was found to be wrong and science decided it had to be God that misplaced dark matter in obscure places in order to have the Universe expand instead of contract as Newton said it does. Science decide on behalf of God that God was wrong when the Universe expanded instead of contracted as Newton said it does because Newton couldn't be wrong so it's God that wrong.

The way science present a star such as the sun is similar to a furnish burning to boil water. The furnace will burn until the coal is finished and then the star will **_die._** This view is carried over from since a time before they new about electricity or internal combustion engines or air-conditioning or nuclear energy. They maintain to present the sun as a coal stove furnace that boils and heats and when the fuel is finished the boiling is done. What a lot of primitive gargle and they pretend to alight with the twenty first century.

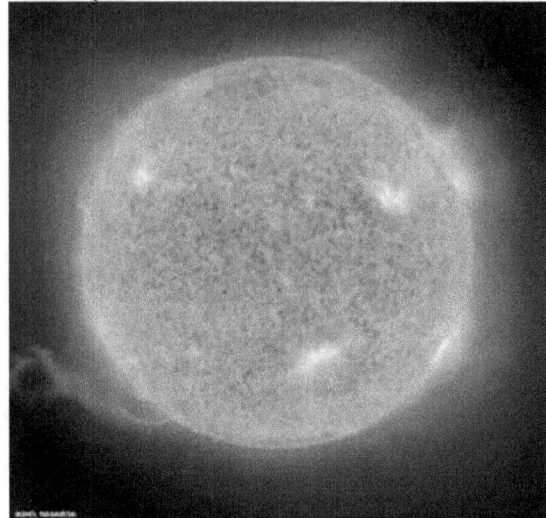

This came from Newton's time when they still thought it was possible to reform lead into gold and Newton in person was such an alchemist. He saw a coal stove within the sun that never was there. Newton saw a furnace boiling steam to drive engines because that was the nuclear science of the day! I say the sun is hydrogen and hydrogen is not a gas but is a substance that is volatile meaning it absorbs heat. I can see the sun squirts liquid out of the sun. The sun serves as a hydrogen pump that pumps liquid hydrogen.

The sun is filled wit a liquid and if it is hydrogen then my eyes tell me that the hydrogen is in a liquid form. Anything that pumps liquid hydrogen is not a coal stove as Newtonian science portrays the sun but with the sun pumping liquid hydrogen then it is an air conditioner and an air conditioner serves one purpose and that is to cool the inside by pumping heat from within to the outside to maintain a cool temperature inside.

That shows that in the star heat serves to keep the atoms cool. That is why we see supernova burst and explode. Without sufficient heat flow the atoms will get hot and expand. In every star that went supernova the star overheated and in overheating it burst by heat expanding and releasing massive heat into space. When the atom expands it not only holds more space depreciating the relevancy of space to material within the star but it spins less and the heat 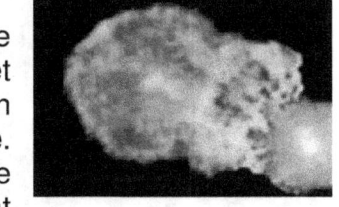 within the star rises even further. This forms more space as heat that expands by overheating produces space and when the space reaches an ultimate critical level the Roche limit is reached and following the Roche limit come the Roche lobe. I explain this process in much better detail elsewhere in this book.

Therefore the fuel that a star runs on is not coal producing fire that can finish when the stoker retires from his job with old age and it leaves the star to *"die"*. No star can ever *"die"* because a star is not a coal stove burning as it cooks oil or burning hydrogen. The hydrogen and other atoms are pumps pumping liquid into the star and the liquid allow the material to grow while the liquid produces the rise of relevant density where the relevant density matches the overall density applying in the Universe at that moment. A current Black Hole was a star a very long time ago and our sun is a future Black Hole a long while from now but it is in the process of developing.

The prominence squirting from the sun can only be liquid heating up as it touches the much more hot cosmic gas. The fact that it rises can only be because it is heating up or getting hotter. If the sun had "pressure" it will release that "pressure" in a cloud of gaseous steam and the sun would go supernova in an instant. The sun spins and this movement contracts the space that by spin or gravity becomes a liquid air that came in as a gas then turns to heat being in a liquid state because of the density increase changing the sun's inner space or atmosphere on the inside from a gas to a fluid that surrounds all the atomic solid particles. Our Earth's atmosphere has all the characteristics of a liquid and is then as such also a liquid and that makes the sun's atmosphere so much denser and therefore so much a denser liquid. At one very specific point gravity compresses outer space from a gas into a liquid and heat in that space then becomes liquid. In the photo's we see the heat returns back to space as it cuts through the sun's curving surface wall bringing about "an explosion".

The main issue to realise is that the pumping produces a density increase and the density increase turns the inside from gas to liquid. It is not the oxygen or the hydrogen or whatever fills the container that is a gas or a liquid but it is the amount of space that turns to liquid heat that turns the container from a "planet" into a "star". Even the earth has already some flimsy liquid atmospheres in comparing to outer space. This is the only difference between planets and stars if you insist on having planets and having stars. The sun has no pressure but the excessive movement freezes the gas in the sun into a liquid because the idea of expressing values in terms of temperature is a Newtonian made custom.

When we go in search of what principles applies to form the building material in the Universe we better look and see what is it that the Universe shows us most graphic and we better stop telling the Universe what it is that we want to see and what the Universe should offer us that we wish to see. We better stop telling the Universe it must get mass and start to see what the Universe tells us what it has to offer us to see. If stars burst by releasing heat then stars are constructions that confine heat or cram heat into a small space. If this is true then gravity must be the process of freezing heat by turning movement and displacing space into compacted heat making gravity a process whereby space freezes as it condenses.

I would suggest we think of stars in the following terms. A star that generates and transmits a lot of light is weak on gravity because their progress started recently. They command a lot of space-time but the demand they have to keep their cooling acceptable is very low. In that they can generate a lot of light but with the demand on cooling low and the gravity in the centre not very developed, those stars cast a lot of light back into outer space. It is just because of the size the stars holds that tell that the stars are still young and have a weak developed governing singularity. The stars will have very prominent hydrogen and helium layers, with the inner core not very prominent. The control of the star is still very much in the individual atoms and in that the motion the atoms have to produce in order to maintain their individual singularity will only come about through motion.

The atom has to make contact with as much space-time through motion as possible since it has a very poor ability in contracting space –time in support of the cooling system. There cannot be something big or small except in the relevancies of perceptions and then the relativity of such perceptions becomes questionable. There cannot be hot as much as there cannot be cold The sun freezes hydrogen to a liquid at 6500 ^{0}C and outer space boils over at 0 K.

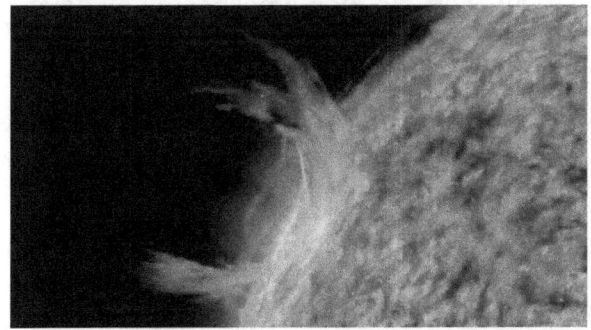

If we humans cannot or will not abandon our human culture driven perceptions and our mankind's pre-programmed perspective we may as well return to astrology for what the future hols. There are so many boundaries out there ready to destroy us because of our lack of insight, as did the challenger disaster. Creation birth started off with one dot so small eternity met infinity within. Then came one more, and another and they continued coming until there were a countless number of dots. The accumulative size of the dots were the same size as one dot because in the true Universe big and small plays no part. The dots were infinitely small and eternally big at the same time because size is a relevancy and without one the other has no size.

So in the true perception, there is no difference in size.

Te sun condenses space into a liquid. In that process it collects hydrogen from outer space where hydrogen is the most abundant element of all materials. By compressing the space into a liquid the hydrogen then becomes a liquid in the atmosphere environment of the sun. However this collecting of hydrogen from space certifies the working principle and indeed the working layout of all stars as an air conditioning unit and not a burning furnace. I prove by formula that electricity and gravity is the very same thing, not the same principle but it is the same process. This makes the sun and indeed all the stars forming gravity a star that is an electrical driven air conditioning unit that can never run out of movement. That too I show because the sun uses comic fuel and cosmic fuel will never run out of supply.

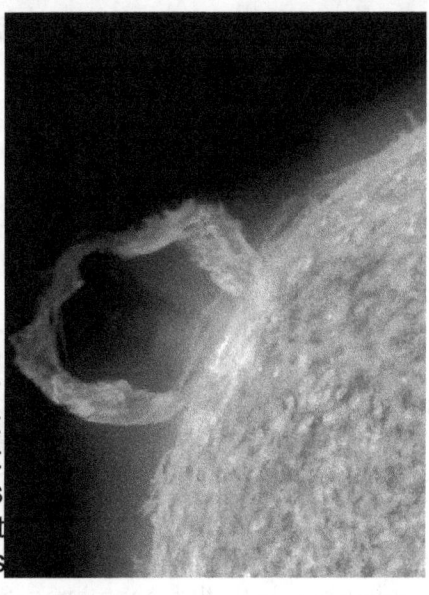

Look closely at the sun and see it releases liquid into space. The sun is not gas but is liquid. The sun spins and this movement contracts the space that by spin or gravity becomes a liquid air that came in as a gas then turns to heat being in a liquid state because of the density increases, which changes the sun's inner space or atmosphere on the inside from a gas to a fluid that surrounds all the atomic solid particles. Our Earth's atmosphere has all the characteristics of a liquid and is then as such also is a liquid and that makes the sun's atmosphere so much denser and therefore so much a denser liquid.

Look very good at every picture of the prominence flowing from inside the sun. By gravity or also is movement it cools or contracts space. The sun freezes hydrogen into liquid. What you see is not vapour forming a cloud of mist. It is liquid squirting from an even cooler inside but then gravity cools the liquid as it falls back into the sun. On the inside the movement of the sun freezes the gas forming cosmic gas, which is what is between the earth and the moon to a liquid filling the sun. Again it comes down to enforcing standards our Newtonians apply to life onto the cosmos and the result is stupidity only a Newtonian wise man could be capable of.

The sun is not a coal stove burning charcoal or wood and becomes a furnace within a boiler to produce steam. In Newton's days when they were not that sure that the earth was truly round steam engines was the cutting edge of science but all parts of science progressed past the Middle Age views to become modern. Science still believes in Newton and Newton was an alchemist that believed in black magic. Still physics wishes to cling onto black magic science just as Newton believed in magical science. Steam and a boiling sun are prehistoric outdated ideas best kept in history and not in physics. That is a concept that is as outdated as steam ships are. We have new technology that teaches us new concepts. The sun us a hydrogen pump that pumps gasses and when there is a device that pumps gasses the gasses cool and freeze the area though which the gas is pumped. Even the rocket engines use the liquid fuel the engines burn to maintain structural integrity and to keep the engine walls cold. The liquid fuel that burns the rocket engines are applying the same process that the sun uses. The sun uses liquid hydrogen to keep cool.

I am the first person ever to explain WHY the Titius Bode is in space functioning as it is. But since those such as Brainwashed students with their pioneering professors and the "Small-brain-brigade" such as he can't fathom any explanation going this far with the way the Titius Bode law works but before I could understand nature I had to accept I was never going to understand Newton." In their smallness they have not the ability to understand that planets are not allocated according to size or mass and this they do not understand notwithstanding all the planets' sizes or individual mass values proving distinctly that Newtonian science is senselessly wrong.

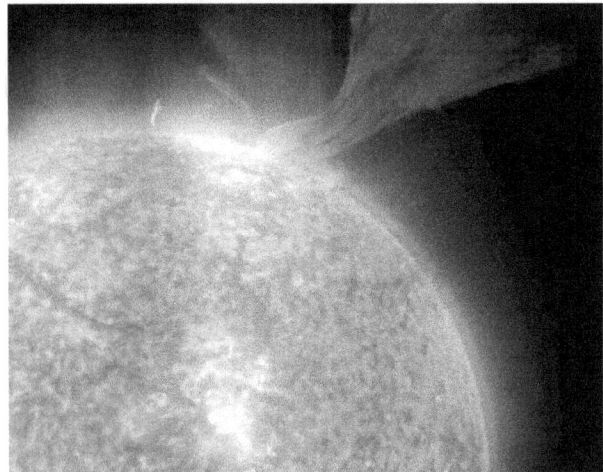

The only form of energy is turning cosmic gas into cosmic liquid and freezing cosmic liquid into cosmic solids. This process aims to keep singularity cold and to avoid another explosion or heat outburst of which the so-called "Big Bang" was only one of many. Also another form of energy forming the cosmos is transferring solids that are frozen heat to cosmic liquid and to cosmic gas. This interaction is the only form of energy consumption in the Universe and all other methods is only materials that are able to hold more heat and then transfer such heat to another source where the process converts heat to movement. For example we take in oxygen and use the heat the oxygen carries to convert that heat into life's ability to move our bodies.

Then to compensate for the loss of heat by the oxygen our bodies connect the heat-deprived oxygen with two part of carbon and release the carbons from our bodies in that manner. In that way the body gets rid of used carbon that it discards in the process of rebuilding new part in our body with new sells. This process we named aging and that makes life as a processes eternally linked to time and disposal of manufactured human bodies. We definitely don't burn oxygen but use oxygen to transport heat in a container to use and when the heat in the oxygen is transferred the oxygen is re-supplied by plants to the oxygen.

The gravity or freezing of the liquid and transforming of heat to a solid by gravity must at all times stay in equilibrium with the containing of heat within the star. When gravity slows down and the star's spinning motion is unable to provide the required level of freezing of heat to contain the heat inside the star, heat becomes space and the release of heat from a semis solid form becomes a cosmic liquid and cosmic gas. The remains of the star can still be seen as frozen vapour

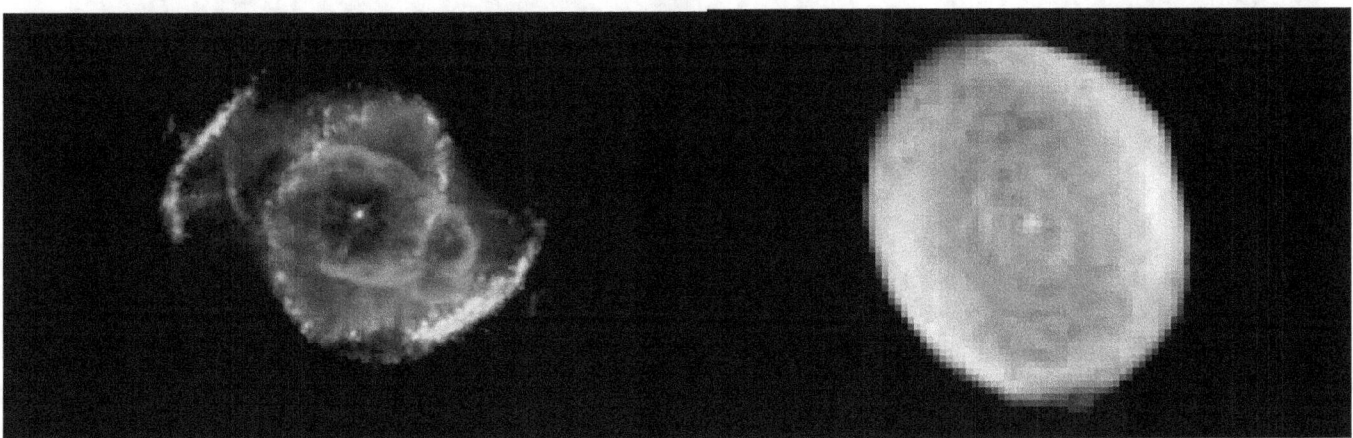

What is inside the star now is outside the star and we can see it contained heat that went out of gravitational control. We can see what the star discarded, that which the star found unacceptable, the rubbish dump we visit with our eyes. The inside, which is precious to the star, remained. The growth the star accomplished it hold contained in the inner nucleus. That precious part we can only reach with what the Creator supplied to the human, distinguishing it from other species, that which I call extelegence.

There is a direct relevancy between heat expanding into space and heat contracting into increase in intensity. By compacting the heat tries to escape and we see the escaping heat as levels rising, but the rising is symptomatic of an increase in cold increasing where a drop in heat intensity means it will take in an ever increase that indicates it will never discard heat but will accept heat unlimited.

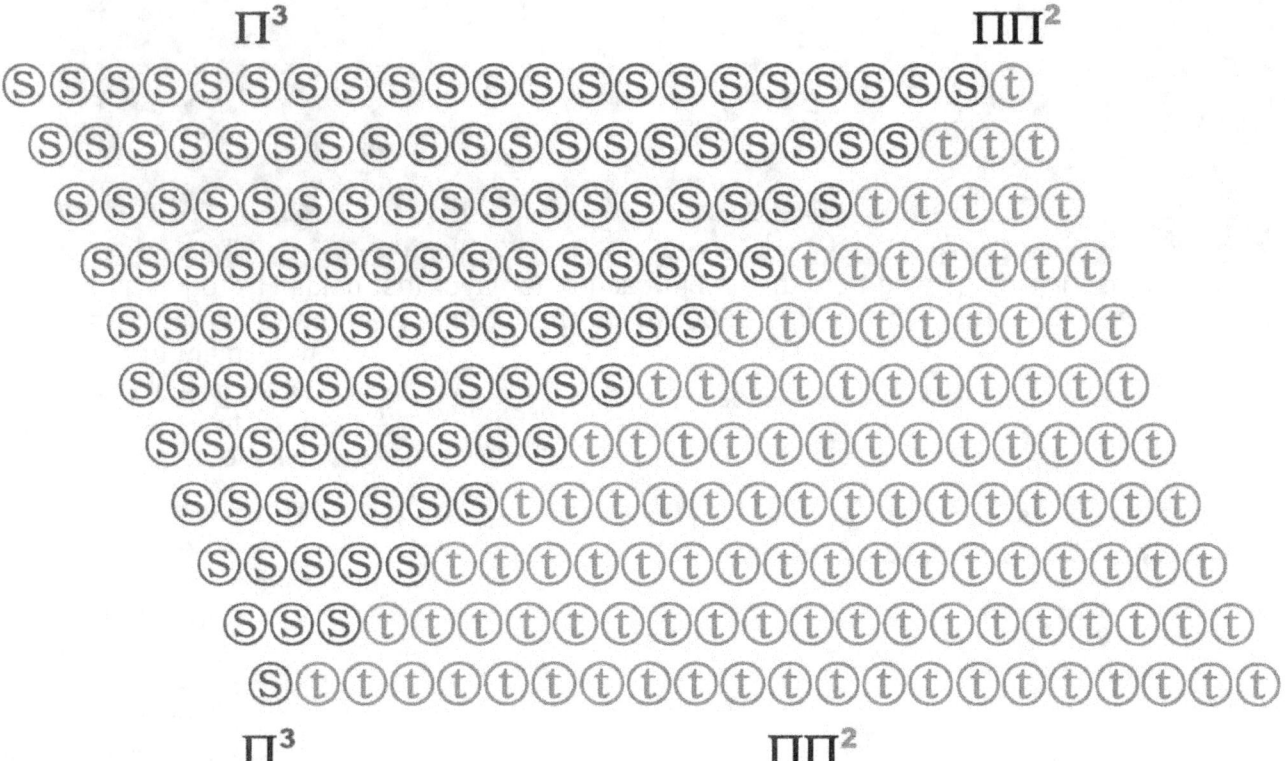

Let us review the dimensions that exist and to which we can identify

Finding the reason behind fusion follows a similar path, as does the sound barrier, because as stated in previously, putting pressure and heat will lead to an outburst not to fusion. The links with the unoccupied space-time to occupied space-time ratio plays an all-crucial role in the process of matter joining in one nucleus. At one point the water vapour stays stable in the air, suspended, and then something change enabling the vapour to form water and fall.

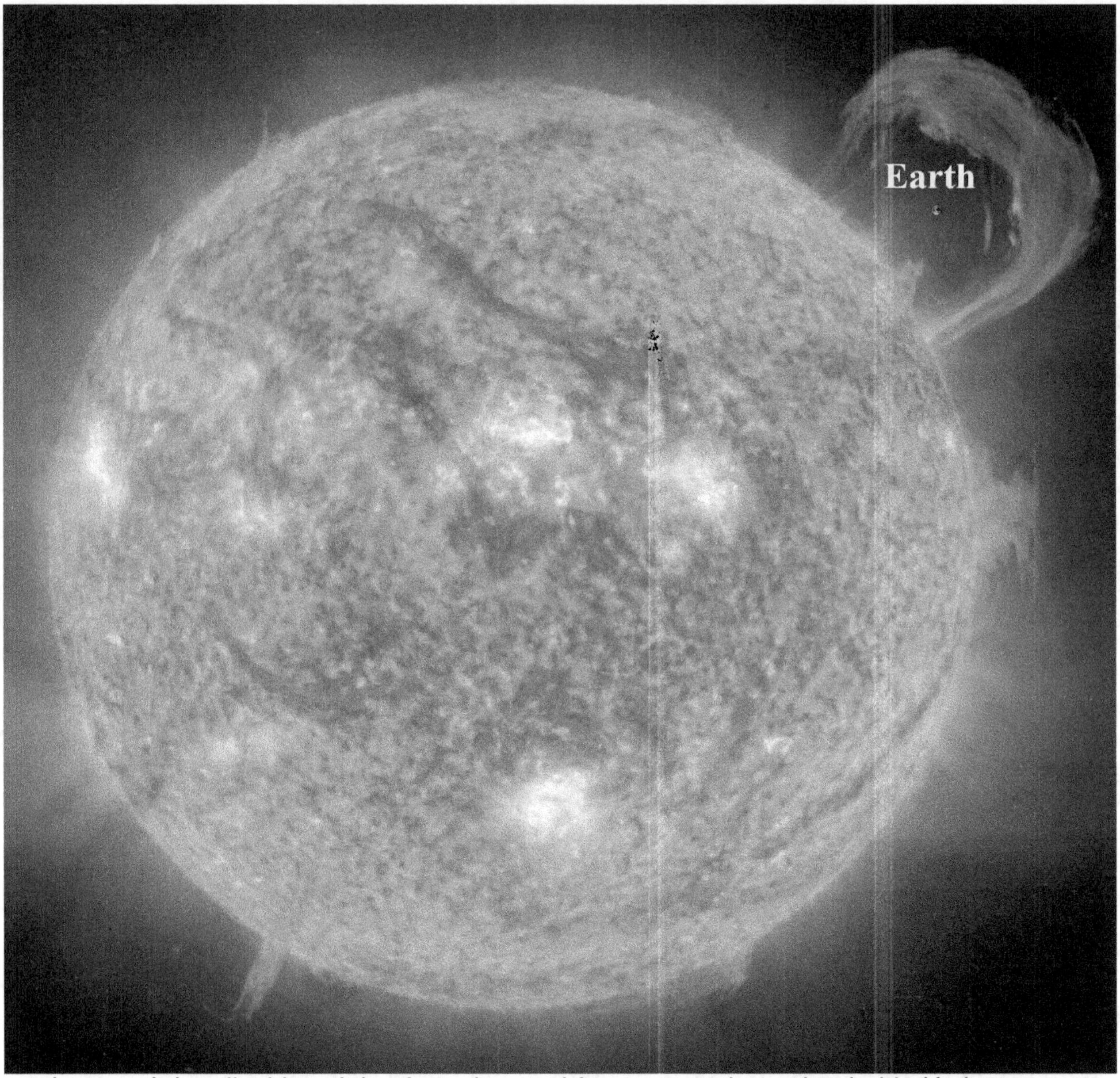

Earth

Look at the sun. It has liquid squirting from the semi frozen outer layer that holds Hydrogen as a semis frozen semi liquid at 6500°. The fact that it is 6500° might be hot in human terms but in terms of the movement of the sun, it forms a cold condition that freezes what we think to be a gas.

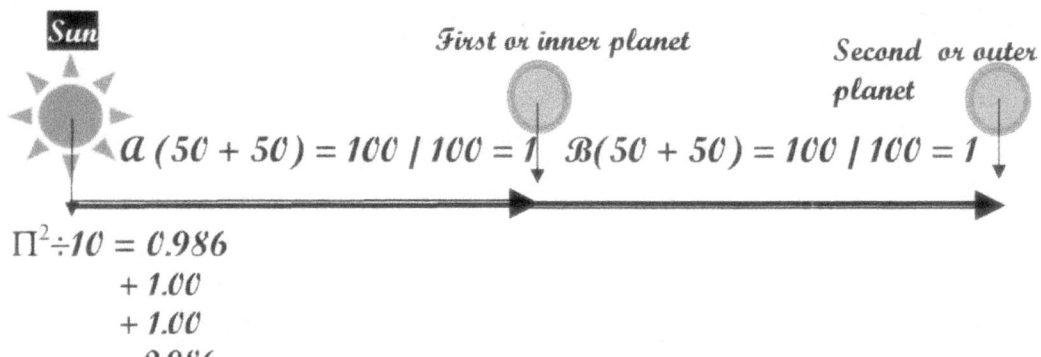

$a (50 + 50) = 100 \mid 100 = 1$ $\mathcal{B}(50 + 50) = 100 \mid 100 = 1$

$\Pi^2 \div 10 = 0.986$
$+ 1.00$
$+ 1.00$
2.986

The movement works in three with the sun rotating as Π^2 in relation to 10 forming outer space. The space between any planets forming the inner relation is 10 or 100

according to Pythagoras and the space formed by the outer planet also valued at either 10 or 100 depending on how you value Pythagoras forms a relation to 100 or the square thereof and that is10. That gives the spin of the sun .986 which is the square of Π and one plus one. That is the relative shift of space flowing towards the sun while the planets show absolute resistance to move with the space. This resistance is known as the Roche limit, which I call the Roche factor. I explain the Roche factor later on. I was asked once what is the Universe and I answered the Universe is light. The Universe is made up of mostly dark invisible light with spots of bright visible light placed at random to break the monotony of the dark invisible light filling the sky. There is light between the earth and the moon and the moon shines as light but the moon as such is not light. The moon I can see is not the moon that I can see because the moon I can see is only light coming from the sun and reflecting off the moon's surface and that light that forms the reflection of the moon's surface is the moon I think I can see. I can see light and therefore everything I can see is light and the only thing I can see is light and therefore what I see albeit dark or bright is light. I see a hologram formed by light because I can't see the moon as I only see light reflected.

If light came as individual streams of photon flurries our visage would translate that as such shown individual stars forming fragmented picture. We would only see small specks of light and very small specks of white and tiny white dots. It would be a picture unconnected bringing across some photons in the manner where every object stands apart not being related in any way and that will be what we see, if it is anything that we see. That we know is not the case but that means geodesic zero is as much rubbish as anything Newtonians regard with simplicity and with careless thought. We would not be able to see the darkness or black that makes up the major part of outer space.

Geodesic zero means outer space is filled with nothing and how can I see nothing as darkness because "nothing" is not darkness, nothing is "nothing" and the darkness I see is darkness showing the darkness is something. Nothing us the absence of whatever is not present while what I see does not represent "nothing" since what I see is preset and is there to see. By me using a spotlight I am able to come close to the animal and witness the animal crawling away while it clearly does not see me with my spotlight. If animals I see at night with my flashlight don't see the flashlight's light or my vehicle lights but sees by the light the darkness provides the darkness just can't be "nothing" and be discarded. What can be "nothing" is what we understand about which we think of as being "nothing". I came to ask where could I find most nothing, between Pluto and the sun with least space for there are more space filled with nothing between Pluto and the sun than Mercury as it is the nearest to the sun and therefore also having the least or nothing between the sun and Mercury. Does space with it being "nothing" from more "nothing" or less space from more "nothing". My book "**A Cosmic Birth…Dismissing "Nothing"**" was rejected because it was condemned by the view as being labelled as incoherent. All I asked is to show where is more nothing to be found? I doubted and questioned the nothing that fills outer space and I was labelled as incoherent.

Newtonians think of outer space as geodesic zero, with nothing in outer space but space. That puts what we don't see as nothing but that is how humans think. I think I know everything there is to know because anything I don't know to me does not exist and therefore I know everything because anything I don't know is equal to nothing I don't know. Everything that exists to me in my Universe I know about because what I don't know does not exist in my Universe and therefore that is darkness and the darkness is "nothing". Therefore I know everything there is to know as far as my Universe goes and there is nothing to me in my Universe that I don't know anything about. Only when you start to grow in things you do know can you realise how much there is that we could never know because the totality we are able to know is so small in comparison to everything there is to know that we forever would be unable to know since we are too small in mind to know. Geodesic zero means the light travels in a straight line from where it originates unhindered all across space to where the light connects the eye. Such an idea by itself is outrages because the stream of photons reduce in space to such a minute quantity that taken the area the photons travel and the space in vastness it covers, the chances of one photon coming across many hundreds of light years through billions upon trillions of cubic kilometres of space and selecting my eye to convey the electricity is less than infinite. Yet such conveying takes place every second of every minute.

If the darkness was the representation of "nothing", then that should be exactly what we must see, nothing but the stars. Taken from a picture of the cosmos we should se some stars and leaving the rest to nothing is what we see in the rest of the picture. A blind person sees nothing but also see no light with the light forming part as "nothing" and not only the darkness forming a selective part of what is nothing. When we look at space, we see something we see as space. One cannot have the ability of sight and see nothing except by closing your eyelids and then you see nothing not even light. But in that case you do not see "nothing" in contrast of "something" you see "nothing" without it contrasting to "something".

Going back to the initial argument of colours being rejected and not accepted while culture enforces acceptance while being rejected, the argument has to go further. When we see white we also see the whole spectrum spinning and therefore the whole spectrum becomes one kaleidoscope of colours mingling. If that is the case we have to see darkness too, because the whole spectrum is not mingling. By spinning the wheel with all the colours, which then forms white will mean all the colours are rejected and with black it will mean none of the colours are rejected as all the colours remain with outer space. Only when you finish this book would you realise what a dramatic conclusion this argument offers about how Creation works. In this concept hides the other half; the half we have not yet realised is there.

That may be one solution except for the fact that only Newtonian's nothing is that simple. The rejecting and / or accepting are correct, but the spinning part just comes across a tad too simple to make sense. To find substantiation one has to find the manner in which light connects to singularity because everything connects to singularity. With the establishing if the value Π and identifying r, one has to distinguish and define each item in order to bring comparison.

I take physics into a Universe that was in place before light was in place. I introduce the Universe when only darkness prevailed because light calls for space and in that era of singularity I introduce, space was not even a thought yet. I show why the Universe goes "flat" and in a "flat" Universe only the value of 1 holds measure since singularity is 1. If you can understand 1 or $5^0 x 7^0 x 3^0 = 1^1$ you have all the mathematical skills required to understand the applying concepts. To reach a value of 1 does not require big mathematical equations but to reach singularity requires 1. The formula such as the following can't

form part of the Universe $\dfrac{\partial f}{\partial t} + \Pi \dfrac{\partial f}{\partial r} + \dfrac{\theta}{r}\dfrac{\partial f}{\partial \theta} + Z\dfrac{\partial f}{\partial z} + \left(\dfrac{\theta^2}{r} - \dfrac{\partial \varphi}{\partial r}\right)\dfrac{\partial f}{\partial \Pi} - \left(\dfrac{\Pi\theta}{r} + \dfrac{1}{r}\dfrac{\partial \varphi}{\partial \theta}\right)\dfrac{\partial f}{\partial \theta} - \dfrac{\partial \varphi}{\partial z}\dfrac{\partial f}{\partial z} = 0.$

There is no 0 in the Universe and finding any number such as zero disqualifies the outcome.

According to Einstein singularity is a mathematical reality within the Black Hole but it is far more a reality of significance in every sphere. Einstein may be the first to name singularity and Galileo (unwittingly) may have been the first to define singularity as Kepler was the first to formulate singularity, but in mathematical terms singularity is the most basic principle. At this point I wish to establish a fact that seems lost in all other grandeurs of cosmology. When tracing the radius down into the sphere the radius stars where all lines start and a straight line cannot begin at zero or nil since it can only start at infinity. Such a statement will hardly seem appropriate but the relevancy of this fact has no limits. If gravity is motion then motion starts with a line. Let us follow the line as motion abides by the rules of the line.

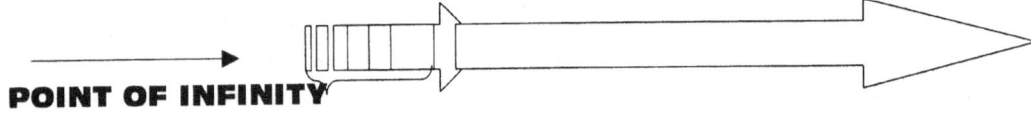

POINT OF INFINITY If the line started at zero there was no line to start because zero multiplied by whatever results in zero as the answer. That must also be the cosmic starting point. Einstein introduced such a point and named that point singularity. When looking at the cosmos from whichever angle the indications seem to be that the fact that the cosmos is entirely in motion. It is forever spinning and it

When spinning too fast the top keeping it upright starts to fights something because the alignment tarnish. The same apply when spinning too slowly but that makes sense. It is the fact that the same affect comes about when spinning too slow that triggers the questions. Why would the top stand upright by spinning. It must be because singularity charges the top into a cosmic independent reality.

is going too as much as it is coming from. Everything is on the move and always encircling something of greater importance. A top can spin but the parameters of its spin are limiting the motion it can apply. By not spinning the top is still spinning as the Earth are doing the spinning on its behalf.

A line starting with zero must repeat zero with every addition of a number. If it started with zero and then moved on to add 1 to form the line then the zero part has to fall away. Fore example: 0+0=0; 0+1=1; 1+1 = 2 Then the 0+0=0 ; 0+1=1 must be disregarded because the actual line will then start from one as flows: 1+1=2+1= 3+1=4+1=5+1=6+1=7+1=8+1=9+1=10 and so on. At this point this seems tedious but when the argument begins on the subject of how the Universe started this is detrimental to be correct. If the Universe started from zero then the adding of zero every time disqualifies a Universe being in place.

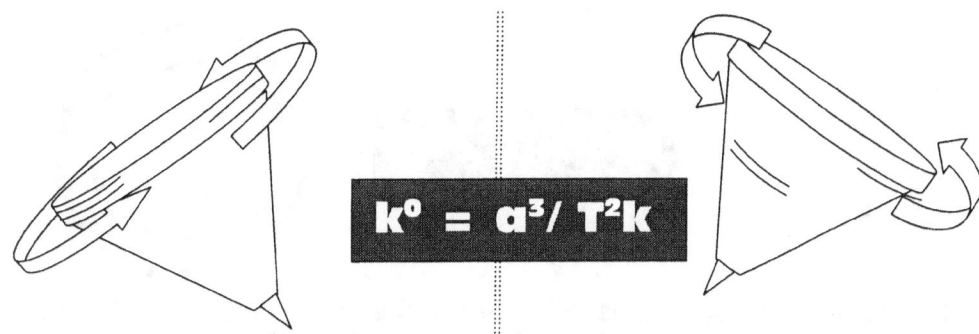

$$k^0 = a^3 / T^2 k$$

The spinning top is all the evidence any one needs to come to such a conclusion. I know probably as much as any graduate about cosmology but lack certificates to prove my knowledge. I am not part of established science. In my developing of knowledge accumulation I came to some conclusions about cosmology that are unique and divert somewhat to drastic form the accepted norm. Most of the work I see the same way as the norm does but in a reverse. Allow me a short explanation

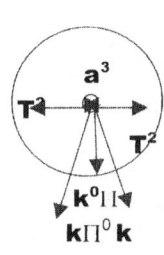

In dimensional terms, which I explain later on the value of **2k** relates to T^2. That relation extends to the next value where T^2 relates to **k**, which relates to T^2. The first space in the circle will then be T^2 **k**. From the centre being in infinity, one can realise by applying mental power the single dimension factor not seen but present all the same. Extending that into the 3D comes six **k** and any one of the six will further extend to form a seventh point as T^2 All this is a multiplying of $k^0 = a^3 / (T^2 k) = 7$

We have to be clear about what we think of when we think of the Universe. Most people think of a picture recalling the black night sky when thinking of the Universe and that thought is most incorrect. Einstein was most correct when he declared the Universe was going flat where gravity is at its utmost, but the concern we should have is not with the mathematics being valid or not but with the vision about the Universe being what we think of and where we place the Universe. The Universe is in the centre of what is spinning and the biggest single particle that is spinning in total independence of the rest of what forms a total Universe is the atom. The atom spins and by the motion the atom evokes the universe forming what must be the group effort of all the atoms then

With the top spinning the Coanda effect steps in and do justice to Kepler's formula.

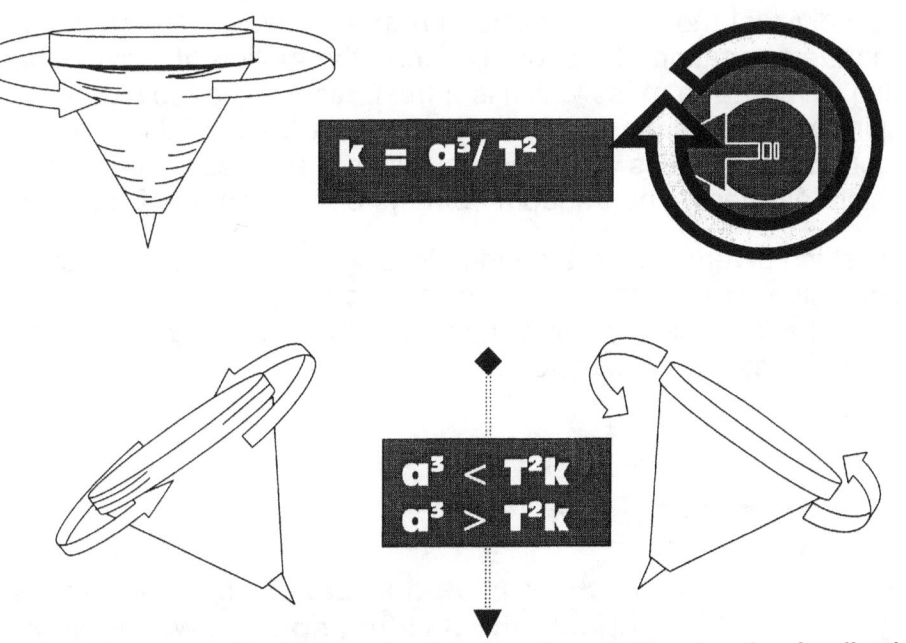

$$k = a^3 / T^2$$

$$a^3 < T^2 k$$
$$a^3 > T^2 k$$

Time is always a displacement of space in relation to the implication of singularity, and comes about between two points in space relating to the centre of singularity as positioned by k, either too the value of k or too k^0.

spin by the motion the atom renders the rest of the larger Universe. The Universe is the part that allow the rest of what the Universe establish to spin. What spin you may ask. Kepler said it without saying it: $k^0 = a^3 / T^2 k$ and not even Einstein with his super human mathematical skills could say it better or more accurately.

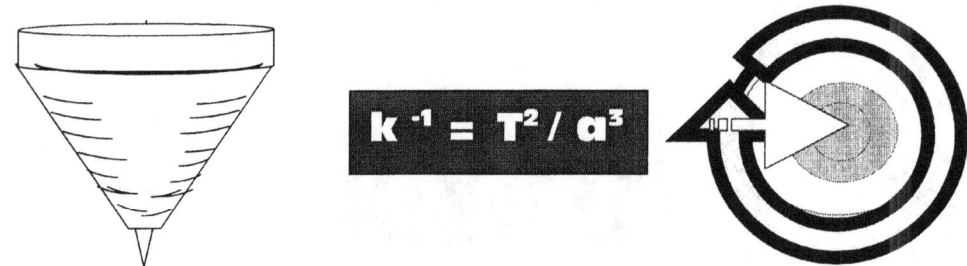

$$k^{-1} = T^2 / a^3$$

The motion established by singularity results in the implicating of the Coanda effect as much as the motion establish the Coanda effect. The spin realises the space limit while the space limits attaches the motion onto the space in the time within the time.

The motion established by singularity results in the implicating of the Coanda effect as much as the motion establish the Coanda effect. The spin realises the space limit while the space limits attaches the motion onto the space in the time within the time.

With the top spinning the Coanda effect steps in and do justice to Kepler's formula.

Time is always a displacement of space in relation to the implication of singularity, and comes about between two points in space relating to the centre of singularity as positioned by **k**, either too the value of **k** or to **k^0**.

 When one takes Kepler's equation into consideration the whole process of motion starts to make sense. It is motion that keeps the top erect and that was accepted from the time of Newton. Now however by close scrutiny as well as considering Kepler's equation where the statement emphatically reeds that the space **a^3** is equal= to the motion of the space in relation to a very specific centre. But this relation works both ways and not only from one side. It has nothing to do with pulling because if the two were pulling the top had no chance of any motion. The top uses the motion of the Earth to the advantage of the top and then on top of the Earth's motion, the top apply individual initiative and claim even more independence than the structural independence it had before. If it were merely gravity of mass and nothing more, then the gravity discrepancy that the Earth produce would prove to be such an imbalance that the top would never stand a chance of committing individual motion by erecting an axis. However should my view be correct about motion and mass being the frustration of motion hindering the motion of gravity, then yes, the top is by motion free the distorting of the Earth mass which is a frustration to the top's gravity motion and that would enable the top to spin even with such slight energy applied. Merely taking into account that the top has to overcome the considerable mass of the Earth brings the Mass theory into dispute because how can the top with such slender energy find freedom from the enormity of the Earth mass.

Every round object has a point establishing a very centre, a middle dividing one side from the other. That division determines the space from one side away from the other side. At one point there must be a point that does not fall on either side of the divide. Such a point forming the space less centre will still be a circle, because from that side the circle divides into four sectors.

In every spinning object does not turn because it such a point becomes a line in the midst of many others. outwards and from that point there is a point of infinity, a point that holds the dividing spin. However when that cannot spin a new Universe is born At the birth that point diverts space the spin is either clockwise or anti clockwise in all directions. As I pointed out no line can start at zero because then there is no line and no rotating point can start at zero because then there is no rotation. Calculating a square involves two aspects that we think of as sides.

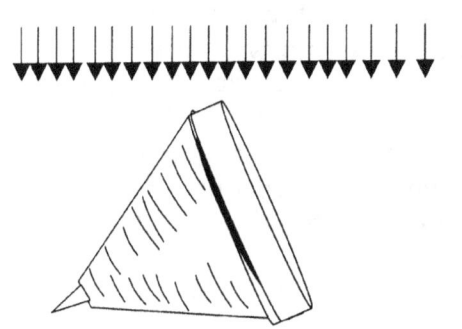

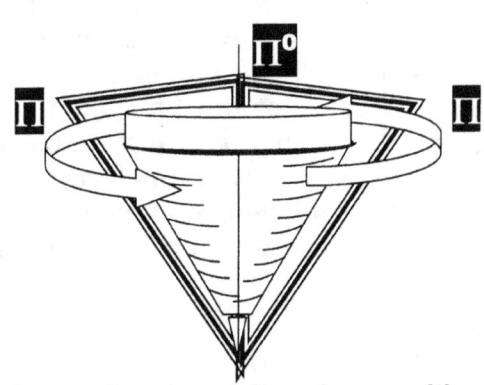

There is a Universe in differences between the top lying down without any individual motion and ostentatiously independent, self assured spinning top that even produce a sound to match the occasion. While without motion the top submits to the contraction lines running as the straight line holding half the value of the square being 180°. The top seems dead as it surrendered its long-term position and would eventually succumb to the Earth's gravity by relinquishing the structural independence it has. Then the motion brings life into the top and gives the top reasons to fight the Earth by fighting for independence. The top just became independent by the motion it received from the combined efforts of all the independent atoms forming the structure of the top.

Lets have a look at singularity in the way of showing form. In singularity there is no value in numbers except the value of 1. The only value is one and in singularity which is the first dimension every number leads to one, for instance $23917402651864872641208765673927561963748^0 = 1$, which proves that all values in singularity is one. Therefore singularity is one in every case and is only applying basic form as a concept.

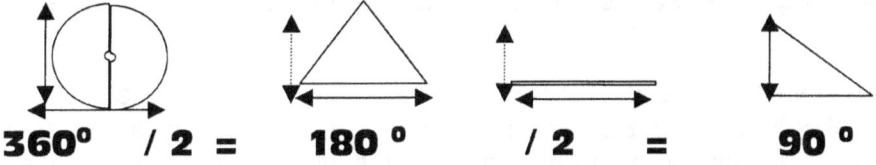

360°   / 2 =   180°   / 2 =   90°

The circle is a square holding a round shape, as the straight line is a square holding one side to infinity. Calculating a circle involves two aspects where the one is either the radius or the diameter that is double the radius. The other is the factor Π. Because gravity works both ways equally and not singularly in one direction, as the Newtonian myth would have us believe. There is then interaction in the neutron's position and the total proton's /electron's forming material $(\Pi^2+\Pi^2)(\Pi^2\Pi)3 = 1836$, where the proton is $(\Pi^2+\Pi^2)$, the neutron is $(\Pi^2\Pi)$ and the electron is **3**. This shows an interacting relation between time formed in space as space and time formed in space in relation to the total of material.

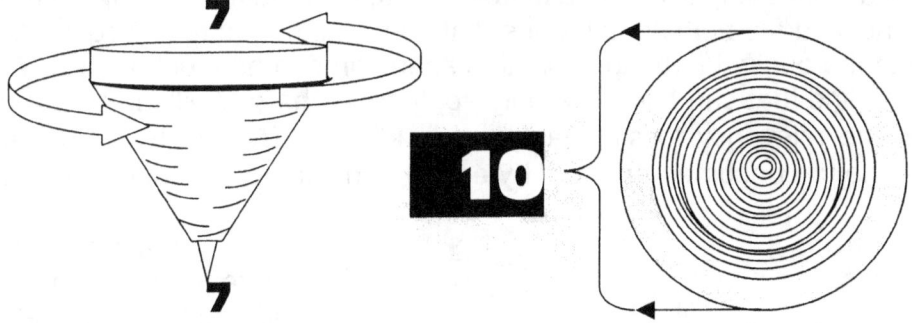

With everything in a cube or a circle or a potential of the two, brings about the implication of eternity in a form of singularity or the point of creation. Removing the radius of a circle does not remove the circle, because the circle is there, securing the ring. If the line (or imaginary line if you wish) holding the value of Π^0 = 1there has to be a point where the circle is no longer in infinity but claims existing outside the imaginary. At that point the radius may be lightly more than infinity, but to all calculating purposes it still remain as infinity.

The spin was going on for eternity because the spin does not apply, it has a value of zero and zero is another expression for eternity.

Having edges where Π^0 duplicate to present the edges singularity lost the value of Π^0 to the value of Π^1 with the same value singularity had being Π^1 to the one side and Π^1 to the other side, Π^0 must be the point splitting singularity into two parts of eternity, the eternal value of the first dimension outside eternity. It was the square of Π^1 being Π^{1+1}. That was the first dimension outside singularity Π^0 where singularity has a value of Π^1 in the form of $\Pi^{1+1=2}$. The first claim to space had a value of Π^2. This applied to both sides of the claim to space outside singularity, and the double proton became the dominant factor on matter.

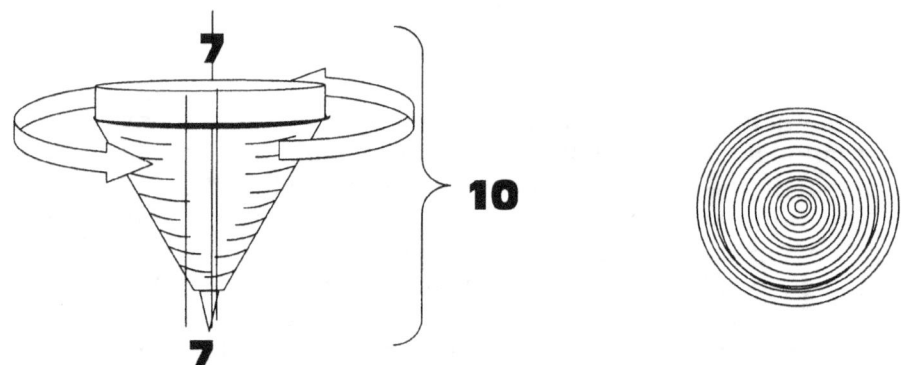

The top has space but there is a space in which the top spins the covers the time part of space. That is the time part Einstein identified (1) as coming from (2) being at and (3) going too and the position holding singularity is represented by the entire body that holds all the space of the spinning top. As soon as the spinning of the top commences, the time aspect releases space in which the top spins from the space holding the time of the Earth and the rest of the Earth within that time. It is this space in time that becomes so hot when the aircraft is speeding because the motion takes the time back to what the time was when time was nearer to the Big Bang.

Using such logic makes science appear foolish. There is just no rational in the time verses events that can explain facts without. Since the time of Newton, the arguments tarnished from being brilliant to clever to fair too poor and a hundred years ago to the point of being stupid.   That is what Kepler's formula is all about? That is what Kepler indicated with his formula $a^3 = T^2 k$. The space of an object (a^3) is equal to the **time (T^2),** which it is in, in every given instant **(k).** If the space becomes smaller, the time duration becomes longer every instant of time's progress. From the graph one can establish the link in the circle's rotation around a conforming unit being singularity.

Saying that one therefore has to admit that the smallest spot has to hold space because the most insignificant dot can transmit light and being able to accomplish that, one must accept it to carry a value of something. If that spot had the value of nothing, it means that spot was not there to begin with. Holding space-time one should return to the original formula indicating space-time in as much as $a^3 = T^2 k$. Being time it has to alternate positions and that can therefore only apply to **k** where **k** will indicate a relation to the space-time in question or the relevancy to singularity being $k^0 = 1$. By receiving **k** on top of the already $k^0 = 1$ that is in place the top becomes an atom by erecting the line of singularity from $k^0 = 1$ to $k^0 = a^3 / T^2 k$

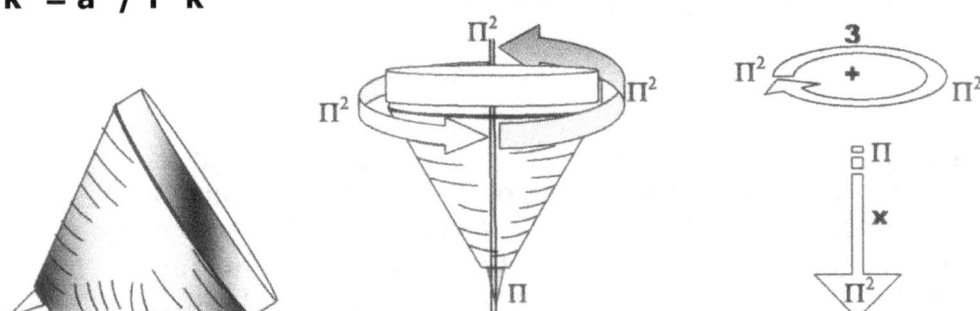

From such a relevancy there then must be four different values relating to singularity and since the atom has a proven relevancy of the atom $(\Pi^2 + \Pi^2)(\Pi^2 \, X \, \Pi \, X \, 3)$. In the motion that replace the motionless, the motion made the motion less an atom by putting the object through the commitment of motion into an independent Universe where every aspect of the Universe becomes an individual atom that maintains its independence as all atoms do. Since the top fills all requirements the top becomes an atom with motion privileges forming $(\Pi^2 + \Pi^2)(\Pi^2 \, X \, \Pi \, X \, 3)$.

When an object is in outer space that object encounters a specific relation with what we presume is space. This comes about by motion and through material volumetric size. The space the object encounter by moving through outer space puts a value of a ratio between the space it moved through and the space moving through which Kepler introduced as $a^3 = T^2/k$. That means there is a contact ratio between space containing and space contained by.

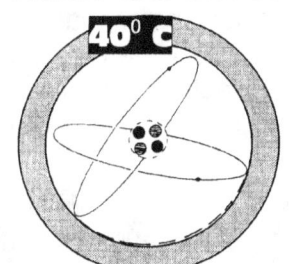

When the atom is in outer space the atom is surrounded by a temperature of zero Kelvin and that is because zero Kelvin is the presumably the coldest any temperature can get. Being zero Kelvin on the outside and with zero Kelvin being the coldest temperature there can, it would make the atom also zero Kelvin on the inside since there can be no colder than that. That would mean the entire atom is then zero Kelvin.

However applying motion going on inside the atom. coldest there can be makes atom is 40^0 C the outside of from the fact of what the that proves that the outside temperature of the atom does influence the inside temperature of the atom.

reduces temperature and there is much motion That means the fact that zero Kelvin produces the a little nonsense of such a statement. When the the atom must affect the inside of the atom because Balmer and the Lyman series would represent and

The normal summer's day temperature on my farm is 40^0 C normally in the shade because at that temperature little in loony enough to venture outside the shade. We consider that the atom must be 40^0 C because that is what the daily temperature is outside the atom. We feel and experience the 40^0 and we presume that all around is suffering from the heat of 40^0 C.

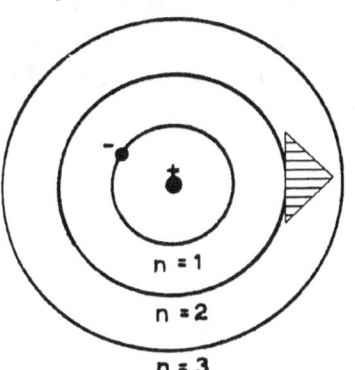

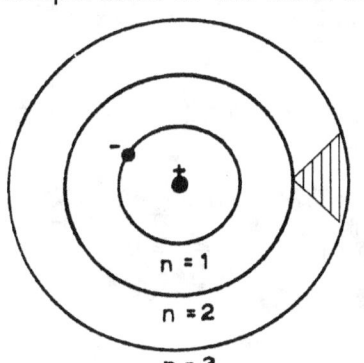

We know that the action brings about a reaction and the actions leads to a response. If the atom heats on the outside by measure that it finds a need to reposition the electron by one band, then also the inside got smaller in relation to the growth by one band. We associate such repositioning with the heat on the outside that amplify or reduce. However the adding of heat brings on a faster flow of liquid, which results in higher motion and it is in the motion that we find the answer to the cosmic principle applying. In the cosmos there is no hot or cold. There is higher or less motion.

The relocating of the electron position where the electron into a new jumps a band is done by implication of the Coanda effect. From the Coanda effect we know that the liquid attach to the solid using the formula $T^2 = a^3 / k$ where as space identify new boundaries by identifying the allocated boundary set by the liquid as $k^{-1} = T^2/ a^3$ where the space then forms the limit at $k = a^3 / T^2$. Every time the motion of the liquid intensifies the motion will attach to the solid by applying a new relation, which alters the relation of the solid by extending the space the solid has differently.

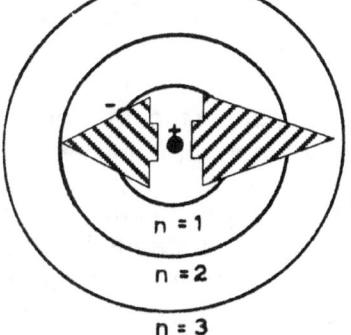

However we must not lock our focus on the heat but we must refocus on the motion that intensify or weakens. It is the motion that produces the new electron allocation and the motion produces a heat that establishes a cold. The focus is on the motion because the motion brings on

accelerated duplication and accelerated duplication produces cooling that brings on a relevant cold within the atom.

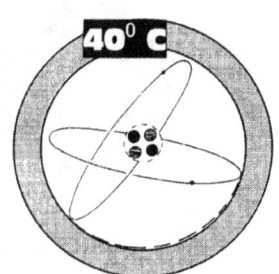

If the temperature on the outside of the atom changes from zero Kelvin to 400 C it is not the temperature that changes but the atom is responding to higher motion. With the atom in outer space the atom is subject to lesser motion since the atom is only in distinct and personal orbital motion in relation to the sun. That is why the atom can be subjected to zero Kelvin. When the atom is within the boundaries of the Earth and circling around the sun in a location set by the singularity of the Earth, the motion is distinctly more than what it would be if the atom were located in outer space.

The outside of the atom calls for a direct response to condition inside the atom since the outside can change very little if the inside does not respond in an opposing manner to what the outside produce. In such a relevancy there are always three factors performing as gravity and in that is the Coanda effect in charge of committing the standards by applying the gravity or the motion in relation to the solid.

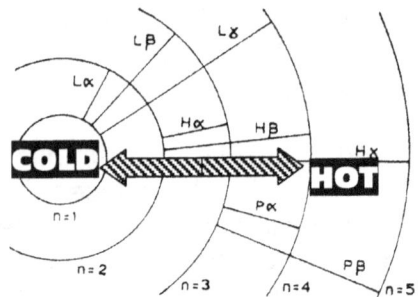

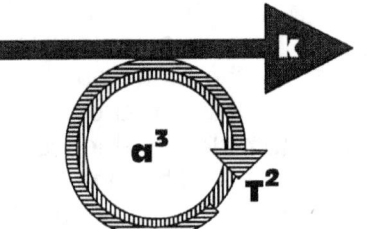

The material revolving through the space holding the material and privilege of motion is in time frame that makes

allowing the material the the amount of material per contact with the space which serves time and that it encounters as the space duplicates its position it holds coming from the past through the present into the future

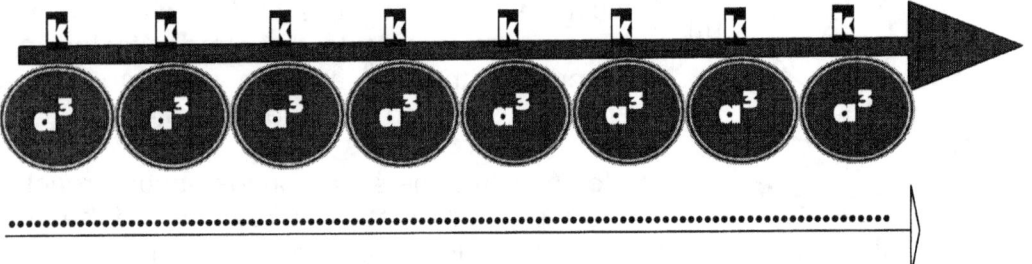

The movement reduces the size the material occupy by duplicating such vat amounts that the duplicating freezes the material into the oblivious.

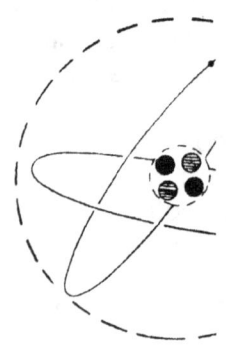

There is a definitive relevancy between the electron and the proton and that factor is what the neutron fills. The neutron is unrestricted gravity or liquid motion whereas the proton as well as the electron is very much restriction of motion of space-time flow, hence the mass. It is proposed that when the atom becomes hotter the electron jumps a band but that statement is not altogether the truth. The proton shrinks as much as the electron jumps a band just as much as the neutron fills the vacant space.

By jumping a band the space within the electron becomes more and the neutron fills that relevancy therefore the neutron becomes more. But if the neutron becomes more the neutron is there to bridge the gap between the electron and the proton and that will have it that the proton needs to respond just as much by becoming colder in the presence of the electron facing more heat. The heat is not the factor but the motion contributed by the heat is what brings about the larger jump in spin.

The neutron facing off the electron as well as the proton will respond on both sides of that which it influences because the response is that of bringing over more motion from the electron to the proton. One cannot gauge the electron behaviour without extending such behaviour to the reaction that the proton would have since the neutron fill the gap and also provide the response on both sides and the changes is what the neutron contributes by suffering the greater discrepancy in changes. However in the ratio or relevancy there will never be any change. The changes come in the form of an amplifying of the motion, which is a relation the space has with time.

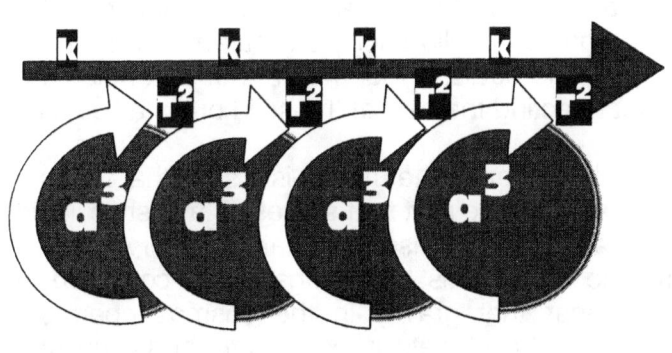

When an object is in a location with little motion the duplication present a lot of heat because the distribution of the heat over the space in duplication has very little possibilities of spreading the overall heat over a wide area. The motion of something as small as the earth will confine the atoms into a relative hot area since the space in duplication does not reduce the extent of the heat by distributing the heat over much space.

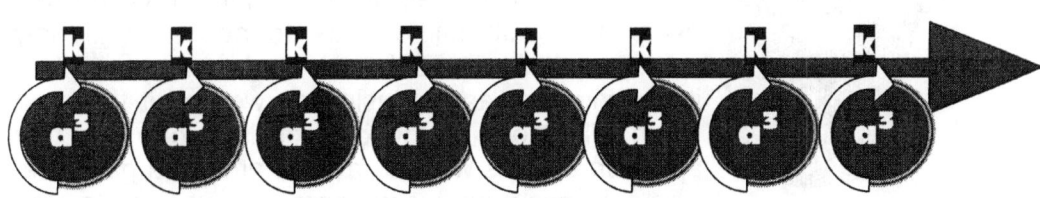

In a structure with the size of the sun the motion of space is enormous by the sure quantity of space in need of duplication. Shifting that volume of space needs duplication that is millions if not billions of times more extensive than what the earth may produce. By duplicating such a vast area in a period reduces the individual atom to a fraction of what the situation on earth would allow. The more the spin of the liquid is in relation to the solid state of space is reduces the space and extends the material in quantifiable

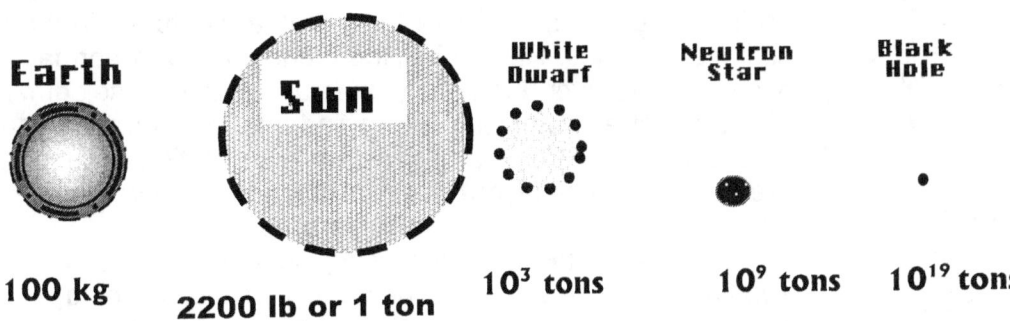

measure many billion times over to what smaller stars are. It is not the space that holds the matter but it is the spin in relation to what the matter holds that puts the relevancy of hot and cold within the star. The more cold there is because of the mot\re liquid heat bringing about motion, the colder would the atomic material be and the higher the relative contracting gravity that the star produces. This we see in the admitting of Mainstream science confessing that the reducing of space produces an increase in mass and because mass is the frustration of material unable to move, it admits to the fact that mass in volumetric size has no influence on gravity.

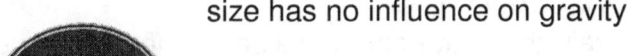

The physics we encounter on Earth allow us to use a common and a constant, a fit all and an all-purpose because we find us captures by the Earth singularity. The Earth provides the space we may claim as well as the time in which such material duplication will take place.

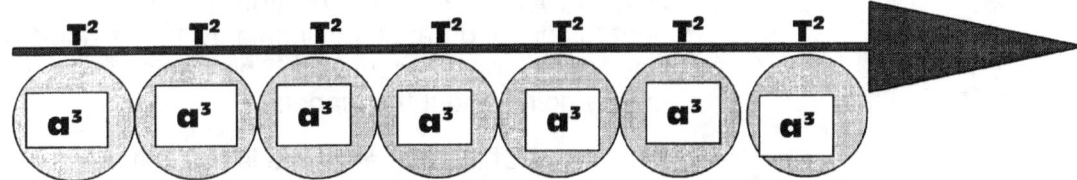

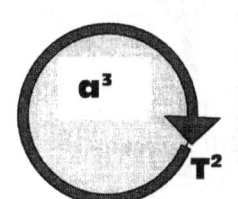

With the enormous motion going on in the Sun the material in ratio shrinks to freeze and this freezing/ shrinking allows the material to accommodate massive heat on the outside while all the shrinking is going on, on the inside.

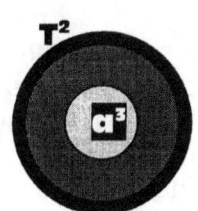

Let us for once and for all accept that Newton's mass activated gravity is inspired by his imagination. The process is when liquid in the form of outer space lines up against a solid such as a palate or the Earth. There is motion in the one department that acts as if it is a solid but is in fact the partner that holds the motion. Then there is the liquid, which by being the stationary acting the part of the solid but is a liquid all the same.

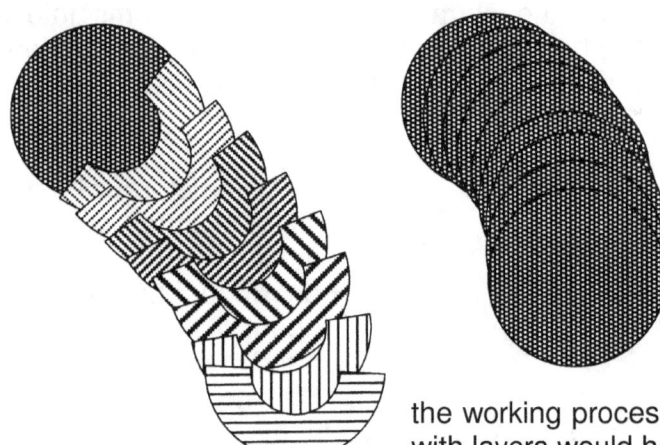

The body delivering the motion is a solid that forms a unit as space. The part that serves as a cosmic liquid is the partner that is also stationary and serves as an immobile liquid. This has things rather confused in the manner that gravity in the cosmos operates. Remember the planets is not a normal set up and even stars with micro stars to attend to is holy unnatural. It is very seldom in combination and when it is things get as confusing as we find it to be on Earth. The norm in the cosmos is a lone star that spins on an axis while in motion around a galactica. The galactica presents the same layout as a star and the working process in the galactica that apply in a similar mode, as stars with layers would have.

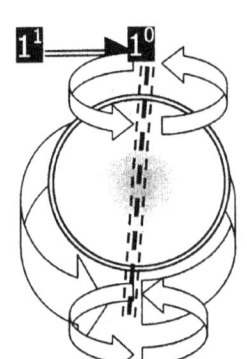

The body delivering the motion is a solid that forms a unit as space. The part that serves as a cosmic liquid is the partner that is also stationary and serves as an immobile liquid. This has things rather confused in the manner that gravity in the cosmos operates. Remember the planets is not a normal set up and even stars with micro stars to attend to is holy unnatural. It is very seldom in combination and when it is things get as confusing as we find it to be on Earth. The norm in the cosmos is a lone star that spins on an axis while in motion around a galactica. The galactica presents the same layout as a star and the working process in the galactica that apply in a similar mode, as stars with layers would have.

Do not look for the pumping going on where one can see time that meets space. The pumping action is going on where the proton pumps time into singularity by expanding and then contracting in the very heart of the atom nucleus. There the duplication present the expanding and the contracting which feeds the star with the motion either in duplication or in contraction that the star requires to comply with the demand space-time insist on as gravity. The reducing of heat by motion is presented as cooling since motion

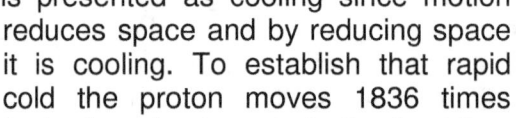

reduces space and by reducing space it is cooling. To establish that rapid cold the proton moves 1836 times faster in order to restrain the heat from the value the heat had when the heat was at the electron relevance. At the electron the heat was already at the speed of light and therefore the atom removes all heat by freezing the heat into the oblivious every atom is a black hole. All this adds up as a general reducing of space by the governing singularity in charge. Every atom in the star is a pump that coverts heat to cold and transfers singularity 1^1 to singularity 1^0 to regain what was lost during moment-Alfa. The gravity in the star is not nearly the gravity going about the planets.

In the case of the planets there is an orbit motion that puts liquids in relation to solids without the much pumping being the dominant factor. The liquid allows the solids space within to move. As the solid pushes against the liquid the liquid bears down on the solid and some liquid give way but the inner liquid increases the density at the point and just above where it touches the solid moving structure. The liquid pushes down the solid while in accordance with the Coanda principle the space expands to a point directly relevant to the motion that the solid provides. That which is without motion is secured by the liquid to be part of the Earth while that which is liquid is secured onto the Earth as an extension of space.

There is an allocated line designated by the extending of the solid that includes the liquid to gather that liquid into the unit forming the solid. We gave that so many names ending with sphere even the thought of all these sphere makes ones head spin. How Newtonians fit the sphere as in stratosphere and atmosphere and what not into gravity is still a puzzle, which is eluding me in the manner that Newton's vision on mass was eluding me. In the end of all this there is a line that is the friction point and it is at that line where liquid tear from solid while the solid is actually intensified liquid.

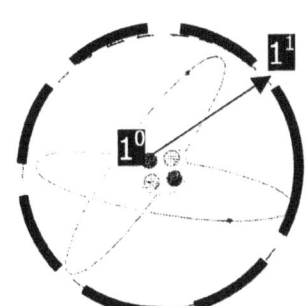

In the case of comets the Sun is the solid that forms stability while the comet is the solidity that moves and outer space is the liquid that does not move. In the case of comets the cosmic law is transgressed. The Sun is an atom. The Sun consists of a unit forming an atom where the Sun is the atom in compiled group but also where the group serves the unit. Every layer in the Sun is a liquid to the top where the bottom serves as a solid to the top layer forming the liquid. The proton puts time at motion where time puts space in demise. Time devour space as eternity meets infinity. The atom is Black Hole with matter in between infinity and eternity and this fills the black Hole with substance that is forming space - time. The final conclusion that any star can arrive at is when it takes

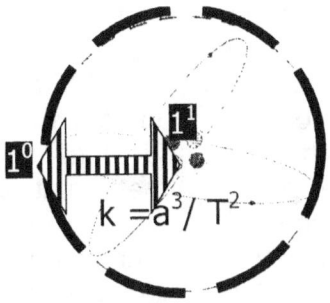

The proton serves as 1^0 to the neutron being 1^1 where the neutron serves as 10 to the electron being 1^1. The atom forms a Universe that hold both eternity and infinity apart by allowing motion to separate time. The atom concludes the Universe because the atom is what concludes the Universe as much as it started the Universe. In the end all star will be one atom in the hydrogen atom but that sis the final conclusion where the last era arrives. The atom is the Universe.

as 1^1. The core is the solid provides the motion electron is located it is the atom is a solid that does because singularity the motion. All atoms being 1^1 while singularity

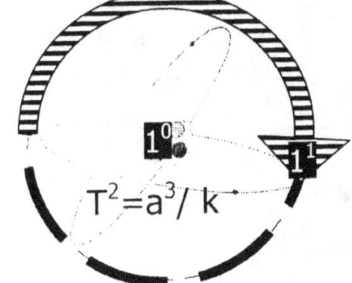

The atom maintains relevancies where the core within the atom serves as 1^0 and the orbit serves and the electron is the liquid. The electron because in relevancy at the point where the electron that is in motion while the core within the not move. The atom serves as movement generated by all atoms forming the unit provides forming the star are allocated the value of motion charged with governing the star is 1^0.

However the only constant in the Universe is that there is no constant applying. Everything is in cyclic shifting as the relevance relocate and alternate positions. In order to get a flow of space - time 1^0 and 1^1 must be forever alternating. The fact of constants are that constants are as Newtonian as mass can ever be and constants are as much a fact that does not apply as mass where then mass has the same position. The planet forms an electron to the Sun becoming the solid and the Sun allow the planet to spin while the planet receives it alternating which forms motion from the Sun that provide the governing singularity not only to the Sun but also everything orbiting the Sun as an electron Because the planet is

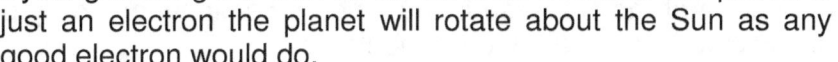

just an electron the planet will rotate about the Sun as any good electron would do.

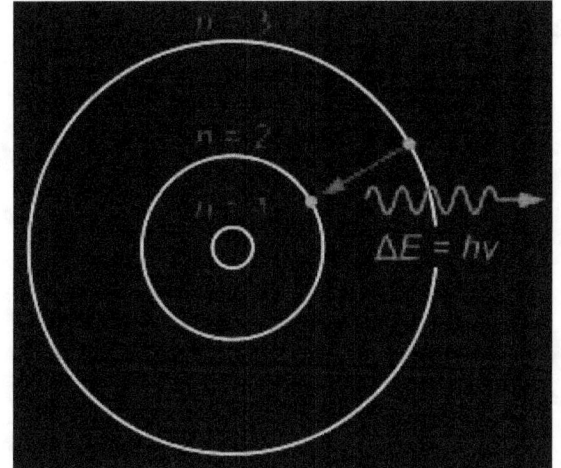

At the time when overheating of singularity brought a position in between that which have no inside and that which have no outside the development first went with time forming three, future present and past, but then one more dimension was included when two went square by adding two. That introduced an entire new dimension into the cosmos as 2 + 2 became 4 and 2^2 also became 4 by forming a rotation as a circle. When this whole lot shifts (duplicating by going square) the fifth point realizes and the process begins from 1 to repeat all over again. This overheating and forming circular growth is still evident in the Balmer process of jumping or growing into another orbit.

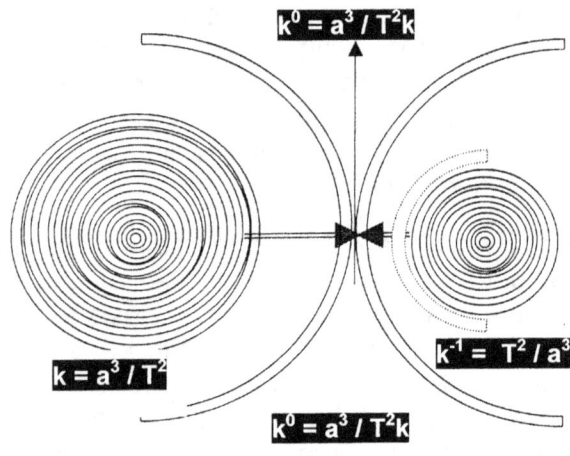

$$k^0 = a^3 / T^2 k$$

$$k = a^3 / T^2$$

$$k^{-1} = T^2 / a^3$$

$$k^0 = a^3 / T^2 k$$

There are changing relevancies as the one cosmic object orbits the other cosmic object and there are forever on cosmic object orbiting another cosmic object except in the case of Black Holes where the cosmos circles around a centre we call a Black Hole. During the cyclic the flow of time T^2 through the four quarters or four seasons the dynamics of the relations **k** change and inter change. The location of positions and the allocation of positions place dynamics in different concepts in ratio to each other.

Creation started with a dot, because that is the only form, size and dimension mathematical logic will

allow our brain to accept. From the one dot had to come a second dot and a third dot and that is a logical mathematical conclusion. The dynamics of such a dot is smaller than we can understand because such a dot is in negative relation to what we see Π to be, and the deeper we delve in finding the smallest fragment where space started, in the spot where time is still eternal as much as we can accept eternity to be.

The reason why we should first locate the spot is because we can only work from that point forward. By working forward we have to work backwards to locate where we are heading in order to find the past. The cosmos started at a point and where such a point is, we will find the Universe. Every one knows where the Universe is, because we can see where the Universe is, but if we can see where the Universe is, then we should find the centre of the Universe in that spot. Einstein theoretically positioned the point of beginning at a place he indicated where singularity should be.

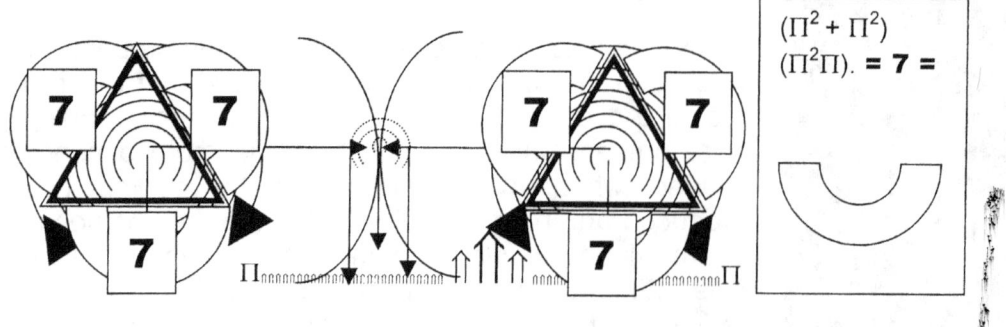

$$(\Pi^2 + \Pi^2)$$
$$(\Pi^2 \Pi). = 7 =$$

$(\Pi^2 + \Pi^2) \ (\Pi^2 \Pi)$ **3.** $= 1836$, which after wards the atom was about and the Big Bang could then proceed.

$(\Pi^2 + \Pi^2)$ $(\Pi^2\Pi)$ 3. =1836, which after wards the atom was about and the Big bang proceeded

With the cosmos the size it is and space so large compared to our smallness we have should no chance in finding the centre of the Universe. The Universe started where singularity is and singularity is the sure indicator of the Universe. With all spinning objects holding singularity we then have located singularity in as much as finding the centre of the Universe. The Universe started with a dot forming. That answer arrive from taking mathematics back to a point of being the smallest possible position, far smaller than we may be able to calculate form. The ten dimensions I named the atomic relevancy is also showing the double value of singularity as singularity extends into form and beyond as space. The atomic relevancy is $(\Pi^2+\Pi^2)(\Pi^2 \text{ X } \Pi \text{ X } 3)$ = **1836** that is the mass relation between the electron **(3)** and the proton. Proton = $(\Pi^2+\Pi^2)$ Neutron $=\Pi^2\Pi$. The atomic relevancy holds the dynamics of singularity control. In the ratio and dimensions we find in the atom, all space-time derives from the atom, whatever the atom is.

Our instincts, our logic and our calculating process all indicate that the sphere holds a centre point from where six evenly positioned point's position matter to be. Using The formula $F = G$ $(M_1.m_2)/$ r^2 it indicates to a force pulling objects closer, where each force is coming from each centre point the body in question has. The contraction must commit the two bodies towards a point in each case being spot on in the middle, not withstanding what direction the force is applying, the body will draw to the centre.

If the Universe spins where singularity particle holds the centre immeasurable many, and every smaller bits, are all not pieces of Universes. Every atom ends the why what we think of is time that matter how small is holding the Π^2 around a centre point that is holding singularity, and confirms the centre of the Universe, then every of the Universe making the number of universal centres atom and sub atom particle presented outside the atom in the Universe but they are a Universe surrounded by many Universe but to explain that I need about 300 pages to explain forms development in space. If every atomic particle no centre of the Universe, then the gravity is coming about from that point because that is where the gravity applying in the Universe are applying contraction.

It then is the atom in the most centre part where space and time meets singularity, that is why Einstein found a Universe collapsing to a single dimension, and every atom at a centre point of the proton where gravity initiates in according with the proton dimensional colas of $(\Pi^2+\Pi^2)(\Pi^2 \text{ X } \Pi \text{ X } 3)$ = 1836

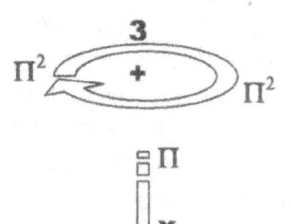

The atom also shares qualities with the sphere and electricity by forming cosmic space in formulating time developing space. The following is the displacement value of each one mentioned. The sphere is 7/10 $(\Pi^6)6$ =112.1, while electricity or gravity starts at 10/7$(4(\Pi^2+\Pi^2)$= **112.8** and electricity or light as form ends at 10/7$(4(\Pi^2)$= **56.4** and gravity ends at 7/10$(4(\Pi^2+\Pi^2)$= **55.2**.

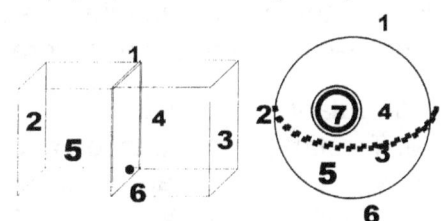

Kepler's formula also indicates that a sphere is within a cube that is holding a sphere

If the Universe did start from one single point and time matter and space flowed from that point, then that point must have a relative connecting base because such a point holding singularity must be eternal as space matter and time link eternal. There therefore must be one point linking the entire universe when regarding the fact of singularity. Then according to the theory off relativity there has to be one exact point holding time in a relevance notwithstanding the fact that time depart from that position and relate differently to all space-time away from such a point. This has to be the centre of the Universe.

Every person I have discussed facts about creation recollects images in the trend depicted in a presentation as one may find to the picture of what creation now represents with material just being formed and that is, is already in place. That would be the most unlikely way Creation came in place. The recalling of pictures representing images about creation must have form, but to mathematics starting with singularity such a point that holds singularity such a point has no form. From this thought the very opposite arises of where Creation came from and to think it is nothing such an idea is mathematically simply not possible. The thought of nothing is just what it is, a thought of nothing and although it is in the human mind common nature to present nothing as a value in the recalling of something, nothing is a presentation of the figment in the human mind. There can be no number such as nothing and that was (possibly) Newton's biggest error. Nothing represent non-existing and that is just what nothing is, it is non-existing. This is why the start of the cosmos in terms of mainstream thinking is still unexplained.

In order to prove my point I wish to ask the reader to define the shortest line there can theoretically be. If he should answer anything but that the shortest line should be at a point where the beginning starts and is the very same spot that it ends. The shortest line that can ever be anywhere must have a start and finish holding the exact same spot. The line will be humanly impossible to create but we humans are capable of very little. The spot has to be invisibly small and hold on this spot the beginning and end of the line. Then the line has a beginning and an end at the very same spot and if it then wishes to extend the position as to further the possibility it has, which direction should it then favour. Humans in the west would naturally think of extending from left to right while in the east humans may want to go from right to left. Some persons will tend to go up or down, but all of the options are about human preference and not mathematical conclusions. Extending the line in any one direction will favour one direction without a conclusion about not extending in other directions. Such a conclusion has no sound mathematical foundation. The only option about extending will be in all directions equally in order to give a meaningful non-bias flow of mathematical equilibrium. From the centre it will extend in six directions equally and evenly. It will grow from the centre outwards in all six directions evenly, much as described what happens to the Universe when witnessing the growth of the cosmos seen from a large telescope.

The shortest line in the realm of possibilities must have a start and finish holding one spot and such a line will also be a dot or a multitude-circles. Not favouring one direction puts all directions at equilibrium meaning that any form what ever may be can develop from such a spot with the end and the start being the same. This reasoning prompted me to look for singularity in such a spot because if the prime spot from which all came was a spot, then the spot must hold the shortest line but more prominent it will hold the smallest form and seeing it started from singularity that includes the smallest circle.

One possibility that the shortest spot can never have is having a starting point on the zero mark. If the mark of zero holds the start it must also hold the end because the end and the beginning has the same position. If the position of zero then is the beginning, the end will also be zero leaving the line without an end as well as without a beginning. I am trying to lay some groundwork later when I explain how the cosmos started. The conclusion from this is that no line can start at zero because that will be a mathematical impossibility. Being a start with zero removes the start and by removing g the start it must then remove the end of the line because the start being zero will illuminate the zero end. A line or spot starting at zero would therefore be shorter than the shortest line possible. A line growing or extending from zero can never leave zero because of the influence of being zero disqualifies any possibility of growth. If the line then had to grow in all directions at the same pace the line must therefore be a circle. The value of the circle is Π, and therefore that is where to search to find where creation started.

That gave me the clue where to start looking for singularity. One would find singularity in the value Π and the value Π will be in all things rotating in a circle. To start my explanation about my cosmic theory I wish to firstly bring some nostalgic and the relevancy will become apparent later on. Such is the importance however that I wish to place this at the very start of the prologue.

Again I whish to repeat how obvious the conclusion is about how the Universe started. When we were boys we played with a top we called the spinning top. I cannot imagine that there is one boy in the western world that did not hold such a devise in his hand. Tying a string securely around the tapered cone started the operation and then with a jerking or pulling throw the devise is launched in a projectile manner and the big knack to success was getting the nail end firmly on the ground and by the realizing jerk the top was rotating. The champion was always the one boy that could throw his top to spin the fastest and that would create a humming sound. The louder the sound produced the bigger champion. This supply of thrust wheeled the top into action and in that we must find the starting process of not only the top but how the Universe started.

When a back braking effort produced a throw of enormity the spinning top would not only produce sound varying in pitch but also create a spin that would seem to have some instability. There are very many limitations about the spin, parameters that determine the slowest and the highest spin rate and spinning is within the parameters of such settings. The question arising is why such parameters are there in the first place?

Singularity by Motion

Singularity by Time

An enormous effort will have the top going oblong while spinning violently and as the pace reduced the top will stabilize by coming to an upright position. In the upright position it wall then spin for the remainder of the period where it will in the end start tilting to the side and in a last effort throw a few wild oblong turns and fall over.

To put the top having motion into the context of dimensional dynamics we will find the top in time holds four positions in which time moves as the top rotate about time or in fact this effort would be an attempt to acknowledge

time. The four positions are the proton dynamics serving singularity in motion. Then in a ninety-degree alignment we have the top using space to move as the space is duplicating material through motion. The duplication is the reproducing of material as Kepler pointed it out when he said space is a dimension using time as a dimension where space is one part and the moving of the space through the time dynamics is the other part of space. This is all evident in the formula $a^3 = T^2 k$.

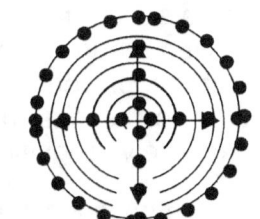

If the alignment is in ninety degrees to each other then Pythagoras has to apply strictly. We have to be able to use Pythagoras to determine the value of time in space, should my argument be sound and which it is sound. When the material rotates or moves the filling of the material is in perspective to the time however, material can only be in one location in one split time-instant. Since material has to cross over to the other side of the Universe in order to duplicate, which is how material move, then the point in which that specific part of that specific material can be only on one side of the Universe. By moving to another point the point also has to rotate and fill another spot holding space in the same circle. In order to rotate the circle it has to move the circle in a directional dimension. When we see anything just rotating we forget that the earth is doing the displacement in direction of the rotation.

Every time singularity complying with matter generated moves, it is a new set of points each holding singularity that is being charged with the motion. The Universe started from allocating singularity charged by heat into positions where such positions contributed to space- time. Every time the spot overheated the spot expanded into four dots and by expanding the spot cooled. In cooling the spot retained heat by which it spawned the dots allocated as time. In overheating the objects expand and by expanding the

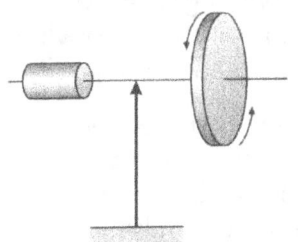

objects cooled as it spread the heat over a larger area. That is gravity. Gravity is the expanding in relation with the cooling which means it is duplicating material in relation to a generated centre that is contracting the motion by cooling. Every inclination of motion is in fact motion and every movement be it contraction or expansion is moving to the other side of the Universe by bridging singularity because singularity as a given is immovable. That is because $1 \times 1 = 1^2 = 1^0$. Mathematically singularity cannot move. Therefore by being immovable, therefore motion has to cross the division singularity apply and by crossing the division the

factor that comes in place is $\Pi^2/4$, which results in the Roche limit. But such motion is three and the square of three in addition to the square of four brings about time in space.

Singularity by Time

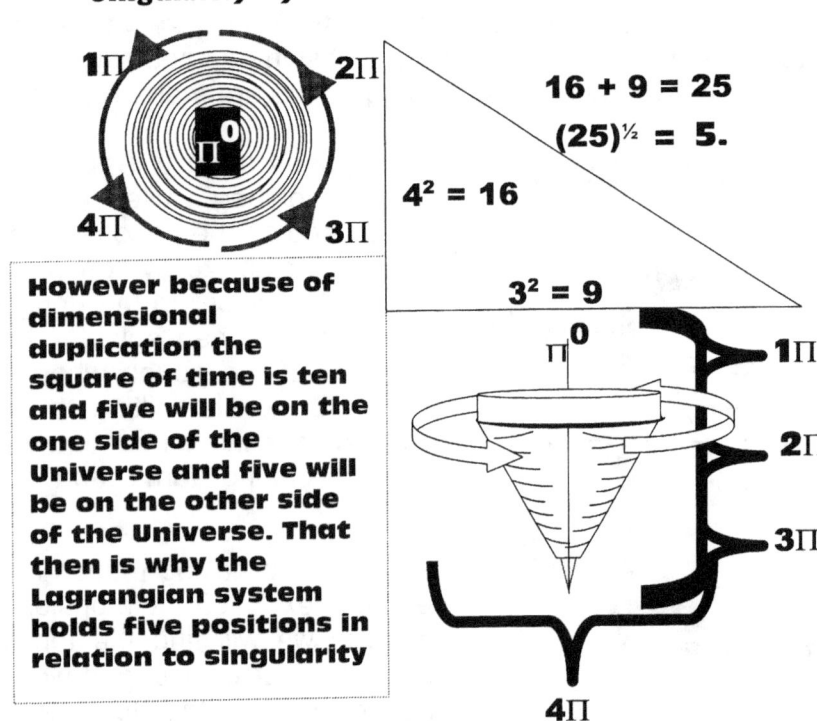

1π   **2π**

π⁰

4π   **3π**

16 + 9 = 25

(25)½ = 5.

4² = 16

3² = 9

π⁰

1π

2π

3π

4π

However because of dimensional duplication the square of time is ten and five will be on the one side of the Universe and five will be on the other side of the Universe. That then is why the Lagrangian system holds five positions in relation to singularity

As a school going youngster I was fascinated by astronomy and in particular the cosmology aspect. In a long and strenuous process of self-education I was completely stunned by the behaviour pattern that the comet had in its relation as it orbits the sun. Please forgive my boyish way of presenting the following but it is important that I bring it across as I saw it as a boy and as a matter of fact still see it today as a middle -aged adult. Push this double standard applied back to before the sun took its position and there was not earth to indicate the year. How small was the year circle at that point in time and space. Take this right down to the:" Big Bang" where "the whole Universe were the size of a man's fist" (To use their words), how far did the circle go to indicate a year then where no earth was and our sun was not a thought yet? The year was immeasurably smaller, shorter and faster than at present. This is logic even the Newtonians must accept. The movement of space marks the duration of time and the more space moves the shorter the time period must be. But back then the space that moved around held a few million galaxies within the space of our thumb now holds that is if the Universe was the size of our fist. There is no space outside insanity to apply time to the past at the value it has at present and far worse, to use something so extremely insignificant as the earth to measure time applying to the Universe by.

Using such logic makes science appear foolish and those in science look like jesters. There is just no rational in the time verses events that can explain facts without correlating the space the time applies to. Since the time of Newton, the arguments tarnished from being brilliant to clever to fair too poor and a hundred years ago to the point of being stupid. That is what Kepler's formula is all about! That is what Kepler indicated with his formula $a^3 = T^2 k$. The space of an object (a^3) is equal to the time (T^2), which it is in, in every given instant (k). If the space becomes smaller, the time duration becomes longer every instant of time's progress. Every planet hold a different time in a different space and that is the Titius bode law. A Pendulum working on Jupiter or on Mars will have a much different swing action in time calculation that it has under the conditions applying on earth. That is because thee gravity is much different that it is on earth. The pendulum uses gravity to measure time on earth. If the pendulum uses gravity to measure time then the fact coming about from this is that gravity is time and time is the moment of gravity. When one applies a pendulum to measure time then the device must measure time. When such a device is applied to work off the flow of gravity it also measures gravity.

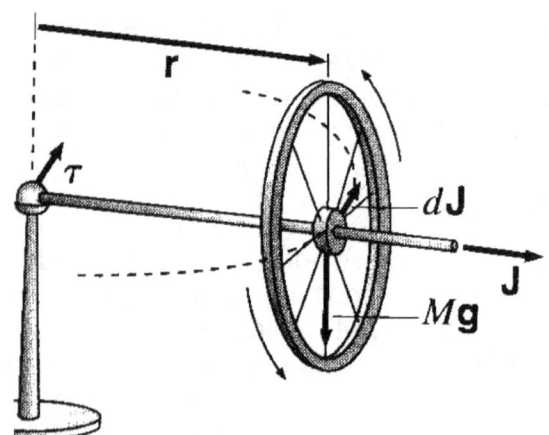

The pendulum not only stops at a precise point that the earth holds as singularity presenting the singularity the earth dictates at that given time. The relation there is in what Kepler discovered and that which Galileo discovered has gone by without many too my knowledge seeing such an extreme direct link. Kepler formulated space-time and Galileo implemented space-time. The space the pendulum swing through is representative of the space captured by the anchor singularity formulated by Kepler as a^3, the pendulum arm becomes the indicator k the swing distance of the arm becomes the time T^2. This is the recopy for time keeping since coming into the light from the dark ages. It is a half circle indicated by a straight line forming two sides where each side is holding a triangle in relevancy. This is in sharp contrast to Newtonian claims that the cyclic repeat the Earth has with the sun and all the numeral Equalities derived from that by Using Kepler's formula still after one year comes to nothing.

The spinning or not spinning is not part of the issue because at the point of absolute singularity the object never spins. Therefore spinning or not spinning does not apply to the point of singularity because singularity never spins in any event. In the whole structure with a pivotal centre as the control to the motion of the space the fact of Π is a natural outflow and any adding of Π is totally incorrect.

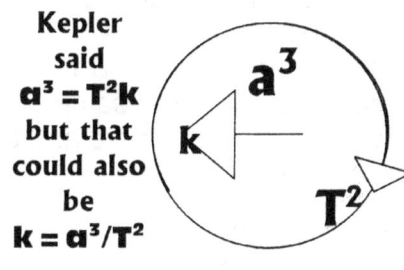

Kepler said $a^3 = T^2k$ but that could also be $k = a^3/T^2$

When translating Kepler's mathematical expression into English we can see what Kepler said also read as $k = a^3 / T^2$ where **k** is one point from a centre point that is space a^3 relating to time T^2. From a centre comes space-time. The centre **k** brings space a^3 in ratio to time T^2, which is space / time a^3 / T^2. Reading this correctly cannot bring any dispute...yet it does...and it has been doing it for centuries on end!

The Universe form by numbers going into dimensions that ends up as dimensional space. However this is also where we have to find the start of mathematics because mathematics started by starting the Universe. Mathematics is not God Almighty that was before the Universe was but grew as the Universe grew. By enlisting the way mathematics started and comparing that to how the Universe began we can trace the beginning back to a point when space was 1 and mathematics was 1 that this formed physics.

Π^0

The cosmos in its entirety started with one point but that is the beginning and not the answer. I show where this point is and that allows me to prove why did one point start. I guess it is easy to say and show it started with one point as Einstein did but there are reasons I give to prove what and how it came about.

The big question we face is to see how the Universe started form one spot. We now know where to look for such a spot because the spot can only be in singularity. It has to be in the centre of the Universe and we know where the centre of the Universe is. The centre of the universe is in the centre of everything that spins, no matter how small or large the object is. It is as much in the centre of the smallest sub-atomic particle as it is in the centre of the largest galaxy and the centres are very much equal as they are very much the same. The only aspect that puts them apart is that the two as being one holds relevancy to the position each one hold in relation to the other but eventually the two are the flip side of the same coin. They hold the value of Π^0 and end up as $\Pi^2\Pi$. It is only because of movement Π^2 that Π forms from Π^0 and with that being evident we can start to look from where this came about.

The Universe never started but it ended. The Universe was perfect with on e dot never changing and there fore it moved from the future through the present to the past with no distinction parting time. Time was as much infinite as it was eternal because the same spot covered the concept. The spot never came about because the spot never went away. The spot was so small it was less than a thought and it was so large that everything that is, still in within the spot from which the lot came about.

From the smallest ever possible dot will grow a line in every imaginable direction relating to a prospect of Π not favouring one direction that puts all directions at equilibrium meaning that any form of what ever might develop from such a spot will have the end and the start being in the same position, which will also have to be a sphere as the flow outward will be equal in all directions. This reasoning prompted me to look for singularity in such a spot because if the prime spot from which all came was a spot holding all, then the spot must hold the shortest line but more prominent it will hold the smallest form including the smallest circle or for that matter the smallest sphere. One possibility that the shortest line or smallest spot can never have is having a starting point on the zero mark. If the mark of zero holds the start it must also hold the end because the end and the beginning has the same position. If the position of zero then is the beginning, the end will also be zero leaving the line or spot without an end as well as without a beginning. Such a spot will constitute all of nothing Any line starting from zero would inevitably start from a point where it ignores the zero mark because the fact of zero does not implicate a start or a size of value, but only the not being there of that position. All lines would form a duplication of another line sharing value since there will always be a possibility of yet another line in the realms of singularity lying between the two lines in question reducing the size infinitely to either side of the divide we humans create. Boundaries therefore are human and as man made substances it does not belong to the cosmos outside the influence of man and must be discarded.

To back this argument that no line can ever start at zero is to ask the simple question: what will the length of the shortest possible line be. It must be a line where the starting point is so close to the ending point the distance parting the two is incalculable yet there is the line therefore the end and the start is apart still sharing the same spot. The length cannot be zero because zero means no line. The starting point and the ending point may be inseparably the same point with virtually no space between the two points but neither of the three points can be zero simply because there is a line (be it infinitely small it is there). If the point is zero, then the line will be shorter than the shortest possible line. If there is a line and the two points starting the shortest possible line and being the continuing of the line it may still be the same point and even by sharing the spot as the point ending the shortest line possible the line must be there and the line holding the start and finish is next to zero but can never be zero otherwise there is no line.

As highly developed we seem to regard science to be, it took a genius like Max Planck all his life to try and reconstruct from what there are to what the universe started with. Once again the human perspective baffled the genius of a Master such as Planck was, because he placed man amongst the first of creation. Where it started from and what we now have is many, many eternities apart and any evidence of an attempt matching what we have now to what first was is foolish.

It started off with one dot so small eternity met infinity within. Then came one more, and another and they continued coming until there were a countless number of dots. The accumulative size of the dots were the same size as one dot because in the true universe big and small plays no part. The dots were infinitely small and eternally big at the same time because size is a relevancy and without one the other has no size. So in the true perception, there is no difference in size.

It started with the fact that there is no place or part in with which one may associate zero or nothing. There are no room for a number such as nothing. Next to the one dot (infinitely close) one will find the next dot, and if nothing was a factor then that is precisely what one will find between the two dots. Nothing of space, a non existing entity, taking up no space, and much more important, no time, therefore the dots are infinitely close to one another, being the same space, eternally big as much as infinitely small. If we as humans cannot find a manner in comprehending this notion, there can be no manner ever understanding the cosmos as much as the start to the cosmos. The number that covers this quantitative is 1^0 where 1 defines the spot and ^{0} sets the spot apart from position in time. The spot moves from 1^0 to 1^1 by being the very same. Only in time moving can the same dot find relevance to form.

Every dot was a Universe in its own and the accumulation was a Universe because every spot was the same spot that time divided apart. Today the earth in itself is a Universe as the moon is a Universe, as every atom is a Universe because rules applying on earth do not apply on the moon and visa versa. When in the ocean another set of rules apply, therefore being in the sea places a body in another Universe. The number of universal entities is still countless, as much as it was in the beginning.

Every dot insignificantly small as it may be, is a part of another Universe as much as it is part of the accumulative universe and every dot in the infinity holds singularity, which we translate as " nothing" being " darkness". There cannot be "nothing" just as much as there cannot be "darkness". There cannot be something big or small, but it is in relevancy of perception, and then the relativity of perception becomes the question. There cannot be hot as much as there cannot be cold. The sun FREEZES hydrogen to a liquid at six and a half thousand degrees Celsius and Universe boils over in the form of the Hubble constant at the temperature (we presume from our vantage point) at minus 273 degrees C. If we Humans cannot or will not abandon our human perception and our manly perspective, we may as well return to astrology for all its worth. The Universe cannot grow because it cannot become "bigger" from at first being "smaller". The Universe has to stay the same because it remains the same Universe that has nowhere to go and therefore nowhere to change. We ca have density shifting position as heat converts from space in eternity to space in definitive where it moves on to infinity to stop infinity becoming eternity once again. It is a shift in density that we perceive as progress but progress then is just a shift in balance that shifts borders.

Every point in the infinity we may observe at is not merely part of the Universe in not being "nothing", but is the point where the Universe started representing singularity. It is the very first point where everything began so many eternities ago, because after all, how can we ever determine where the first point was, as they were very much equal and alike at the beginning. Every aspect of the Universe started with the fundamental fact that no point in the Universe can represent "nothing" as a number, because every aspect in the Universe represents singularity in what ever form it may hold in that specific spot forming space-time. If man does not reach a conclusion where that conclusion is matching the Universe as a reality and stop to match the Universe with man (and man's incapability), we may all go back to caves and become starving hunter-gatherers again, because we will never find a way to progress to the ultimate understanding of the Universe.

We now sit with the question as to why does space begin singularity and not the other way around. When we look at what we have we will find what we have and that will point to where we came from. First the top has to spin to allow the top to form the non-existing line and it is the non-existing line that places the top in the Universe as part of the Universe. Space establishes time but time confirms space in its position and allocation. That is what we have and now we have to move from there.

With this evidence we have all we need to determine how this lot began. The main thing to remember is that once something is within the Universe it remains in the Universe and it may change in relevance but it can never go away or leave the Universe again. This is rule number one. This means every time movement in the Universe comes about the Universe starts all over again from a beginning that is only different in relevance through space that holds a new position to time in time. The density applies differently but the concept and the quantities remain the very same. This is rule two. The Universe changes every instant in order to remain the same. The Universe stops to start again and never start. Infinity by which the Universe starts is twice in duration compared to eternity in which the Universe starts. Infinity has to accept the way eternity was to change by not moving (infinity can't move) and reposition eternity into a new location where eternity then moved and not infinity.

The Universe consists of singularity and singularity holds relevance in heat. Heat changes density.

This was the era of distinction, when separation brought an all-possible new Universe

The spot becoming the Dot

The Spot

The spot was in position eternally because the spot is with us now eternally and so there is not denial in the fact that the spot is eternally present and part of everything we think is present. The spot is smaller than a thought while being so big the spot holds the lot inside the spot.

Then came the first instant that set it apart from the last instant. This was when the perfect ended to begin the imperfect. This was when that which was eternal became instant. I do not have space in this book to show the entire concept of what was not to become what is everything but it is a huge concept to understand.

What was, was heat because what is, is heat. Heat became more as what was overheated to become more. Hot parted from cold. Heat grew into space The movement that this partition established created a growth in time as space parted point from point but the recognition of the point was that time moved the point. Time moves but in this instant movement was required that was independent of time moving. Time moves eternity that what was required was movement outside eternity to bring change to what was already in pace.

Time was I eternity which means it was not yet time to move but movement was required to break the moment of overheating. The overheating brought destruction to what was present and what destructed was the perfection, which was in place. With something overheating the immovable to move by establishing what never was before. Movement had to come that was outside time because only time moved by never moving. Everything stayed the same forever and therefore change had to come about to bring movement to what never moved before.

We can witness today what is in place to place what was in a concept we can understand. Today we have Π^0 becoming Π through the movement of Π^2 forming the moment Π^3. This still is in place and this is everything that is and therefore this is everything that was. Still today we have the four of the top establishing the three of time to become the moment we witness as space. I put this in place to show that everything did start from one point where space had to supply an alternative movement to time.

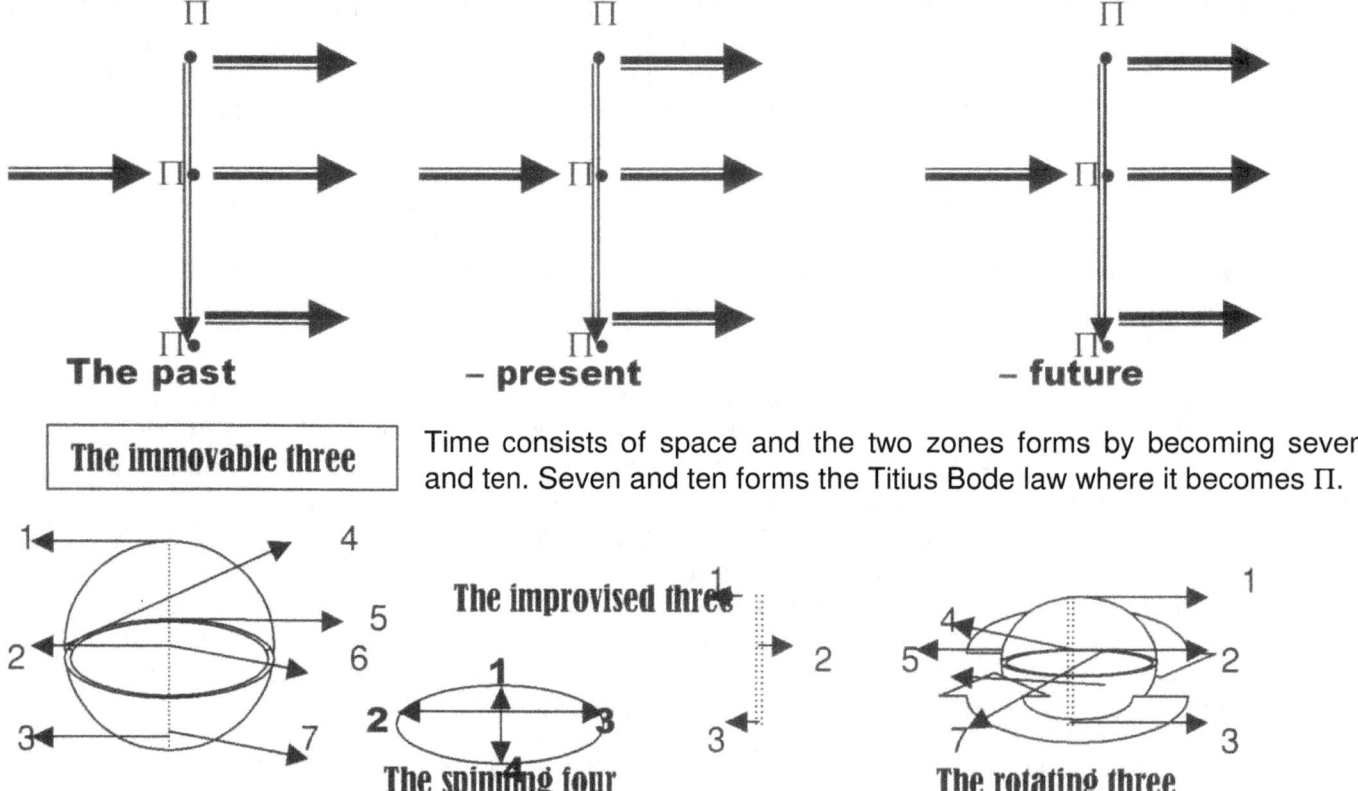

The past   – present   – future

The immovable three

Time consists of space and the two zones forms by becoming seven and ten. Seven and ten forms the Titius Bode law where it becomes Π.

The improvised three

The spinning four   The rotating three

The question these laws answer is when the Universe started with one point where did that one point come from. Yes, Einstein proved that one point was in place but what then came before that one point and where did that one point come from.

1^0 1^1

Then came two but how did one become two. I'll give the purpose of two and then later on as the book develops I'll relay that to how the lot started. The sun is one and the two is I being with and on the earth and together the earth and I as a unit is two. Because the sun that is one secured my past being the two it at the same instant provide me with a future because my past secures my future. There is a relevancy brought about by these 4 laws in which the sun and the earth secure each others future because the past places two positions in relation. From each other's perspective the one holding a place is the other's past and future.

1^0 1^1 1^2

If ever one of the two would disappear it would leave a hole in space wherein the rest of space would fall. That is why we see Black Holes devouring space. It is because one of the two and which ever of the two does not matter, lost a position because of lack of movement sustaining the position the rest of space that secures two loses the value of two because one lost the place one held.

This is where the

$\bullet + \bullet + \bullet = 3$

the   three comes in. Since the light moves from one (the Sun) to two (my place on to replace the   earth) the movement of light forms three by securing the time in place it takes past falls away and   light that brought the past in the instant to secure a future. If one falls away the that drops the future into the past and that is what takes place in a Black Hole. As long as one secures two and two holds a place forming three the Universe sustains space. Now we have to take this very argument back to when the first 3 came in place.

Then because the Universe in 1••▶2••▶3 double as well as the Universe in space formed a place and a position to be.

three secure one and two by providing one and two a reason to fill the Uniqueness both has, and it is for that reason four forming a a square of two became the first dimension outside one and the Universe in space formed a place and a position to be. This is how the Universe is filled today but why did all this brought about a start. It is so easy to show senseless mathematical equating that says nothing and proves even less and it then runs away from any understanding.

Taking the argument further the argument gets more and more logical but also the argument consumes more pages and that is why I leave the rest to explain as the book develops. It is so simple the reasoning is quite understandable to those that wish to understand. The sun is one and the earth is two and the space is three and the movement and relocates these object in movement brings four, which proves that movement, is the square root of everything that forms the Universe.

The rest is as simple to explain but science pretend to search for planets far outside our solar system and all the while science has no idea why our solar system functions in the manner that it does. Would it not be much better to first find out how nature applies in forming our solar system before trying to pretend to know about other thought to be planets? Science ignores nature and pretends to know much more while all of science know so little they have not figured out how gravity works…and that I prove! When you read this book you will know much more about the solar system than what the Super-Educated ever knew about science. In this book for the first time ever the true nature of gravity is revealed.

If we wish to find the place where the Universe started the last place to look would be in the overall picture. That is where we now are and that is where the story developed. Therefore this picture of the place where it is now is worthless. My finding the way to decipher these four laws put in me the place where I could locate the beginnings of the Universe because without realising whey these laws form gravity we will never know how gravity starts and started. By deciphering how these four combine in principle we can locate how everything started because these four principles take the Universe into its smallest detail and that is singularity.

That is where Einstein proved the Universe started because that is singularity.

 To talk about singularity one has to refer to the ultimate one. It is 1 or 10 going on to 11 that begin to form as 12 and so on. That is singularity and that is where the Universe started. It started in singularity that holds a dimensional space of 1 or 2567892^0, which also forms 1. There is the single dimension also called singularity or the Ultimate 1 and that has no dimensional value.

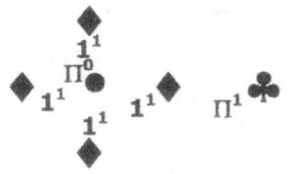

However the Universes formed by relevancy holding positions as many and that is where five begins. If it is true that four holds movement to three forming space then five secures allocation by bringing stability to whatever moves. If we take Jupiter and its Trojan system as an example we see Jupiter spins around the sun (4) while also providing moment for the Trojan asteroids each moving (2^2) but giving location stability (4 + 1 = 5) and that forms the Lagrangian law.

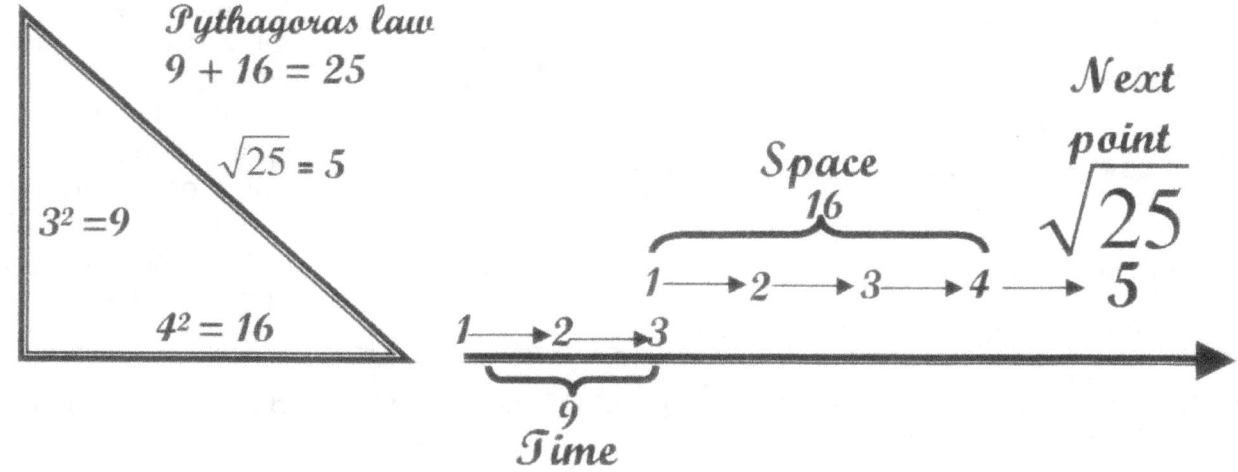

Pythagoras law
$9 + 16 = 25$
$\sqrt{25} = 5$
$3^2 = 9$
$4^2 = 16$

Space
16
1 ⟶ 2 ⟶ 3 ⟶ 4 ⟶ 5

Next point
$\sqrt{25}$
5

1 ⟶ 2 ⟶ 3
9
Time

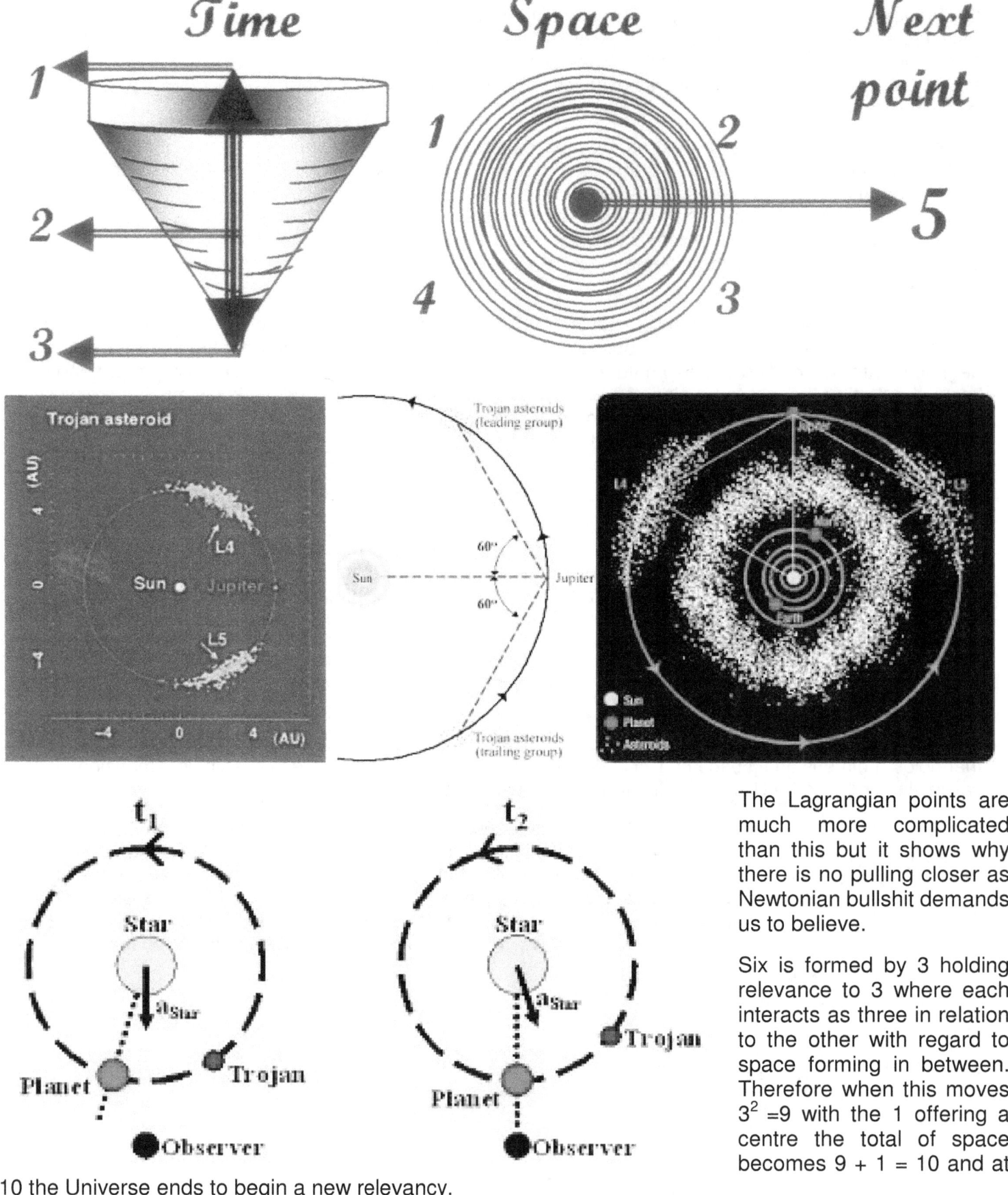

Time

Space

Next point

The Lagrangian points are much more complicated than this but it shows why there is no pulling closer as Newtonian bullshit demands us to believe.

Six is formed by 3 holding relevance to 3 where each interacts as three in relation to the other with regard to space forming in between. Therefore when this moves $3^2 = 9$ with the 1 offering a centre the total of space becomes $9 + 1 = 10$ and at 10 the Universe ends to begin a new relevancy.

After six we have seven and seven in relation to ten is the Titius Bode law and the Titius bode law is how time forms space by movement thereof.

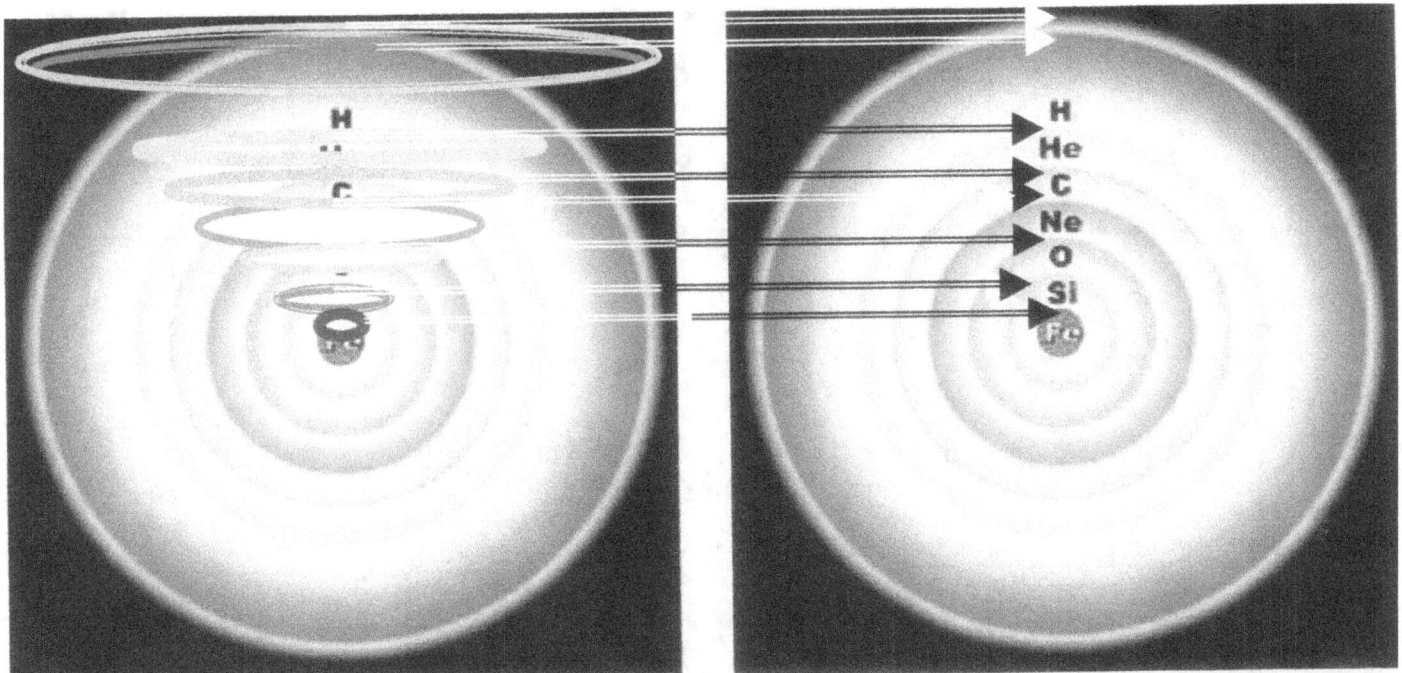

Every layer within every star represents a new era of development that the star has to progress through. Every layer forms a star within a star and acts independent of the star while also producing a united development of the star. Each layer has a purpose that the development has to fulfil before progressing.

> **The sun has a liquid exterior that freezes space – time to hydrogen at a temperature of 6500 ^{0} C**

> **Inside the sun the temperature rises but it does not get hotter, the space becomes vacant and the eat rises.**

> **Within the star hot and cold becomes meaningless as space-time approaches singularity.**

At such a point that is within all cosmic atoms, space and time, heat and space becomes irrelevant because the conditions applying is such that we will never understand the cosmic principles ruling there.

However the Titus Bode law is not only about developing the solar system but it is also proving why stars and galactica form layers while each form by a different selection of material. All this I explain in **Nature Annihilating Newton**, which is available on the Internet.

The idea that nature and Newton fit hand in glove is a myth carried on by mainstream science for more than three hundred years but it holds as much truth as Newtonian approach to cosmology present in accuracy and that is nothing. Newton's views and claims on physics and what nature applies do not even share a Universe in comparison. I bring you facts about what is true in nature and not what you think is true or science thinks is true but what nature uses as the truth. When you argue about what I say you argue with nature and what I bring you then you also falsify nature in favour of Newton as science do.

All in science has this notion they project that nature can't be trusted because nature is a freak show and nature has one purpose and that is to mislead the human species. The person that trusts nature or trust in nature will be a fool when nature bites back to make him foolish. When science split paths with nature science never went back to nature and see where science was wrong but then propagated that nature is wrong and untrustworthy. Science never went back to see why it seems as if the sun revolves around the earth but just blew nature of as fooling humans. Regrettably it was then that science went on a fool's detour about the truth. Science does not wish to acknowledge that nature can't b e wrong because nature presents what applies as reality

We have those mathematical equating geniuses reconstructing the Universe with formulas that subdue our senses and speaks in languages we mortals my never guess like this example:

$$A = \left(\frac{E_T}{\rho_0}\right)^{\frac{1}{2}} \left[\frac{3(\gamma-1)(\gamma+1)^2}{4\pi(3\gamma-1)}\right]^{\frac{1}{2}} \equiv \frac{2}{5}\left(\frac{E_T}{\rho_0}\right)^{\frac{1}{2}} \xi_0^{\frac{5}{2}}$$ or going even much more impressive as

$$\frac{\partial f}{\partial t} + \Pi \frac{\partial f}{\partial r} + \frac{\theta}{r}\frac{\partial f}{\partial \theta} + Z\frac{\partial f}{\partial z} + \left(\frac{\theta^2}{r} - \frac{\partial \phi}{\partial r}\right)\frac{\partial f}{\partial \Pi} - \left(\frac{\Pi\theta}{r} + \frac{1}{r}\frac{\partial \phi}{\partial \theta}\right)\frac{\partial f}{\partial \theta} - \frac{\partial \phi}{\partial z}\frac{\partial f}{\partial z} = 0.$$ To them these formulas

above is the entire Universe. No it does not represent the Universe but to them these equations become the Universe. However I want to give them the real challenge: Determine the size differentiation that the Universe has to undergo when shrinking the largest view of the Universe into the size of an electron with which we see.

The ratio is astonishing, but more-over what is truly astonishing is the arrogance of man to think of his position, as being important while the space man holds is beyond any comparison in ratio to everything we see in the Universe we see. Think how small we are when we are able to see the entirety out there! Even if there was other life out there, what is the worth of it in comparison to what there is that we see? Tell your physics professor not to reinvent the Universe will his brilliant mathematics, but just too mathematically formulate how all the light forming a Universe, the visible Universe can fit into my eye. This is physics and this is where the cosmos starts. It starts by fitting the Universe into something so small that it touches only one nerve ending and still carries all the information there is to my brain.

Albert Einstein formulated a concept in 1905 he called **The Special Theory of Relativity** and in 1915 he introduced his assessment on the principle of **The General Theory on Relativity**. I do not quite agree with his findings. What I discovered goes far beyond the discovery that Albert Einstein formulated. I have discovered that the Universe is not employing a general relevance of singularity, but throughout the Universe there is a fixed overall state of *The Absolute Relevancy of Singularity* that is not only **controlling the Universe**, but is what the Universe **constitutes of**...**it forms the Universe**...**it is the Universe**. However, notwithstanding the magnitude in significance *The Absolute Relevancy of Singularity* presents as a breakthrough in science, the influential members of the scientific establishment will not recognise my theory on **The Absolute Relevancy of Singularity**. Past encounters taught me that mainstream science in physics will again ignore the ideas that I formulated as *The Absolute Relevancy of Singularity* and I don't believe it would be well received, it will be seriously considered and much less be accepted by those with the authority to change physics principles. I think the theory I introduce would never be accepted during my lifetime because science is fixated on Newtonian ideas, which makes them bent on believing in the outrageously marvellous, and the unexplainable magical powers with gravity working by mass supplying a pulling power, which is a fact never proven and accepted only on Newton's word and Newtonian cultural bias, although they claim to only use proven facts.

What I ask of readers is to beforehand forfeit the culture of Newtonian bias when reading this by paying attention to what I say and not about the degree in which I stray from mainstream science's thinking. This way the exercise will present many new ideas and explaining my new concept will become clear. There is so much to benefit from. Science has no idea what a Black Hole is while I can prove what a Black Hole is. I formulate mathematically what "the sound barrier" is. I prove what gravity is. By using the four cosmic phenomena, which is what the cosmos uses to form gravity, I show what "the sound barrier" is and I go much further than that. I show that gravity forms from using the **Roche limit**, the **Lagrangian system**, the **Titius Bode law** and the **Coanda effect**. I uncover these principles by placing Π within the formulating of gravity and when using Π I bring clarity to the misunderstood cosmic principles. The list of the unknowns I can then explain is almost endless.

Gravity forms by movement that establishes singularity initiating a circle in using Π. I show why gravity is there, how gravity forms and what role stars play in forming gravity. There is no difference between how gravity and electricity forms and that I prove mathematically by decoding the cosmos. I prove mathematically when atoms spin they establish Π that forms the Universe. Whatever forms gravity has to link closely to Π since everything that has anything to do with gravity forms a circle that is Π by the value of the square radius. If mass has anything to do with generating gravity, then mass has to apply Π or otherwise mass has nothing to do with the forming of gravity. Everything using gravity forms a circle of sorts, which forms the curvature of space-time, which is Π and which curves light. The way the planets orbit the Sun and how stars spin has all to do with Π. In spinning in a circle, Π forms gravity as a centrifugal force that condenses space.

I researched the work of Kepler and found science doesn't even recognise his work, while it is his formula that forms the basis of all physics. Everyone thinks that Kepler found planets rotating, with Newton being able to explain Kepler, which makes everyone more concerned about how Newton saw **Kepler's work.** **The formula used in physics as a principle is** $F=mV$, which should be $F^3=mV^2$. $F^3=mV$ is replicating Kepler's formula in detail as $a^3=T^2k$. By using Kepler's formula we have $F^3=mV$ that is a precise replica of $a^3=T^2k$. The duplication is so obvious that we have (**F^3 becoming a^3**) while (**m is k**) and (**V^2 is T^2**). Einstein also only duplicated Kepler's formula by putting $E=mC$, which also should read $E^3=mC$. Again that is precisely Kepler's formula $a^3=T^2k$. (**E^3 is a^3**), (**m is k**) and (**C^2 is T^2**). In $E^3=mC$ Einstein mimicked $a^3=T^2k$, Kepler's formula. (**E^3 is F^3 is a^3**), (**m is k**) and (**C^2 is V^2 is T^2**). So what is so brilliant about Einstein's formula if Kepler had it centuries before? $E^3=mC^2$ is $F^3=mV^2$ which is $a^3=T^2k$. Newton corrupted the formula when he added $4\Pi^2$ to the formula and removed k that Kepler introduced while $a^3=T^2k$ Newton ignored. Newton changed $a^3=T^2k$ by using the symbols G (m + m$_p$) to replace k and then declared $a^3 = T^2$. I still wish to see the proof confirming Newton's changes as being correct notwithstanding that everyone thinks physics is entirely based on this conception. Whether the formula used is $F^3=mV^2$ or is $E^3=mC$, it still remains duplicating what Kepler introduced as $a^3=T^2k$. So I changed it back to Kepler's version of $a^3=T^2k$ as to better the understanding of the foundation of astrophysics and mainstream physics. The entirety of physics is not based on Newton. Physics precisely duplicates Kepler's findings while science doesn't even recognise Kepler's formula. By giving Kepler the credit due, the entire Universe becomes completely understandable...but then for my audacity to show mistakes in physics I am ignored flat! All I ever ask is prove the truthfulness of G(Mxm)÷r² because it is **F³=mV²** that forms the basis of physics and that accuracy comes from Kepler's view **of a³=T²k** that became Einstein's **E³=mC²**.

By re-implementing Kepler's full formula **a³=T²k** and using Π I was able to prove what I discovered as follows:

1) The location, the position and the value of singularity as a factor forming space-time
2) Finding space-time by dissecting Kepler's formula in relation to valuing singularity
3) Finding space-time, proving space-time and aligning space-time with gravity
4) The working principals behind and manifesting of gravity as a cosmic occurrence.
5) The Roche limit and explaining the resulting of a law coming about from singularity.
6) The Lagrangian system, how and why that becomes the building form of the Universe.
7) The Titius Bode law and I show mathematically how gravity comes about from that
8) The Coanda effect and the producing of gravity through reproducing space-time
9) The sound barrier by proving it is gravity generated by motion in space becoming independent motion.

This I conclude because Kepler said **a³=T²k** but that could also be **k=a³/T²** and could be **k⁻¹ = T²/a³** and that is the Coanda effect. Mathematics says a sphere is **a³ = 4/3 Π r³**, which is mathematically correct. However, Kepler said the cosmos told him a cosmic sphere is **a³ = k T²** where that puts the cosmos in completely different mathematical dynamics altogether. There are the two distinct possibilities of **a³**, which Newton saw and which Kepler saw and both are most valid, but are altogether unequal. Between Newton's **a³ = 4/3 Π r³** and Kepler's **a³ = k T²** concepts there is one Universal difference.

It is true that when measuring the sphere, Newton's method or formula a³ = 4/3 Π r³ is used in calculating, but **Kepler received his code of calculation a³ = T² k from a very high authority,** which **is none other than the Universe** and therefore Newton can't discard **k**. Kepler saw singularity forming relevancies and Newton knew nothing about that. It is the duty of the cosmologist not to reject Kepler's findings, or as Newton did, try to transform it into something that Newton could understand, because it then strays from the original meaning...but science should dutifully search for the meaning as Kepler received the formula **a³=T²k** from the cosmos. We can test any of the following symbolic values in the mathematical expression and also test the principal behind the expression in which Kepler stated them. By such testing **a³=T²k** repeatedly we find that the translations of Kepler's formula into English never required any corrections in translation because Kepler never presented it incorrectly. By taking the formula on face value it can change as follows: **a³ = T² k** can become **k = a³ / T²** or become **k⁻¹ = T²/a³**. When translating Kepler's mathematical expression into English we can see what Kepler said also could read as **k = a³ / T²** where **k** is indicating one point from a centre point that is space **a³** relating to time **T²**.

From a centre comes space-time. The centre **k** brings space **a³** in ratio to time **T²**, which is space **a³** / time **T²k.** Reading this correctly can't bring any dispute...yet it does...and it's been doing it for centuries! Kepler said **a³ = T²k** and that correctly translates to a mathematical expression **k⁰ =a³ / T²k** which in the English verbal statement translates that Kepler said that there is a **space a³** which is **equal =** to the motion in **the time duration T²** thereof between two specific points which holds a relation onto a centre **k⁰** where from there forms **a straight line k** that is centred on the spot where space begins from **k⁰ that produces k** as well as producing the circle. Therefore that spot where the specific point is at **k⁰ =a³ / T²k,** that allocated spot holds **k⁰** at a value of having the least space there could ever form. The line **k** is centred onto a spot where space begins specifically at **k⁰**.

This point not only produces the line **k** coming from a point **k⁰** but represents also the space **a³** that forms the eventual circle by the rotation of **T²**. Therefore from the centre holding **k⁰, k⁰** leads to **k** that forms the revolving space **a³**, which is rotating **T²** at a distance **k** where **T²** forms the outer limit of **k⁰**. Mathematically **a³ = T² k** will also be **k⁰ =a³ / (T²k)** because **k⁰ = 1.** But **k⁰ = 1** also presents the single dimension where all factors are a product of one. If anyone can locate **k⁰** then also that person will find singularity. That is where gravity is because gravity is strongest where space is least. Then that suggests that gravity is strongest at **k⁰** because there space is least. That is gravity because that is what keeps the orbiting objects in orbit but also that is what Newton completely missed when he changed Kepler's work. Newton failed to recognise gravity as the only ingredient in Kepler's formula. He admitted that he, Newton missed this because he admitted he did not know what gravity is while Kepler explicitly showed what gravity is. Gravity is what keeps the orbiting objects in rotation while orbiting. **k =a³ / T²** is **distance¹ =**

space ^{3}/ time^2 forming from a pivoting centre k^0. That is a cycle and moreover it is a cycle formed **by space/time**. What Kepler said is that space is a^3 **being in motion T^2 k.**

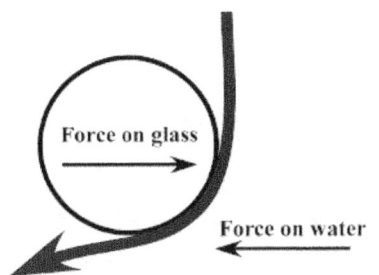

Force on glass

Force on water

As Kepler said $a^3 = k\,T^2$ **and therefore** $k^0 = a^3 / k\,T^2$ **and therefore we have to find** k^0. As a result of examining this proposition, I located two principle positions both holding singularity. The cosmos is made up of one type (1^0) that is in two categories where one type moves and the other type does not move. The one is a liquid and the other is a solid.

The condition for the presence of this singularity that forms everything, controls everything and is everything is centralised to a centre singularity $k^0 = a^3 / (T^2\,k)$ that forms by movement $T^2 = a^3 / k$ of space $a^3 = k\,T^2$ placed in relevancy $k = a^3 / T^2$ that is centrifugally going both ways $k^{-1} = T^2 / a^3$ thereof (Newton's 3rd law). This explains the Coanda effect and the Coanda effect is gravity and gravity "glues" the water to the glass by implementing Π to form singularity! *What is in the Universe is spinning*. **The entirety of everything forming the Universe is spinning inside the Universe** and such spinning is always in the centre of one specific point, wherever such a point might be. In the **precise middle** of all **objects in rotation** is a precise centre where this pre-designated centre is dividing the object in rotation into sectors that will **start the spinning initiation** from that centre point. This is what Kepler's formula confirms in $a^3 = T^2 k$. By spinning, the one side is coming towards while the opposing side at that time is going away. Thus, the spinning object **will have a middle point**, a very specific **centre point that does not spin** and only holds Π as a specific value because within that centre being that small, no radius can apply. We have named this position or line the axis, but the true meaning of this line has eluded us since the concept was realised. This line that forms holds no space although it directs all the space that it controls by spin. When going toward the centre where the axis forms at the very centre of rotation, the space on the one side has to end and the space at the other side has to begin with the line unable to hold space.

On the one side space turns in a completely oppositional direction from the direction in which the space spin on the other side and in between the opposing movement a line forms without the ability to contribute space. But also within the one value forming, such a line cannot have a value of zero because the line **is there** and holds contact to the rest of the material bringing about that **zero does not start any line** and therefore the **value of the line must be infinite,** just as described **in accordance** and by the **definition of singularity**. In dimensional terms, which I explain later on, the value of **2k** relates to T^2. That relation extends to the next value where T^2 relates to **k,** which positions T^2. The first space in the circle T^2 will then be located at point **k.** From the centre being in infinity, one can realise by thought that the single dimension factor is not visible, but is present all the same. Extending that into the 3D comes six **k** and any one of the six will further extend to form a seventh point as T^2. All this forms a point that finally refers to the location of one spot holding singularity attached to space by the measure of $k^0 = a^3/(T^2\,k) = 7$

Let's find $k^0 = a^3 / (T^2\,k)$ and see where it is hidden. The sphere is a circle in many facets and therefore we will approach the sphere as one multi dimensional circle. However, the sphere as such remains one circle to the power of many. When investigating a circle, one would draw a line from one edge running through a centre all the way to the other edge. In doing that we would find the measure of the diameter, which is most important when trying to establish the volumetric worth of the sphere. The circle has Π to indicate form and uses r^2 to establish the worth of such a circle by using the radius symbolised as r in drawing a straight line. In any circle or sphere the size only depends on the fluctuation of r in the square as a component to the circle or sphere but that does not affect the form, which comes by indication of Π in any way there may be. The conclusion from this is that no line can start at zero because that will be a mathematical impossibility.

Lines mathematically cannot start at zero because there is no evidence of zero as a factor in mathematics. Should you disagree with my statement, the question in need of answering is this: What will the length of the shortest hypothetical line imaginable be and moreover, what would the total overall length be in that case? A line or spot starting at zero would therefore be shorter than the shortest line possible. For obvious reasons can no line, or any line grow or extend from zero because such a line must then quit zero and become something, thus abandon its original value by the adding of the first value. Mathematically said it would be as follows $0+0=0$ whereas if it started with something infinitively small it would be $1^0 + 1^0 = 2$ and then from using something infinitively small it will grow into something immense such as the Universe. In any circle or sphere the size only depend on the fluctuation of r in the square as a component to the circle or sphere but that does not affect the form by indication of Π in any way there may be. The conclusion from this is that no line can start at zero because that will be a mathematical impossibility. If a line started with zero, that would nullify Π ($0^2 \times \Pi = 0$) and that would leave the form without having any form because $\Pi \times 0 = 0$.

This statement by itself excludes zero and with zero excluded one then begins to appreciate all the rest of the concepts governing corrected cosmology. If there is a distance, it holds a measured one of whatever norm or value, which is a specific length that applies and that zero or nothing then could never fill. By saying the distance constitutes of nothing we have to substitute the one factor with a factor of zero to find what mainstream says fills the Universe. Including nothing as to state the presence of that part contained by the calculation delivers the total of zero. It seems as if science has ignored this mathematical principle that $1 \times 0 = 0$ as an issue by simply not thinking about the fact of the matter and therefore simply ignoring that which is measured forming the sole value of space. It is somehow more convenient to put the value of nothing as part of the distance in calculation because that is what is understood.

Measure zero and then see how one can multiply when using zero in mathematics to reach a distance holding a value other than zero when multiplying with zero. I agree that what is filling outer space is invisible, but also it is there, it is present and being present and there while being invisible disqualifies whatever is there from being zero because being zero will mean it is not there and we cannot deny whatever is there of being there. Then what is there will be there, while being invisibly small, but it will still be possible to form a line because every aspect of the Universe forms lines while also it will have the potential to fill space and can still form a measurable unit. That then must be 1 because while $1 \times 1 = 1$, $1 + 1 = 2$ and that qualifies that invisible thing to be present ($1 + 1 = 2$) but at the same time be completely invisible ($1^3 = 1$). When realising this I knew what conclusion coming from this had to be true about that which I was looking for and that it had to be singularity because singularity can only have one value and that is 1.

To find the invisible I had to locate singularity. I realised that my effort to locate the point holding singularity enabled me to backtrack the exploding Universe to its origins. The Universe is a sphere because it is filled with spheres filling the void spaces (not the nothings) and in that I first had to investigate the visible. Newton's mathematics says a sphere is $\underline{a^3 = 4/3\Pi r^3}$ while Kepler said a sphere is $\mathbf{a^3 = T^2 k}$, and both are equally correct because the cosmos gave numbers to support its statement. Where Kepler says $\mathbf{a^3 = T^2 k}$ and with mathematics saying that $\underline{a^3 = 4/3\Pi r^3}$, we think of volumetric size of space in terms of using normal mathematic formulations. We think if it is volume then it has three sides and in the case of a sphere the measure is $\underline{a^3 = 4/3\Pi r^3}$.

Comparing $\mathbf{a^3 = T^2 k}$ to $\underline{a^3 = 4/3\Pi r^3}$ is like comparing the equal ness of a triangle and half circle and line to numerical values. $\mathbf{a^3 = T^2 k}$ predates mathematics, where $\mathbf{a^3 = T^2 k}$ determines positions at a period in cosmic development when only form was used going before when numbers as value were in place. It shows the half circle $= 180°$ is equal to the triangle $= 180°$ and both are equal to the straight line $= 180°$ notwithstanding the obvious differences used in form. However, the starting point of these forms has to be equal and also has to be not zero to have the end be equal and result in all being equal in value in the end.

Kepler said a sphere is $a^3 = T^2 k$, which also mathematically is $a^3 \div (T^2 k) = 1 = k^0$.

In honesty we have to realise that we cannot dismiss the whole formula that Kepler produced just because it doesn't match the scenario set to determine volumetric size as the Newtonian version does. Kepler's version holds a foundation based on movement and it is in the movement we find the measure and not in the size as Newton's mathematical formula does. In Kepler's formula the entire formula is formulating a circle being motion. However, with the correct interpretation we find so much more than just motion. The correct formula is $\mathbf{a^3 = k / T^2}$: That is what Kepler brought into civilization for all time to come. He saw space $\mathbf{a^3}$ being in isolation due to the time it uses to move $\mathbf{T^2}$ claiming such space forming independence according to what the line $\mathbf{k}$ indicates. Let us look at the factors in more detail before we proceed with the rest.

Space $\mathbf{a^3}$ will always be circling around as $\mathbf{T^2}$ is in a position referring $\mathbf{k}$ to the centre $\mathbf{k^0}$. That is what Kepler said when he said $\mathbf{a^3 = T^2 k}$. Kepler indicated space $\mathbf{a^3}$ will forever fight for independence and show separate individuality in remaining apart as identifiable cosmic components by means of motion. Every space will cling to independence indicated by **k** through fighting off the integrating of another overall unifying unit by applying the motion of T^2! The problem we have to solve is what will the cosmos use to secure such independence between all particles? What sets space apart from the rest of space? First we have to admit that Kepler was the one that introduced the following: Kepler gave us the answer to the following but no one ever took notice!

Kepler was the one who discovered **space / time** as **space** a^3 = **time** T^2 **k**
Kepler was the one who discovered **singularity** as $k^0 = a^3/T^2 k$
Kepler was the one who discovered **gravity** is holding **space-time** relative by the measure of distancing **k** as $k = a^3/T^2$ and $k^{-1} = T^2/a^3$

Kepler said gravity in space is about the area a^3 that would always keep equilibrium with the time T^2 it takes to travel the distance of the full circle position placed by the indicator **k,** therefore adjusting **k** as the need arrives. With **k** shifting in length a^3 will have to readjust and therefore T^2 will find a new relating value each time. This was the finding of Kepler and came after his intense study of orbiting planets. Translating Kepler's mathematical expression $a^3 = T^2 k$ correctly to the verbal statement in English Kepler said that there is a space a^3 which is equal = to the motion in the time duration T^2 thereof between two specific points which is a straight line **k that holds a relation from a centre** k^0 **to an end** k **where the two ends run from the beginning of** k^0 to connect at the end of **k.** I might not be the smartest boy on the block but I'm not that stupid either. I know how to translate mathematics

a^3 must have a volumetric interpretation because the third dimension is sure evidence of multiple conjunctions of dimensions put together in three sides opposing three sides having the third dimension in place. The fact that any symbol uses a value to the **third power** a^3 indicates **space** or a volumetric established and separate unit. Using a cube by three dimensions symbolises a cube, a room, a space to be filled, a unit able to hold other ingredients on the inside when empty or partly filled. It is space because it is volume using the third dimension.

T^2 is an indication of something having a cubic nature other than the square forming motion that is provided by the motion the square indicates, which is where the moving object is representing a third dimensional object that is moving from point to point and it is this point to point that multiplies into the square. The space is moving as a unit from one point to another point and the moving between the points are represented by a flat square or following a flat distance between two points. The cubic space was in one instant in one place and then the second instant in the other and because time can never stand still or become single dimensional (this I am about to prove) insisting that time must always support the motion it consist of or space as well as time in time cannot be. It is motion that is taking time, which is motion in the second dimension moving the space in the cube.

k is the symbol used to indicate a straight line between two points with a definite beginning and a specific end position. It is the location where the form in question is holding space running from where the space was to where the space will be the very next split instant that follows while time by movement repositions the allocations. This indicates points of representing **k** in different time positions to which the points will then be multiplying to form the square that forms between k_1 and k_2. The movement indicates not a square surface but it indicates movement by the square. This indicates the time the journey took to move the space from one point where **k** is to where **k** will be. It indicates the location of the space where from to the point where the next indication of **k** runs. T^2 will shift **k** where **k** indicates the position of the space a^3 that forms as a result of the movement T^2 of being the space a^3 indicated by the point at the end of **k.** Since time represents the square T^2 and with **k** being the distance, this fact proves that the **k** represents the distance of the ending of the space a^3, which represents the form relative to the circle that T^2 forms. It is obvious that T^2 represents the time that represents the space a^3 in the square T^2 through the motion. It is the distance moving space a^3 in the cube to complete time in duration in the square of motion T^2; therefore **k** is permitted to be in the single dimension.

Let us find the smallest possible line first. We have already reached the conclusion that by reducing the line, the reduced line will eventually leave all sides on the same spot on the condition that the circle spins. Such a spot must be round in form since it still holds Π as a factor next to r^0. We now are entering the domain of singularity where the visible is no longer traceable and only intellect can bring understanding of the scenario. With the line being the smallest line, such a line will start off as a dot Π that moved away from a spot Π^0. With all possible sides being in precisely the same spot we have all possible sides onto one spot. I chose to differentiate the dot and the spot by giving the spot a value of Π^0 while the dot holds Π next to r^0. Mathematically the spot is placing form evenly spread being Π coming from the single dimension Π^0 where the space is one (1) and holding exponentially zero (1^0). There the space moved over to form the spot Π^0 and by introducing form the movement changed Π^0 to the dot Πr^0 forming a circle as a dot. Again I must draw the attention to the fact that we now are reaching into areas only the human mind can venture by understanding and seeing nothing more than with the eye of intelligence. The understanding of this concept demands our reaching the point where the mind of the animal cannot reach. If it starts with a line it then is there where that line only represents two sides being one and as such that is representing rather a flat Universe. At the dot Π we have roundness because we have Πr^0 while at the spot there is not yet any round form because of Π^0 and only when Π being round forms, it then is requiring a shape or form and this lies beyond or before space at a point where any form of shape comes into the cosmos scenario. This part of the Universe comes in a place at a point in a location where shape and form is a part of the distant space hidden in and beyond where eternity develops. The spot is located at a point where entering the domain of the spot also at the same time is crossing the spot and landing on the other side of the spot where entering the spot is crossing the spot. Nothing can enter the allocated position the spot holds because entering the spot is crossing over to the other side of the spot. It serves us well to realise that the entire Universe was that small at a point where everything started forming because the spot that developed into the dot is still with every spinning circle...and the Universe is a multitude of spinning circles. It is also very wise to remember that once anything becomes a part of the Universe, it can never leave the Universe since it then has no place to go or no gate to pass through in order to leave the Universe. With the spot becoming a dot, there must have been a time when everything in the entire Universe was that big as the spot is, and that then moved on to form the dot and in that it went on growing in relevance. The point around whichever spins becomes the centre of the Universe by singularity. In establishing such a centre containing singularity we find the reason why bullets travel more straight when they are fired circling and circling is what gives the bullet the accuracy in its trajectory that then established a cartelise singularity that establishes a value forming Π in relation to the centre singularity being 1 or as I named it as singularity Π°.

When a rocket is fired and the spin is not present there will be no stable trajectory. The only way to secure the stability of the trajectory is to allow spin (Π^2) that enables a point holding (Π) as this will locate and establish singularity (Π°). Establishing singularity is the most fundamental principle about gravity we can ever find. This is the one part that is most important when we go in search of gravity secured by singularity that forms the absolute relevance of everything filling Universe we have. Everything is a rotating object that holds any point allocated in Universe to form the centre of the Universe because everything in the entire Universe spins around any given point and that then forms the centre of the Universe. Every centre of every atom forms the centre of the Universe by spin! Again I indicate the precise location of such a point. What is in the Universe, is spinning and therefore what I am referring to, applies to everything holding a place in the Universe and therefore this which I mention directly links everything holding any space whatsoever in the entire Universe to one single point around which all spin. In the **precise middle** of all **objects in rotation** is a precise centre dividing the object in sectors that will **start the spinning initiation** from that centre point. Thus, the spinning object **will have a middle point**, a very specific **centre point that does not spin** and only holds Π as a specific value because no radius can apply. But also the one value such a line **cannot have is zero** because the line **is there and holds contact** with the rest of the material bringing about that **zero does not start any** line and therefore the **value of the line must be infinite**, just as described in **accordance** and by **the definition of singularity.** As I am introducing a very new idea, I wish to explain in better detail what I try to convey. While the toy top is spinning one will find singularity by moving the rotating line or radius progressively to the middle by reducing the length the line has from the edge to the middle. At one point all further reducing must end but the ending cannot include zero or nothing because the rest of the line is still attached to the rest of the top. As the rotating direction moves inwards, the rings will become smaller and smaller. Then we reach a point everyone thinks of as being the axis around which everything rotates. The line only forms when everything around the line spins by establishing a circle to the value of Π.

Everyone calls this line that forms the axis. Everyone knows about the axis and yet through so many thousands of years of using an axis, no person ever thought to scrutinise the principle behind the axis. Yet in all the millennia everyone was aware of the line that forms called the axis, no one took time to see it holds singularity at Π° presenting Π. The only conclusive value singularity can have is 1 or Π°. The axis controls all particles spinning around the line being the axis while the axis in itself forming the line represents no particles because the axis represents no space. If there was space within the axis, the space had to spin in some or other direction. Having no space would mean occupying no space which means forming no part of the Universe filled with space and yet it controls all the space as wide as the mind can imagine. Without space it does not form a part of the cosmos, but forms the cosmos as wide and as deep as the cosmos goes. The axis could not be seen but with applying intelligence the axis could be witnessed. Having no part in the cosmos in space, the axis could only be understood and never be seen. The axis could be proven but never be shown. The axis is what controls the Universe from end to end because when there is no end there the axis provides one end to what never can have another end and the axis governs whatever spins in relation to such a line. Again I wish to press this issue to form clarity. The line forming the axis is without space and only holds form, and therefore the line represents a point not having any dimensions while it still is there without ever being there. If ever there is a concept I have to introduce, then it is the concept of how important the axis is and how science up to now missed the biggest issue that is responsible for all movement within the cosmos. The line forming the axis is there but only intelligence will ever form the concept whereby one can realise where the line is without ever seeing the line. Anyone unable to understand this concept can never see the validity of space-time. In the axis line there is a something that is there but only intelligence can bring understanding to the understanding thereof. Only motion of space can resurrect the line coming from the point it holds as a dot. Everything in the cosmos spins and everything that spins has to form a line that doesn't exist but yet the line controls everything that spins around this line that never can hold any space or be part of the Universe. Without having space to fill, the line can never form any viable part of what forms the cosmos, which is space.

The point in reference is the line forming the axis and the axis must be a line that never forms in space because if it did, it would have to rotate in either one of the directions space spins in and by not spinning, it has no space. **That point** albeit hypothetical, is also as much a reality none the less and is placed where that point **must be standing still** because every line **running from that point** in **opposing directions** is also **in opposing directional spin to the other or opposing side**. In considering the spinning motion in the fraction of time in the detailed instant every aspect of rotation will turn in every instant of change in time. Although the points had the same characteristics only one instant before, they oppose the characteristics it had just before and just after the very instant in which they are and to which they relate by similar points also in rotation. Looking at the graph unfold will explain my point about quarterly opposing dimensions and values unfolding.

The circle can reduce one step more when the circle eliminates r completely by returning r to a point of singularity r^0, but the elimination of r as the factor reduces the major factor to the single dimension in Π^0. That will not reduce the cosmos to zero, but it will only eliminate all potential lines r^0 to potential circles $\Pi^0\Pi r^0$ and from there the circle Πr^0 will come about by manifesting as a line but that manifesting can firstly only establish a circle Πr^2. The only value that singularity can have although the single dimension may host the entire Universe is Π^0. Pick a number and elevate it to the power of zero and in the process one may have established another point holding all points in singularity because that is the value of singularity. Only Π^0 or any other value holding one accompanied by zero as an exponential value can ever be the accurate value of singularity while singularity will then host the rest of all the possibilities in the Universe. This means that the entire Universe composes of and is made up of singularity... this much I am going to prove. Every point occupied or otherwise constitutes of singularity either under control by movement in a form we call atoms or being passive in a location we call outer space. This position one can derive from Kepler's formula $a^3 = T^2k$. It is just a question of how to fit this sensibly into Kepler's formula $a^3 = T^2k$ and find a way that will bring much understanding to cosmology and the way that singularity connects one Universe to form cosmology. The top spinning is what connects space to form the Universe. The top being still on the ground and not spinning holds singularity at a value of the dot forming Π° while putting the relevancy on the Earth's roundness by Π. When the top spins the relevancy changes to the line from forming as a dot Π^0 becoming a line Π. The line Π forms as a result of the top forming space Π^3, which is in place as a result of the movement that the top acquires Π^2. It is singularity without space so being a line or a dot makes no difference. The top no longer holds only a dot Π^0 in the centre, but generates the relevance Π by forming $\Pi^0\Pi r^0$. The top, by moving adjusts Π to form space by movement which is $\Pi = \Pi^3 \div \Pi^2$. All of this is what makes gravity be what it is and all of that Newton missed and Newtonians never saw since all of that is covered by a blanket called mass being responsible for gravity.

Reading this mathematically encrypted coded formula of the cosmos given to Kepler and keeping it removed from Newton it reads as being that the space a^3 is equal to $=$ **the motion T^2 of the space a^3 in ratio k to a centre k^0**, which is relevant to the positioning of k. If we bring in the full equation it will be $k^0 = a^3 \div (T^2k)$ which means half of space is solid $k = a^3 \div T^2$ and half of space is liquid $k^{-1} = T^2 \div a^3$ where liquid is moving. However, it is also true that everything through movement defines a value in relation to one point holding singularity k^0 and that is what the formula $k^0 = a^3 \div (T^2k)$ underwrites. What this proves is that gravity is the motion of space provided by time being the liquid. Please allow me to explain. In the formula $a^3 = T^2 k$ the space forms as the space is in motion. Newton suggested that $\frac{dJ}{dt} = 0$ where he

stopped time to have the motion of the circle demolish the work that the circle does. That means he got time standing still or being T^1 and the motion $T= 0$. Let us ponder on that thought for a while, while we remain with the formula Kepler suggested $a^3 = T^2 k$ and then it will seem that according to Newton $a^3 = T^2$ and in that k then becomes 0. Should that be the case then we have space going flat because $a^3 = T^2 k$ where $a^3 = T^2 \times k = 0$ forming a square instead of a cube, and the Universe we have is a three dimensional system in every aspect there is. The concept Newton brought about that $a^3 = T^2$ is putting a person that looks at a mirror equal to the possibility of walking in

and   out   of the mirror by becoming the reflection in the mirror T^2 and then himself a^3 again. It is rediculous to say the very least.

It is quite apparent that Newton saw no difference between the top spinning while the top was standing in an upright position and the top lying down on the Earth. This is a crucial mistake that has such a wide implication that on the one hand it either values the Universe to the value of singularity or on the other side dismisses everything about the Universe to the value of zero.

I am of a very   different opinion about Newton's point of view where he declared that forming a circle moving $\frac{dJ}{dt} = 0$, and by doing such the movement then removes Kepler's relevancy factor. This

places a value of empty space in which a top would spin and Newton missed the difference there is between a top spinning and a top laying on its side on the Earth. There can be no such a thing as empty space. The fact that space is valid removes an empty connection because space can be anything there is in space except empty space that is filled with nothing. The Universe is time contained in space, which makes it space-time. Space has only one value, and this is to contain time and time provides space with a definite value. **I do not disagree** for one instant **with Newton**'s calculations whereby he came to the conclusion that $\frac{dJ}{dt} = 0$ and therefore I am not going into repeating the entire calculating process. All of

the calculations Newton made are very correct except the eventual and final conclusion Newton came to. Newton never understood the mathematical concept of time playing a part in physics. In the time of Newton singularity and the relevance thereof had no feasibility in any concept regarding physics. Newton had the concept that time could stand still and that is impossible in physics or any other place. Time can never stand still because time is forever moving by establishing space in a three dimensional environment.

Being the mathematical genius as Newton is so often portrayed as; Newton had very little insight into mathematical possibilities, because when he suggested that $\frac{dJ}{dt} = 0$ he made one huge mathematical blunder. No person (including Newton) may place any two objects in a direct relation where the two factors divide and have an outcome that forms zero. Much surprising is that not one mathematical genius that came after Newton drew the correct conclusion that forming $\frac{dJ}{dt} = 0$ is mathematically not acceptable.

Newton saw that dividing something into something else could bring about zero and that is impossible. In concluding that $\frac{dJ}{dt} = 0$ bringing in zero as a legitimate value Newton found a way to replace Kepler's symbolic relevancy value of k with using the symbols G $(m + m_p)$. In doing that Newton painted a picture that has no real meaning except where Newton tried and succeeded to put mass into an argument that has no true validity in cosmic principles. This is just a longer and probably a more detailed manner of indicating k and better defining of k but it symbolises precisely to the point what k stands for nonetheless. I wish to draw your attention to the matter of Johannes Kepler's findings that Mainstream science considers as resolved and closed for many a century while it is not. My investigating Kepler helped me to resolve other unresolved matters but it was only possible by using Kepler's work.

Newton never considered why the spinning top stood erect and the top not spinning lay flat and still.

Newton did not think that as soon as the gyroscope started spinning, the balance shifted in favour of a position wherein the gyroscope stands upright. He never thought about what then comes about which has the ability in keeping the gyroscope upright. This is rotational movement and in my other books on the **Absolute Relevancy of Singularity** I explain how rotational by the square of the double seven forms Π and Π is forming the curvature of space-time and in that bending of space-time is what we call the atmosphere that keeps the gyroscope square with the Earth and through that the gyroscope stays upright. The gyroscope is acting in accordance with the Coanda effect where the Coanda effect is gravity. By spinning it establishes a solid forming as $k = a^3 \div T^2$ and a liquid forming as $k^{-1} = T^2 \div a^3$. By spinning $T^2 = a^3 \div k$. That is evoking singularity which forms as $k^0 = a^3 \div T^2 k$ that establishes gravity $a^3 = T^2 k$ in relation to the Earth evoking gravity through also spinning.

Newton found mathematically that the movement of the top by spin removed the value of the radius $\frac{dJ}{dt} = 0$ where quite the opposite applies. The spin of the top $T^2 = a^3 \div k$ positions the relevancy that **k** as a factor produces by initiating singularity k^0 on both sides of the relevancy forming $k^0 = a^3 \div T^2 k$ as well as placing singularity in relation to the spinning top $\frac{dJ}{dt} = 1^0$ because that is the correct mathematical principle coming

from the equation. The smallest any dividing can be is one and one is the form that brings the value producing of singularity. The spin of the circle does not eliminate the relevance of **k** but institutionalise the measure of **k** by confirming the space a^3 in terms of singularity k^0. However **k** has no confirmed and specifically applying value but puts a relevancy of space a^3 forming in relation **k** to movement T^2 applying. By trying to find a measured value applying to **k** such a person is showing no understanding about what **k** is. The value of **k** is finding the space that **k** indicates in terms of what moves. The indicator **k** identifies the space a^3 that the circle claims in terms of singularity k^0 that the movement T^2 isolates from the rest of singularity $\frac{dJ}{dt} = 1^0$. The value of **k** is dictated by T^2 as the movement that isolates the space a^3 but also **k** dictates the value of T^2 to form space a^3. The measure of **k** is the relevance **k** is claiming on behalf of the space a^3, which uses the relevance of **k** to put a limit on the space a^3 by spinning in accordance with T^2.

What Newton suggested while never realising he did suggest it is the following, and that is that the rotary movement of objects put singularity $\frac{dJ}{dt} = 1^0$ in position on the outside of the moving circle. However, by using $\frac{dJ}{dt} = 1^0$ Newton placed emphasis on the turning movement of the circle and saw this as a destroying of the circle while in fact the turning is putting the space that identifies the circle on the cosmic map. That Kepler also found without ever realising what he found. Kepler said $a^3 = T^2 k$ which is $k^0 = a^3 \div T^2 k$ which is the spin $T^2 = a^3 \div k$ which is the circular movement T^2 that validates the space a^3 in relation **k** to a centre k^0 which is exactly and precisely what Newton said when Newton said $\frac{dJ}{dt} = 0$ that actually should read $\dfrac{dJ}{dt} = 1^0$. The location where Newton placed singularity as being singularity established by the movement of space $\dfrac{dJ}{dt} = 1^0$

This indicates four factors forming singularity that absolutely dictates the cosmos in terms of movement. Holding that in mind, I therefore had to name the four positions that equally form singularity by dictating gravity. To argue this concept of singularity guiding movement, let's take the Sun that provides a centre k^0 for the Earth a^3 forming a centre where k points a line that forms the orbital circle T^2 wherefrom the edge of the line k is pointing at the position of whichever planet a^3 forms a circle T^2 in relation to a line coming from a centre of the Sun k^0. The line k indicates the distance from the Sun's centre to the planet that orbits and this forms the circle as the planet a^3 orbits T^2 around the Sun. The line k will provide a line from the Sun's centre k^0 and the line k will provide a spot where T^2 produces a circle holding space a^3 in a located position by running around the centre of the Sun k^0. In this view the space a^3 of the Earth rotates and in that forms the **controlling singularity** that holds the value as Π indicated by k forming between k and k^0 being singularity Π°. The Sun holds singularity in the centre, which is forming the **governing singularity** Π° and from that point the circle T^2 comes that forms the orbit Π^2. That means every single point that k indicates there are positions forming space a^3 implicating sides of a double dimension. In the same manner is k not limited to distance or is T^2 lesser by size. If Kepler said $a^3 = T^2k$ then $k = a^3 / T^2$ is also what Kepler said. There are three dimensions a^3 between any two points T^2 flowing as time from the centre of the Sun, which is indicated by the line k. However in the next scenario the Earth holds the **governing singularity** Π° running from the centre k^0 to k forming the edge while the circling rotation T^2 then forms the **controlling singularity** Π indicating the point in rotation. There are also two other points holding **the mutual singularity** and **the primary singularity**, both which I do not explain in this presentation but without which the four phenomena would not form gravity.

The value of k is not to be put in place as a measured value, but is there to bring a reference to the location of singularity $k^0=a^3/(T^2k)$ applying as to place a specific singularity in as the **governing singularity** and acknowledge the position of another singularity in place as the **controlling singularity** because there always has to be a **controlling singularity** determining the orbit while there has to be a **governing singularity** determining the spin of the body in relevance performing as the space a^3 in question in the formula $a^3=T^2k$ where in that formula k determines the relevance of k^0 as in $k^0=a^3/(T^2k)$. However, this burdens k forever with the responsibility of forming a line and a line is what places the Universe in place while the circle T^2 is forming the Universe a^3 at the same time. Every space a^3 in question puts singularity k^0 in position by the motion T^2 in relation k to the position allocated to k in the Universe a^3. Nothing in the Universe can move without moving straight k that is also going in a circle T^2 to form space a^3 in relation to a centre k^0 while in orbit around another centre k^0. In this point k^0 time forms space and space develops as the history of time running from k^0.

a^3 symbolises in a mathematical interpretation of implicating the three-dimensional space holding a specific centre in relation to another specific centre indicated by k that could apply to either centre points in question. This is always a straight-line k representing the position of the **controlling singularity** moving in a circle T^2. The space forming a^3 is a **positional validity** of the space indicated by $k^0 = a^3 / (T^2k)$.

T^2 is representing the circle that goes around the **governing singularity** k^0 or Π° that forms in relation to the line k pointing to the controlling singularity or Π in reference to the centre k^0. The space that forms holds the orbiting planet a^3 in direct circular contact with the space in relation to a very specific centre k^0 moving from point T_1 to T_2 that then forms Π^2 in relation to a precisely placed centre k^0. The circle coming about from T^2 is the **controlling singularity** Π, which is always a circle Π relating to the centre Π° that is positioned by the line k in relation to the centre k^0 and by forming a circle Π it holds reference to the **governing singularity** Π°. Where **the governing singularity** is the centre of a spinning object such as the Earth, the centre of every atom holds **mutual singularity** Π^3 that collectively puts a mutual value of all the atoms' singularity as a combined equal to the **governing singularity** Π^0. The solar system will provide a **primary singularity** $\Pi^3 = \Pi\Pi^2$. The one would represent T^2 the other forms **k** that then produces the third singularity forming space a^3.

k indicates **controlling singularity** from the centre k^0 ending at the line **k.** This line shows the location around which a planet circles. The specific value about the centre is most important because from the specific centre gravity indicates a positional worth. The line forming **k** is pointing the circle or the **governing singularity** formed from a line that ends at a circle T^2 running from the centre k^0 to where the space a^3 is indicated.

The turning T^2 of any circle holding space a^3 is valid only if forming a reference **k** to a centre k^0. $k^0 = a^3 /$ (T^2k). This depicts a position the domineering singularity k^0 fills in relation to another point serving subordinate singularity **k**. There are always a dominant and a serving singularity interacting. If **k** indicates the centre of the Earth then T^2 rotates initiating the **governing singularity** k^0 where then the centre of the Sun **k** will form the **controlling singularity.** When the Sun rotates, the Sun's centre k^0 forms the **governing singularity** giving the Earth in orbit **k** holds the **controlling singularity**. The measure of **k** is not a specific value but serves only as an indicator to which space rotates or applies by the space rotating in a circle. This role of singularity being **controlling** or **governing** is playing part in movement of gravity forming and is very important when trying to understand the role that the four phenomena play in forming gravity. It is important to understand what happens in the event of an object going through the "sound barrier" or when escaping from the Earth's atmosphere.

Where the object is standing still holding a position that allows the object to have mass, the object is part of the Earth while the Earth has the **governing singularity** and the Sun has the **controlling singularity**. As soon as any object moves on Earth, the movement switches singularity by allowing the object to obtain the **governing singularity** while the Earth then for fills the directional circular control in forming the **controlling singularity.** All four phenomena interacts in a manner forming this role where for instance in the solar system the Sun holds the **controlling singularity** and Milky Way forms the **governing singularity.** To find validity in my argument one must draw this statement of motion back to the point where singularity is getting sides or said mathematically Π^0 is going Π. Π is the **controlling singularity** and Π forming Π^2 is in relation to the **governing singularity** Π^0. When there is singularity there can be no sides. The one forming singularity Π^0 by measure fills no space while form Π develops Π^2 into space. The space that even the dot fills being Πr^0 does not really exist in the manner we humans see space to exist. It is a spot that is there without being there. It does not visually exist because it is not filling any substance and it cannot be recognised since it is not three-dimensional. The spot and the dot have no dimensional worth of any measure but holds relevance. The Universe, to which I am referring, has never been unveiled. Einstein loosely mentioned it but no one could point to the flat Universe. This Universe holds a line in time made up of dots and spots forming no space but holds a Universe relevant.

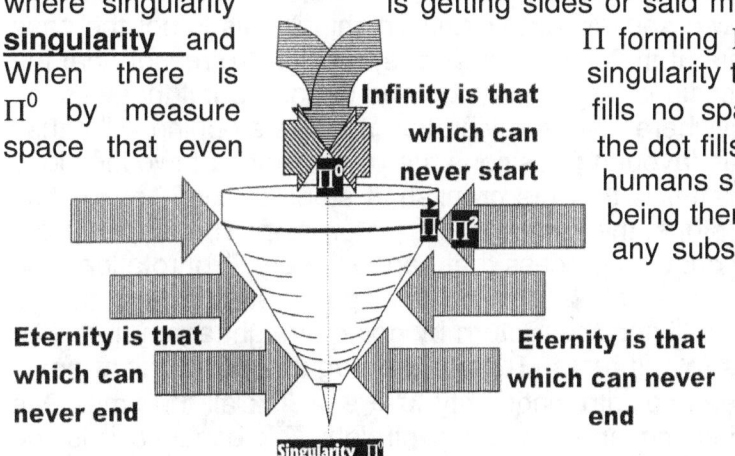

Infinity is that which can never start

Eternity is that which can never end

Eternity is that which can never end

Singularity Π^1

It is the point forming the very centre that plays the part as the **controlling singularity** within the Universe I have named as **Infinity,** which is better known as the axis. It is where nothing can go smaller and anything within that point can never reduce. That point is where the entirety called the Universe begins and where everything holding substance begins. Once one accepts the fact of singularity being present in that location, that accepting of singularity then is contradicting all the things we know and we can measure and we recognise that point being present by merit of the fact that the point referred to is not being formed by any of the things we can recognise. It is made up of everything we don't know and constitutes of everything we are unable to recognise or visualise. In that spot there is no space. That spot holds **Infinity.** In that space there can be no motion because there can be no space to have the motion within. It is formed as a line that is so small that our human reality by perception declare that point as not being there and the only reason why we know it is there is because of the results it left as an imprint of its not being there. We cannot detect it but notwithstanding our failure to note it we can recognise the dot on the merits of its absence and while in our Universe it is always absent, reality disallows the dot ever to be absent, because it is never absent. It cannot be absent. It cannot go absent but it can never be there where it should be in a place from where the third dimension forms and it is always present should I wish to locate it. It is **infinity** that can never go away.

I named the other part of singularity forming space **eternity** because that area never become bigger, or become more or find an end to the outside. Whatever was and is and will ever be is locked in that space I named **eternity** and it is **eternity** that never ends because **eternity** can never end moving. What we think of, as expanding is never ending movement giving eternity the eternal motion that will go on forever. The "so called expanding" of the Universe $T^2 = a^3 \div k$ is where singularity is shifting relevance **k** from liquid $k^{-1} = T^2 \div a^3$ to solid formulated as $k = a^3 \div T^2$ and the process whereby this happens is precisely the same as the Coanda effect. Getting back to my first argument about a line and that no line can start at zero but has to use singularity as a starting point, this is all the proof I require. The line **k** coming from the centre (singularity k^0) forms by forming an initial spot Π^0 becoming the dot Πr^0. However, I went on to say that whatever the line used to start with has to continue in order to repeat the same that began the line. Therefore the line started with Π^0 and it has to continue with Π^0 until such a point, as it must end withΠ. Whether the line is Π^0 or is r^0, or uses 1^0 the outcome all refers to singularity being used. By reducing the line we come to the end of the mathematical equation of the circle but the circle does not end there. That is what Newton did not recognise from the figures the cosmos represented to Kepler. The circle only secures the final cosmic figure and the value to singularity where all things have equal value. The movement of the circle splits singularity in two sectors. By forming Π the circle has to form Π^2 due to the movement coming about in securing the spaceΠ^3. Kepler chose to use different symbols to those being valid, but the concept remains the same. Kepler said that $a^3 = T^2k$ while I show that $\Pi^3 = \Pi^2\Pi$. It still confirms that movement $\Pi^2 =$ is the forming of space by three dimensions Π^3 in relation with the movement Π^2 being relevantΠ to singularity Π^0.

I shall try and explain what this concept holds in terms of a piston moving while working inside an internal combustion engine. The piston goes up to a point we call top dead centre where the piston stops and according to the crank the piston halts in directional movement. Then the piston starts to accelerate to a point we call bottom dead centre where, again it comes to a dead halt. The piston stops directional movement at T.D.C. and at B.D.C. or that is what we see without seeing anything. This is not the case because if this was the case the engine must vibrate at those two points of stopping. We reason that the piston stops twice and starts moving on the two occasions (at the very top and bottom) but if that was the case of stopping at two points without stopping anywhere else, the vibration that the stopping will cause will have the engine disintegrating completely. To us favouring positions the piston stops at two locations

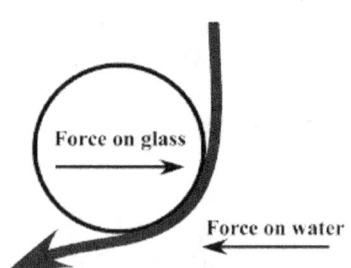

Force on glass

Force on water

but the fact of the matter is that the crankshaft stops every 7^0 of rotation and if the crankshaft stops, then so does the piston stop. The stopping is a continuous and is an ongoing process that happen every 7^0 of rotation. The crankshaft moves in a straight-ahead position going straight and then it stops and redirects by 7^0 and then it turn by going straight again. It is $a^3 = T^2k$ and then it stops (a^3), it turns (T^2) and then again goes straight again (**k**) while holding reference with singularity $k^0 = a^3 \div T^2k$ all the time. One cannot part the redirecting and the going straight T^2k because it is the same movement since the space forming a^3 is equal = to the turning T^2 and the going straight **k**. This is evident when dissecting Kepler's formula $a^3 = T^2k$ that $T^2 = a^3 \div k$ and $k = a^3 \div T^2$ while honouring Newton's 3rd law $k^{-1} = T^2 \div a^3$. Please believe me that this puts movement in such a perspective that it must be the most complicated dimension because this has the material $a^3 = T^2k$ moving $T^2 = a^3 \div k$ in terms of ($k = a^3 \div T^2$ as well as forming $k^{-1} = T^2 \div a^3$) while always referring to singularity $k^0 = a^3 \div T^2k$.

Kepler gave his formula symbols $a^3 = T^2k$ that do not quite represent gravity in its true symbolic nature and that then was the reason why I came on the idea that gravity has to link to Π more than any other value or symbol. It is because everything holding gravity or representing gravity (not mass) is round. Gravity connects by the use ofΠ. We have to part what mass does and what gravity does. Mass is where the object connects to one point on Earth and being at that point with mass the Earth does the moving by spinning. The spinning of the Earth then represents the movement or the intention to move because the Earth spins byΠ. This movement gives mass its qualities because mass does not possess the influential value of Π since mass is a quantity representative of the amount of atoms and not the spin of the atoms within the mass quantity. If we look at the way the Moon connects to the Earth, committing movement in a circle does it. That represents Π. When we look at the way the solar system connects to the Sun in circles every planet holds an individual symbolic value to Π that circles in relation to the Sun. If we look at the roundness of galactica, the formation represents Π. Every cosmic star holds roundness and roundness only represents one value, which is Π. The connection gravity has is not by mass but it is by Π. When we go in search of a cosmic resolve to find gravity, we better start looking for the influence Π has on the subject or leave the entire subject alone because the gateway in understanding gravity goes by the meaning of Π relating to Π^0.

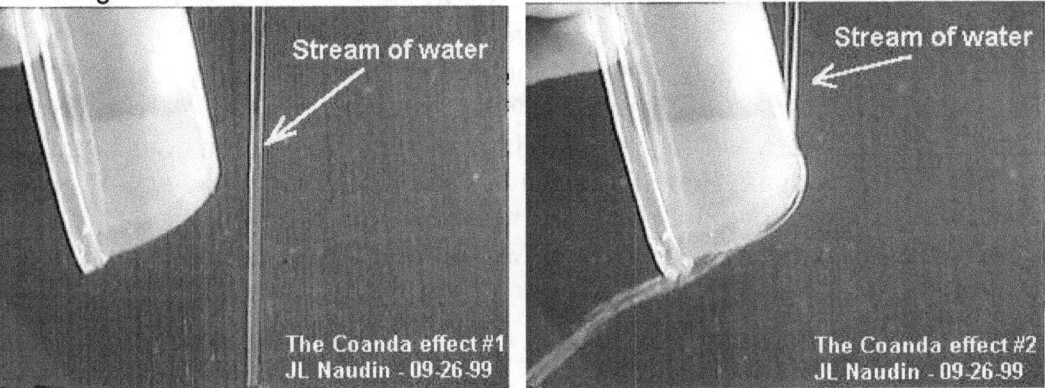

The condition for the presence of this singularity that forms everything, controls everything and is everything is the centralised $k^0 = a^3 / (T^2 k$ singularity that forms by movement $T^2 = a^3 /$ of the space $a^3 = k$ in relevancy $k = a^3 / T^2$ going both ways $k^{-1} = T^2 /$ thereof (Newton's 3rd law). Now put this formula in terms of gravity and we can see the gravitational picture of the Coanda effect come to life.

The condition for the presence of this singularity that forms everything, controls everything and is everything is the centralised $\Pi^0 = \Pi^3 / (\Pi^2$ singularity that forms by movement $\Pi^2 = \Pi^3 / \Pi$ of the space $\Pi^3 = \Pi\Pi$ in relevancy $\Pi = \Pi^3 / \Pi^2$ going both ways $\Pi^{-1} = \Pi^2 / \Pi^3$ thereof (Newton's 3rd law).

This explains the Coanda effect and the **Coanda effect is gravity** and gravity "glues" the water to the glass! The water forms a value of $\Pi^{-1} = \Pi^2 / \Pi^3$ while the glass forms a value of $\Pi = \Pi^3 / \Pi^2$. This process happens to all spinning things and as much as it happens to a piston connected to a crankshaft, just as much this will happen to an atom spinning an electron in a similar manner as the crankshaft is spinning holding a piston connected. This proves that gravity is the Coanda effect and in another book I prove that the Coanda effect has its origins in Π forming a value and that value forms gravity. In order to understand physics applying in cosmology I had to start by dissecting the set-up forming pi. Using this argument I can introduce my theory on the *Absolute Relevancy of Singularity.*

At the point in the centre of the circle a line must start. In the beginning when I explained the way I figured how the line starts I said a lot of dots has to continue in order to form a line. It would be 1 + 1 + 1 etc. because the line must form by holding singularity. After that point does mathematics begin but in the line that forms representing space as all other factors, then time holds 1. The line can only form when all the points forming the line have the value of 1 being 1^0. In that conclusion one realises something must separate singularity from all other factors because singularity hosts all other factors but is by own initiative Π^0. Only when singularity meets the end value can the end value have Π where the final ring of the spinning circle forms $\Pi\Pi^2$. That will be the spot of origin forming the relevance in Π.

That will hold the eternal spot…the smallest spot ever because all spots that ever can be were secured in a position in the centre of that spot that must continue as a line that forms. Because of the progress singularity follows from the single dimension singularity only allows mathematics a start at Π^0 progressing further onto Πr^0 and from there the line is born as $\Pi^0\Pi^0\Pi^0$ and to $\Pi^0\Pi^0\Pi^0\Pi^0$ etc. where Π^0 then may form the concept and value of r. But the line starts at $\Pi^0 = r^0$. This forms because cosmology is singularity based and the value is $\Pi\Pi^0$. This line $\Pi^0\Pi^0\Pi^0$ of singularity can only continue because every spinning atom preserves Π^0 in the very centre and since $\Pi^0 = \Pi^0 = \Pi^0$ and is represented in the circle of every atom spinning, the line is the same without finding conclusion except at the end where it forms mass at Π. At the point where Π forms, the movement Π^2 of the circle defines the space Π^3 of the circle and it confirms the centre Π^0 of the circle through the rotation going through the atoms. Let's call this the solid forming or if you wish, let's call it Kepler's singularity. After that singularity forms a line $\Pi^0 = \Pi^0 = \Pi^0$ where this forms another line again as Newton stipulated it by $\dfrac{dJ}{dt} = 1^0$. Let's call that the liquid singularity or Newton's singularity and the relevance of singularity having a solid base compared to the singularity holding a liquid base comes about by the movement of gravity. From these conclusions I prove that gravity is the result of four cosmic phenomena interacting to form the value of Π which by movement becomes the value of gravity Π^2 and gravity is equal to cosmic time applying. In order to understand the development of the cosmos and moreover the start of the cosmos and the progress in the cosmos as the cosmos formed, one has to understand the measure of Π. One has to see that Π is not merely 22 over 7 or that Π is a ratio that no one ever bothered to clarify, but Π is the key that unlocks every lock that hides a secret in origins of the Universe. One has to microscopically dissect the measure of Π to find the cosmos in measure. One has to understand where 7 fits in Π. The fact that Π is 7 at the bottom and that 7 relates to a double value of 10 is a key issue. Furthermore, it is very important to see why Π is 10 times two by adding 1.991 on the top part of the equation. In this measured value is what holds the building blocks of the entirety we call the Universe. It is behind Π that we will find the four phenomena, which I named the four pillars performing as gravity as they form gravity. It is by the actions of Π that the Universe develops.

The Hubble expanding goes by implementing gravity as Π in the square through the four pillars on which gravity and time rests. It is behind Π we discover the meaning of singularity and how singularity forms the absolute and only building block as a form that forms the Universe. It is in Π we find the Cosmic Code unlocking the meaning of the Universe. Time is centralised in Π^0 that forms Π as space's limit that becomes space by gravity being Π^2.

<u>Space is time gone to the past</u> in which time confirms its presence it had in the cosmos by moving from the present time into space and then onto the future leaving space behind as the past. The proof of this is again the top standing upright. Time places the governing singularity Π° in the centre and then the very next instant the governing singularity of the previous instant moves outward to form Πr° which then keeps on moving outward as time goes on the finally establish $\Pi\Pi^2 = \Pi^3$. That is how the top keeps erect and that is how the Coanda principle "glues" the liquid to the round solid surface by guiding the liquid as it flows in relation to the round solid. The spinning top is the manifestation of the Coanda effect, which is the coming together of the Roche limit, the Titius Bode law and the Lagrangian points. By forming a presentΠ° , time is in infinity forming singularity that then has to move onΠr° and in doing so it leaves a legacy behind being space which will form as $\Pi = \Pi^3 \div \Pi^2$. Time is the movement of everything forming the Universe where in time the movement of time relocates everything in space by moving from the present onto the past leaving behind space as a history of time gone by. That way the top can stay erect.

As time becomes the past by going to the future it forms space as it confirms the past, and in that space is what time forms by going to the past leaving space behind. Space becomes what time was at the point where time formed the particular space in relation to Π. As time becomes the present coming from the past, time has to move on to the future by replacing the past with the present at the same time and as time moved on it left space that represents that instant in time in relation to other space that was in some position at a specific location at such a point in time wherever that point in relevancy might be. The fact of Π not only refers to form but also validates the Universe by splitting infinity from eternity. By forming space when creating Π, time is using Π^0 in establishing movement Π^2. It is in the process of relocating Π to new positions by establishing Π^2 and connecting this as it forms a network consisting of Π° by forming space Π^3 in relation Π that establishes infinity Π° that always stays motionless. If not for movement, the Universe would be one line holding time by repeating singularity Π° uninterrupted and it is in the diverting of eternity to a position away from infinity that the Universe comes about. This is what happens in a Black Hole where no movement within the Black Hole places eternity that always moves in a standing position to infinity that never moves. Without movement the entire Universe will fall back into and onto one point and everything we thought is real and solid will disappear into that one point holding infinity onto eternity where infinity and eternity then reunites without holding space by any measure thereof. This proves the Universe to be an unreal concept with space being no reality at all but for the movement of space in relation to singularity Π° whereby Π confirms everything in a location in relevancy to all other things in a specific time slot or space.

When I, as a person forms a part of the Earth by the virtue of having mass that connects me Π to the Earth Π^2, stands on the Earth Π^3, my position in relation to the Earth gives me a specific positional relation to time Π^0 and the Earth. That gives the Moon a future of say one point five seconds being the past in relation to the Earth and that gives the Earth a past in reference to the Moon's future of one point five seconds. Where I am at any specific point in the present, that point I am holding is that which secures my present point in time.

The Sun is eight and a half minutes into my past with all the space being in-between the Earth and the Sun and by my view of the Sun I have a present time slot, as it also gives me a past of eight and a half minutes in relation to the Sun since the light travelled eight and a half minutes through space to confirm my past during that present instant. That secures my past by eight and a half minutes at the point of giving me a present location in time. However, that also secures my future I have from the point I now have in the present by the margin of eight and a half minutes because that establishes a flow of light that would last another eight and a half minutes of filling a presence worth eight and a half minutes while travelling through space by moving with time and every spot filled on the way would secure a position that I will have in a future presence for the next eight and a half minutes, which then becomes my future as it fills my past.

Looking at this scenario in a view from Alfa Centauri the allocated position Alfa Centauri holds in space relating to the Earth, gives the Earth a past of say four point six years while this secures the present and having that present secure the Earth to a future of say four point six years by forming time as space between Alfa Centauri and the Earth and this is confirming time to the tune of four point six years. By securing movement it forms time in having a past in relation to the present that by the same margin also secures a future in relation to a definite past. This is how the Universe builds space in establishing time. This applies to all allocated positions of rotating objects throughout the Universe. This means that every point away from Π^0 serving as Π, wherever that might be, secures the past the cosmos and I have by giving the cosmos and me a future in terms of the present Π^0.

Take this in relation to Kepler's formula we then find the Earth (a^3), which is in relation as viewed from Alfa Centauri (k) four point six years (T^2). That secures the three dimensional status the Earth has ($a^3 = T^2 k$) within the space from the Earth to Alfa Centauri (a^3) forming the Universe in terms of a present (k^0) being in the Earth centre which then depends on a location (k) secured by a future (T^2) that will come by movement where the future also doubles as a past ($k = a^3 \div T^2$ and $k^{-1} = T^2 \div a^3$). That is time and that is how time forms space and that is how space-time forms the Universe and that is the ***Absolute Relevancy of Singularity***. That then forms time in the centre in infinity in relation to space in eternity in singularity where time that moves forms space by holding time that does not move secured in positions in relevance to where every point that previously formed was in space which is time that has gone by. $\Pi^0\Pi$ **divides** **infinity**Π^0 from **eternity**Π where **infinity**Π^0 can't **move**Π^2 and **eternity**Π eternally moves as time$\Pi^0\Pi$ that establishes space Π^3 in motion$\Pi\Pi\Pi^2$.

If we put this in terms of singularity (Π^0) we find the Earth (Π^3) is in relation as viewed from Alfa Centauri (Π) four point six years (Π^2) while moving in that space that is time that has gone by. That secures the three dimensional status the Earth has (Π^3) in terms of a present (Π^0) that depends on a location (Π) secured by a future (Π^2) that will come by movement where the future ($\Pi = \Pi^3 \div \Pi^2$) moving forward that also doubles as a past ($\Pi^{-1} = \Pi^2 \div \Pi^3$) by the light coming from and thereby confirming the past. That is space formed three dimensionally by keeping time in infinity apart from time in eternity. The relevance (Π) that forms in relation to the present (Π^0) will relate to movement (Π^2) and the movement is circular which ensures that the relevancy forming is circular (Π) by securing that the movement is circular (Π^2) in terms of one specific point (Π^0) in infinity which then secures a roundness (Π^3) that forms an everlasting eternity ($\Pi\Pi^2$) which validates a never ending circleΠ^3. In this time in infinity (Π^0) that secures that there is an everlasting eternity ($\Pi\Pi^2$) in space (Π^3), it is not the space that is everlasting but the movement of time by the line ($\Pi\Pi^2$) that is everlasting. The **governing singularity** (Π^0) holds a **positional validity** (Π^3) of three dimensions $\Pi^3 = (\Pi\Pi^2)$in terms of any **relevance** (Π) formed by the **controlling singularity** ($\Pi\Pi^2$) thus mathematically it equates to $\Pi^0 = \Pi^3 \div (\Pi\Pi^2)$. If a **relevance** ($\Pi$) did not validate a **positional validity** (Π^3) securing a **governing singularity** (Π^0) in terms of movement formed by **the gravity** (Π^2) that produces the **controlling singularity** ($\Pi\Pi^2$) in space, with a three dimensional status Π^3, then space (Π^3) would not be obtained and thereby the Universe would not be secured. That is why space-time is $\Pi^0 = \Pi^3 \div (\Pi\Pi^2)$. However this must be seen where it applies. It applies where singularity as time meets space, which means it applies at a point in the Universe where time still grows and that is at the position that predates the Big Bang. It is where material forms before material forms. It is where the visual will never come. It is where singularity Π^0 forms space Π^3 by singularity (Π) moving (Π^2).

Time is the movement of space in relation to any one centralised point not spinning securing such movement. Everything in the Universe moves in relation to any one single point and every one single point that forms in any location everywhere that then has to stand still to form the centre of the Universe wherefrom that point must be motionlessness to allow everything else movement. The point not moving is anywhere and the rest that moves is everything excluding that one specific point that is motionless. In that manner the Universe is constructed and with every point being confirmed only by the movement of all other points around any specific point that means there is no valid solid Universe because the Universe is constructed from singularity (Π^0) that holds no valid space (Π^3) other than being in position (Π) at a specific point ($\Pi^0\Pi$) while having gravity (Π^2) that forms the time (Π^2), which is also the movement (Π^2) which is gravity of space (Π^3).

The flow of time being the present in singularity forms space by moving time in relation to space as much as relocating the present in terms of a past that is determined by the movement of time whereby that action of time moving by the same token is establishing space that confirms the past as it secures the future as time moves on to leave a positional legacy, a footprint of time gone by presented in terms of light which is the presentation of space. From this we can deduct that the Universe in a three-dimensional form starts at $7/10(\Pi^6) \div 6 = 112$, which is a value forming the start of the element table and that I explain in the Cosmic Code. In the **Cosmic Code** there are numerous values consisting of Π forming the relevancy by which certain rules comply throughout the cosmos.

One is 7/10 which is the Titius Bode law which is the interaction of gravity spinning and by spinning is forming a sphere (Π^6) within a cube ($\div 6$) and that is how the cosmos forms usingΠ. The dimension of $\Pi^0\Pi$ is flat but by spinning $\Pi^3 = (\Pi\Pi^2)$ the Universe goes in a sphere (Π^6) spinning in a cube 6. In this I prove that for instance amongst so many other things that electricity and gravity is the same thing. By ticking $\Pi^0\Pi$ time forms space by becoming space as time moves into the future leaving the past behind as space. Time is a substance and the only renewable substance with the ability to come into the Universe because from the start it came into the Universe to form the Universe as space. As time moves on space grows by the margin of singularity $\Pi^0\Pi$ leaving spots that form dots. The proof of this is in the value of Π being 3.14159 where 3.14159 -3 = 0.14159 x 7 = 0.9911, which is singularity as the spot (0.9911) becoming singularity 1 as the dot. In other work I explain this in much better detail.

There are two definitions we can use when looking at such a growth. We can look at the space not holding material that grew in size in which the stars finally froze their development to end as Black Holes and the growth was in terms of reducing space by remaining behind in terms of the expanding Universe all because of a lesser developing singularity within material compacting singularity $\Pi^0\Pi$. Or we can focus on the stars growing $\Pi^1 = \Pi^3 \div \Pi^2$ and with that push the outer space much more into expanding by reducing the density of outer space $\Pi^{-1} = \Pi^2 \div \Pi^3$. As the cosmos grows in space, the cosmos in expanding progresses just as much as the star was reducing in space and the space in the star that became less is the same space as that with which the cosmos expands.

This ratio is the ultimate relevancy.

This comes from the manner that the star manages to destroy space or dismisses space or compacts space and redirects the space to go from a gas and become more compact and denser by forming a fluid where the fluid is light or heat or the solidity of frozen space as matter really is. In the Black Hole it reduces much further as it claims the singularity, which the object had, and destroys all space and all time there ever was. As the star condenses space on the inside making the star to appear as if it is shrinking away, the space in outer space seemingly becomes more as it seems to be expanding but in real terms this is just a relevancy of one becoming denser and the other losing density. That stars get hotter towards the centre is not the pushing of mass, but is about space condensing.

The entire truth about the cosmos is that the Universe is within the atom that forms a cosmic unit holding singularity as much as it secures singularity and every atom forms a Universe standing apart, parted by time from all other atoms by the spin produced. Every Universe formed by every atom starts in infinity and ends where each atom's spin is forming relevancy between where that Universe starts and ends. All atoms are a Universe formed within the space that time puts between infinity and eternity. All atoms are stitched together by an invisible, unseen singularity - string that is present while also being absent and this invisible string links everything that the Universe is throughout the entirety.

The entirety rests on relevancy. As time moves on forming a line by implementing more dots in relation to the dots that are already there forming the history of time, which is what we call space, the area we call outer space receives many dots that time leaves as a footprint while the dots time leaves within material are less, just because the space is concentrated and thereby is less. The dot that time leaves holds no space but in terms of space moving with time increasing the adding of space-less dots brings about more space which then reduces the concentration of space and the more the dots are, the more the concentration reduces.

That is why the top can stand erect when spinning. Its because time forms a governing singularity Π^0 that then shifts in the next instant outwards to form Π as the controlling singularity in terms of the movement Π^2 that then controls the space spinning Π^3. It is time leaving Π^0 that then the next moment forms Π and in the movement of gravity Π^2 the space forms Π^3.

With more dots landing in outer space since there are more space, the space density reduces as the expanding in outer space seems to be more than what is applying to material where space is at a premium, being condensed. With time duplicating to form dots in singularity, every instant that it produces spots forming dots as the present, the space that outer space gains supersedes the space that material gains and that makes material more compact or more and more dense in relation to outer space. The space gained by the space occupied by the moving of material receives fewer dots than the space forming outer space or that part which we see as outer space and the space material holds advances more in density through the loss of density in the space called outer space.

This leaves material more compact in relevancy that seems to hold less space and this is moreover because of the relative loss of density in outer space is there because of outer space gaining space by time leaving more dots. The density in outer space is thereby lost and in that the density in material is gained by the loss of the density in space in outer space being more because it is a greater recipient of time.

The dot also leaves one point every time on the dot forming the governing singularity and that confirms the point holding governing singularity in terms of many dots received by the spin of the controlling singularity in terms of the gain of endless space in outer space. In that material always grows as outer space declines in density and that forms the "Hubble Constant" that is no constant. The Hubble constant is gravity expanding, which contradicts Newton's gravity contracting. **In a nutshell that is gravity**. **It does not even mention mass because mass has nothing to do with Π while gravity is Π in more forms than what humans are able to imagine. The cosmos grows by gravity which is $\Pi^0 = \Pi^3 \div \Pi^2\Pi$.**

That is why the distance between the Earth and the Moon becomes more. That is why the circumference of the Earth becomes bigger. That is why there are Earthquakes and hurricanes. That is why a human grows and heals and that is why hair and nails grow. That also is why there is aging and eventual unavoidable death to material holding life. The body never stops growing, which brings about the inevitable decline of life's body structure, as time becomes more that the body endures. The ever growing of the body makes the body collapse on itself with aging.

As time goes by everything on the Earth including the Earth and everything in the Universe around the Earth is gaining in space because that is what time leaves. That is why everything in fossils seems to be bigger the further back the fossil goes in the history of the Earth. Newtonians show millipedes that once roamed the Earth that were one metre wide…and Newtonians not only believe that but also advocate this information as the truth! Everything holding material grows by time leaving space as the history of time that went by. The history or the space of the millipede became bigger as time moved on but the millipede never was one metre wide. That is why we can see galactica so far way. It is through time progressing in space that it carries light to move from there to where we are capable to see where the light came from. Time brings light all the way by progressing in space that carries light through space.

There is and there can be no such a thing as "dark matter" What would make matter "dark"? If the material is "light" it then has a higher concentration of light than where we are at present. This puts the object we see in a denser area than where we are. There is much more movement in that area that concentrates the space in that area and thereby we can see the area because the space released from that area expands as it comes towards us and that light expanding is what we visually see. On the other hand areas that seem dark are more expanded with our light flowing outwards to those areas. Being darker is having light flowing to that area from where we are.

That puts that object in a more expanded environment and in higher expanded surrounding than where we are. If the material is "dark" our light is moving towards that position and that makes that area move slower than what we do. That area is therefore less concentrated and more expanded than where we are. Then again if we see the area as light, the area is more concentrated in density having light poring out towards us by the measure of releasing density. The light flowing towards us will make the region seem as if it is lighter. It again is about relevancies.

The part that seems to have brighter light moves faster as the light moves at a greater pace and moves towards us. The area that seems to be darker has light moving much slower because the light is moving away from us as it is the light from our area expanding into that larger area that leaves us with the concept that that area is darker. As the light moves into an expanded area it will seem to slow down. It is a question of relevancies applying by movement in relation to "standing still" or "moving faster" and "moving slower". If we look at the Earth from the Sun the Earth where we now are would be so dark where the earth is located that Earth would be invisible from the Sun because our space where the Earth is, is so much more expanded than the space is that surrounds the Sun.

Again seen from the Earth when looking at Pluto but by only using the naked eye Pluto is so dark it is "invisible" to the normal human perception. It is because "space" is much more expanded out there than it is where we are and if it is more expanded it is moving in relation to the space being available in which it can move making movement seem slower and making that area seem bigger. Then we have Mercury of the approximate same size but are very visible because it is more compressed in that area and therefore more visible than where we are and with the larger density the reflection of the Sun seems to light up the planet.

My question coming from this is why there is this hunt to find dark matter. Dark matter there is because dark matter is only more expanded in terms of denser matter which has light flowing to us which makes us able to see the light coming to us.

I am the first to admit that there is no substantiating proof presented in this article alone and I don't even begin to claim that I deliver any proof in this article. There is no room to present even the least bit of proof in any form possible in the space given to this article. With the limited space available to publish information in a journal by way of a small article such as this and having so much information at a premium I decided to release some vital information and the required proof about my claims in other small but comprehensive works that can be obtained. **For more information visit** http://www.singularityrelavancy.com/

In short the Titus Bode law is the way we find the solar system forms, this is a fact beyond dispute! There is a sequence in which the planets are spaced and this spacing has no relation whatsoever to mass of any description albeit planets, rocks or distances in formation.

The Titius Bode law named after Johann Elerty Bode (1747- 1826), who in 1772 published the law, formulated by Johann Titius in 1766 as Astronomy an empirical rule relating the distance of planets from the sun, based on the numerical sequence 0, 3, 12, 24,… By adding 4 to the sequence each time after the numbers doubled and afterwards the dividing the resulting number by 10 results in a sequence of 0.4, 0.7, 1, 1.6, 2.8, which is a reasonable representation of the actual distances in astronomical units for most planets but that is if the minor or inner planets are counted as a single entity at 2.8

In a sketch it explains as follows:

The first planet will adopt a value in ratio of 3. The 3 then adds 4 and the result divide by 10 in order to locate a distance in ratio in order of the Titius Bode law.

In the first 3 inner planets the ratio does not fit this explanation exactly but the reason why this does not apply has to do with gravitational singularity, which is a major new concept I bring to science. Then to complete this ratio at every planet's worth in distance 4 is added and 10 divide the number. This is how the solar system forms and there is no other way. The reason why it starts with 3 that I explain by proving the concept... The reason why 4 gets added that I explain by introducing the correct laws... The reason why the distance doubles every time becomes clear with using very simple mathematics.

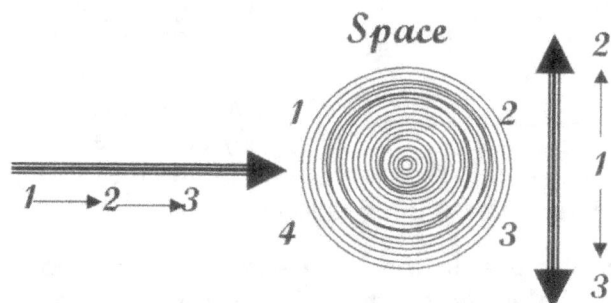

Titius Bode law 3 (time) plus 4 (space) = 7 (definite)

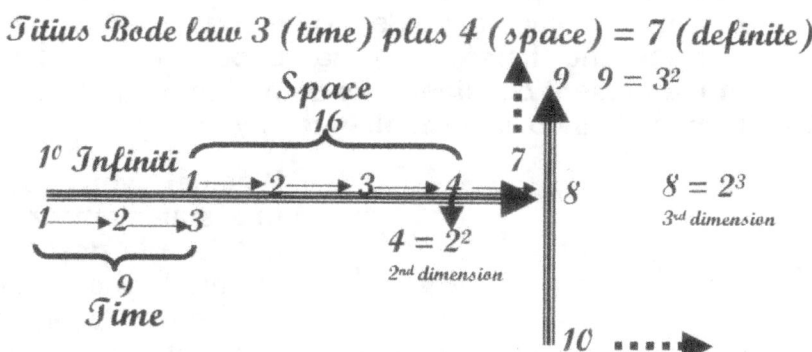

Titius Bode law 3 (time) plus 4 (space) plus 3 (the Universe) = 10 (eternity)

This formula that forms the Titius Bode law still holds its roots in the manner in which the Universe formed from a dot to a spot and into space-time.

Time forms the first 3, which has no place in space or in the Universe. This value composing of three points still holds a place that warrants no space because space starts at 3.1416 and to form movement of space within space can only come after the Titus Bode law forms a 10 interacting with 7.

The entire formation is exactly how time kick-started space by founding space and then form singularity controlling space.

Time control starts with singularity at 1^0. Then by turning four establishes 3 as a secondary position the first planet holds the pivotal position of 3 and with the space forming as a rotation added this value combining and doubling by going square the result is 10 and 10 is the value of space in eternity.

SPACE DIVIDED INTO TIME

(7/10) / (10/7) = 0.49

10 / 7	$(\Pi/2)^2$ **The Roche influence on Titius Bode**
7/10	$2.04 \times (\Pi/2)^2$ = 5.033
$(\Pi/2)^2$	$2.04 \times (\Pi/2)^2$ = 5.033
10 / 7	5.033 +5.033 = **10.066 from both objects**
10 / 7	10 / 7
7/10 =.49	7/10 = .49
.49 +	.49 = .98

.98 X 10.066 = 9.8 =Π^2 **TIME SPACE = Π^2 = 9.869 TIME**

Mercury is at a distance of more or less 50×10^6 but the positional relevancy is 0.40.

Venus has a distance of more or less 100×10^6 but the positional relevancy it has is 0.70.

Earth is at a distance of more or less 150×10^6 but it holds the positional relevancy of 1.

In these numbers hide the mystery of how space form and from this we can read how the Universe formed when the Universe began with one dot, the original spot from where the lot started.

Then from there we find that the distance doubles as the solar system grows. This has a mathematical explaining but such an explaining brings proof that the Universe forms gravity along a completely different method than what science at present would accept.

By using these numbers and values I use the numbers to prove that gravity grows by the measure of forming Π. Gravity comes about as singularity forms the relation of $\Pi^0\Pi$.

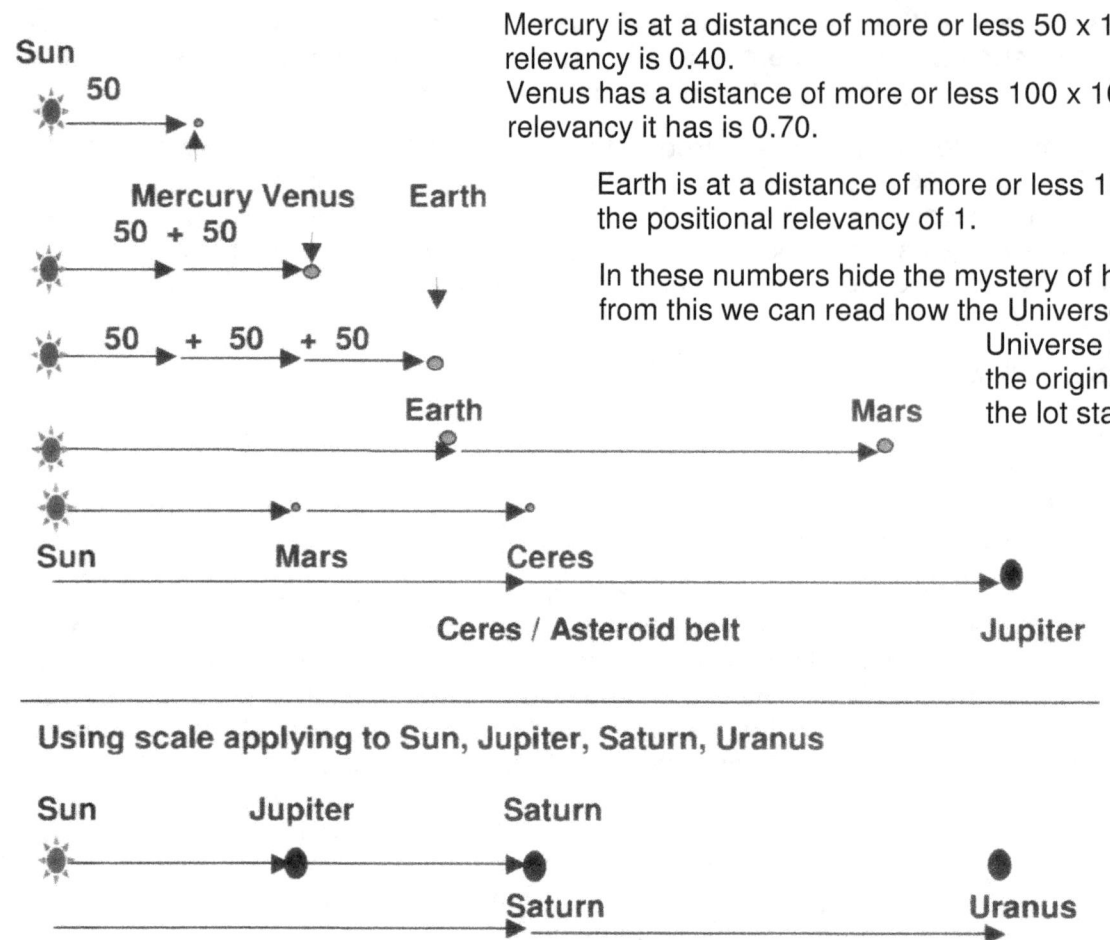

It is extremely clear that "mass" plays no part in the solar system forming or the allocating of planets in their natural positions. The density of planets holds no bearing on the planet position and therefore all the various factors science in the present considerer as being important to planet forming has no value at all in reality.

Whether you accept this statement or not, but for the past three hundred years there has never been any proof to confirm the idea that mass produces gravity and to the contrary, when combing through detail it is nature that proves Newton wrong.

PLANET	Mean Distance from the Sun (AU)	Equatorial Radius (km)	Mass of planet (Earth=1)	Mean density (grams/centimeter³)	
Mercury	0.3871	2439	0.06	5.43	
Venus	0.7233	6052	0.82	5.25	
Earth	1.000	6378	1.000	5.52	
Mars	1.524	3397	0.11	3.95	
Jupiter	5.203	71490	317.89	1.33	
Saturn	9.539	60268	95.18	0.69	
Uranus	19.19	25559	14.53	1.29	
Neptune	30.06	25269	17.14	1.64	
Pluto	39.48	1160	0.002	2.03	

Nothing about what Newton says vaguely corresponds with what applies in nature! Scientists forcefully present science as a copy of nature but that misrepresents the truth.

If you read only this website you will come to see what Newtonian science could never explain.

It's called the Titius Bode law. For the first time ever I explain this law.

The solar system does not use mass to form gravity as Newtonian science declares but this information is never often and openly revealed to the public as detrimentally important. Nature does not apply Newton and Newtonian science but uses another application going by the name of the **Titius Bode law**.

Go on and look it up on the Internet and verify what I say about nature using the **Titius Bode law** being applied instead of Newton. This is fact not widely promoted but nevertheless it is true.

In four books on different levels of intensity in each I show and explain just how I cracked the principle named the **Titius Bode law** and show why it applies and how nature works and in each one of four books on a different understanding level I reveal the concept that **nature** (<u>not Newton or science</u>) uses as the building blocks of space. However… the concept disproves Newtonian science and rubbishes Newton's ideas about what he thought gravity is as unworkable and it is that part that science will never accept as a factual scientific fact.

Introduction part explaining

In the case of...

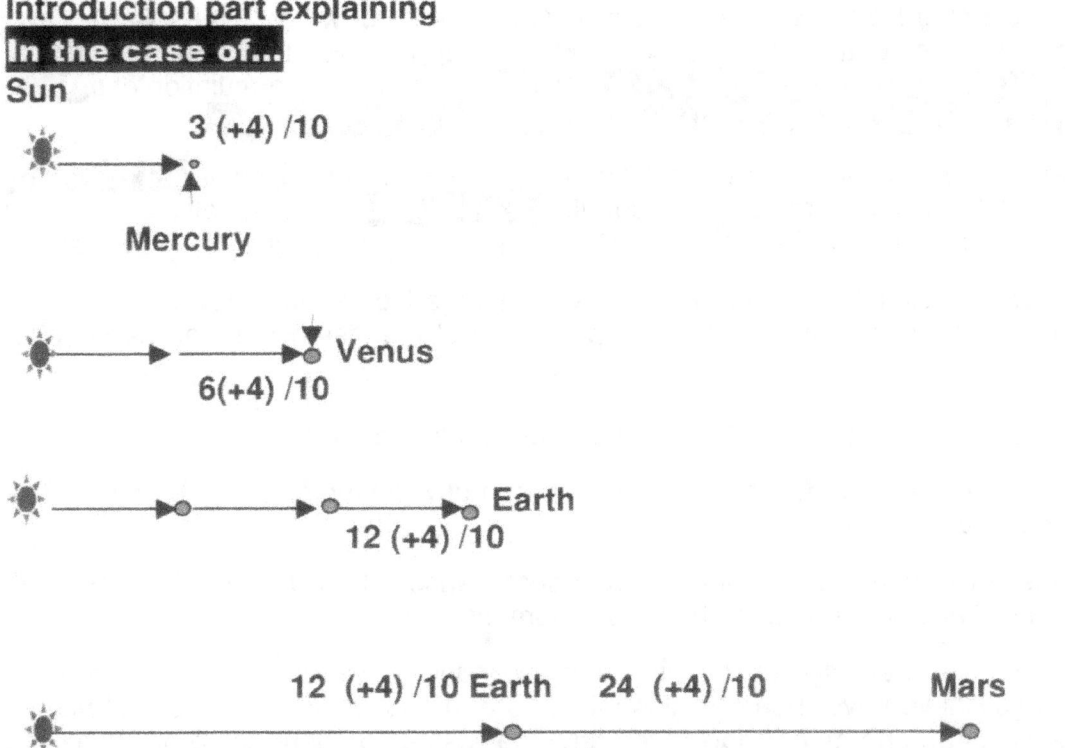

This is the formula nature uses to build the solar system and therefore the Universe.

Where the solar system does not follow the Titus Bode law to a very precise detail we can see how and where development diverted and how history of the solar system changed the normal development and furthermore we can quantify these deviations by finding proof of other laws coming into play.

One thing stands out like little else: Nature dismisses Newton's mass ideas in the cosmos at large because the cosmos applies gravity in relation to gravitational density.

The fact that nature uses the Titius bode law in the solar layout is known for centuries but as usual science tells the Universe what it is and ignores what the Universe really is. Science ignores nature since science has no idea why the Universe is what it is. The solar system and therefore the entire Universe uses a ratio that two men Titius and Bode simultaneously but being well apart at the time discovered. The solar system and indeed the Universe is built not by "mass" as Newton suggested incorrectly but employing another unknown system but is also a very well hidden system called the Titius Bode law. This law is what forms the solar system proving it is not mass.

The following table compares the law's predictions with the actual distances, where the addition of Pluto is a modern modification.

Planet	n	Titius-Bode Law	Semi-Major Axis
Mercury		0.40	0.39
Venus	0	0.70	0.72
Earth	1	1.00	1.00
Mars	2	1.60	1.52
asteroid belt	3	2.80	2.8
Jupiter	4	5.20	5.20
Saturn	5	10.0	9.54
Uranus	6	19.6	19.2
Neptune	-	-	30.1
Pluto	7	38.8	39.4

This is what is there: It is the **Titius Bode Law** and I show why the **Bode's Law"** or **"Titius-Bode Law"** forms this formation ratio as it does but the best science could come up with was to change the original formulation that was $a = (n + 4) / 10$ where $n = 0, 3, 6, 12, 24, 48...$ to the modern formulation of (AU_{earth} $= 147.597 * 10^6$ km): $a = 0.4 + 0.3 \times k$ where "k'= 0, 1, 2, 4, 8, 16, 32, 64, 128 (sequence)

The ratio is so impressively periodic or cyclic correct that it can be put to a formula such as $a = (n + 4)/10$ and the outcome is so predictable that according to the formula $a = (n + 4) / 10$ it led to all the discovery of all the missing planets that was discovered after Galileo Galilee used the first spyglass to look at stars.

This ratio doubles the distance a planet has every time a new planet is located in a new position. The distance of Saturn doubles the distance between it and Jupiter to what the distance is between Jupiter and the sun.

The distance ratio doubles to Mars to what the distance is from the sun to the earth.

This ratio of doubling the relevancy in distance from the sun applies notwithstanding the size or the mass or any specific value that any planet might have.

This shows that the ratio is not vested in material but it is in accordance with the space that holds the material and that throws "mass" and Newton out of the window forever.

This law has been and is known to science for almost if not two hundred years and if you are not a professional astrophysicist you have never heard of this law before in your entire life. It is well hidden under Newton's misconceptions and no matter how much any one wishes to share some brainwashing therapy with me the fact is that Newton's $F = G \dfrac{M_1 M_2}{r^2}$ and $4\pi^2 a^3 = P^2 G(M + m)$ does not wash because it is not applying in any form within the Universe anywhere.

The Titius Bode law is what the solar system uses and if that burns your brainwashing and you mind control inflicts excruciating pain on your brain and you get an overwhelming urge to becomes violent with disapproval towards me it would be best to kneel and pray for it is not me that put the Titius Bode law in application within the cosmos but it is the way God Almighty designed the Universe. When you argue with me you have an issue with God or nature or the cosmos, not with me. It is used in nature and not by me 'cause I show it's there and that's all. What I achieved is to find out why the Titius Bode law applies and how does it come about to form the ratio that it does. I was the one than made the connection (not discovery) how gravity comes about by the implication of the Titius Bode law but also always in conjunction with three other phenomena called
1) The Coanda effect, which is the way, the atmosphere forms.
2) The Titius Bode law is how planets use a ratio to arrange their allocated positions.
3) The Roche limit is the law that applies to what we call as the "sound barrier" and how stars explode.
4) The Lagrangian points are why atmospheric layers form around the earth and some planets has rings.

It is the manner in which the solar system is presented that is completely inaccurate and just as much confusing. The solar system presented with this layout is even less accurate than what the Ptolemaic presentation is because the Ptolemaic have the numeric order correct and the rest is completely mythical. Having the solar system is so close in proximity gives the impression of being cramped for space and huge in material and there is no excuse to explain away the inaccuracy except to help Newton cheat by cheating some more.

The distances apart are more important to science than what sizes are. Newtonian science dismisses that the sun spins around the earth but that I observe every day of my life. In this time I can see the stars in constellations determining which month it is and what season we are in. The stars turn around the earth just as the moon and the sun and denying this is misrepresenting truth.

You are going to say this is wrong to believe the sun is turning around the earth but yes although this is incompatible with science it is not wrong. To say it is not true that the sun spins around the earth is untrue because I can see it happen! It is as far off the mark as Newtonian science is off the mark. The way Newtonian science form your picture you form your concept by the solar system is as much a comedy made up of misunderstanding as this is.

There is not enough space to go into even slight detail but for that I offer four books that is on the market. I show you in detail how far Newtonian science is wrong!

In those four I present you with the truth about science in all of human history and I do not exaggerate. I mathematically prove that according to the Titius Bode law gravity is pi.

This produces time and according to that the earth is divided into time zones where when it is morning at one point it is evening at another point and when it is dark on one side of the earth, it is light on the other side of the earth.

You might begin to feel bored with my rambling about what you know but what you are so very aware of, science says does not happen and your eyes are telling you lies. There are four cosmic phenomena that form the Universe according to principles nature applies and all four principles reject and annihilate Newton and Newtonian science in every way imaginable. The four proves Newton's ideas are completely wrong. The function of gravity is a relevance between that which moves and that which contains movement!

If you don't believe me then just read on and find out the truth for yourself. Find out why you can see the sun rises and crosses the earth's sky while science says this never happens! What you can see everyday of your life happens, science says never happens because science is so far apart from nature they don't even share the same Universe. Newtonian science and nature is un-reconcilably different and Newton's approach to science is different beyond compromise. The two concepts, nature and Newton can't even associate.

By proving the four cosmic principles I prove how Nature forms the Universe and why Newtonian science is so very wrong and by proving nature I show why we see the sun as spinning over our sky and it does happen.

This part of science no one in science ever tells because they want you to believe everything they and Newton says about science is confirmed by nature and nature apposes EVERYTHING science stands for. This you did not know? Now I bring you nature and you will find why science rejects me. It is because science rejects nature. I present you the way nature works.

Please believe me that this is your introduction to the cosmos and every aspect that brings about explaining the cosmos. The cosmos is not mathematical equating of more rubbish. It is about ideas mathematics can't prove. This picture shows the Roche limit and the Roche limit annihilates Mainstream physics' concept about the cosmos Therefore it is not popular. Mainstream physicists might portray science to diminishing nature in presenting nature as "a freak (of nature)". In that they actually show their small understanding about the cosmos. To belittle nature does not diminish nature but reduce scientists' credibility and knowledge. These pictures prove Newton is wrong and that mass don't pull mass as Newton said. This is called the Roche limit and the Roche limit proves stars can't collide.

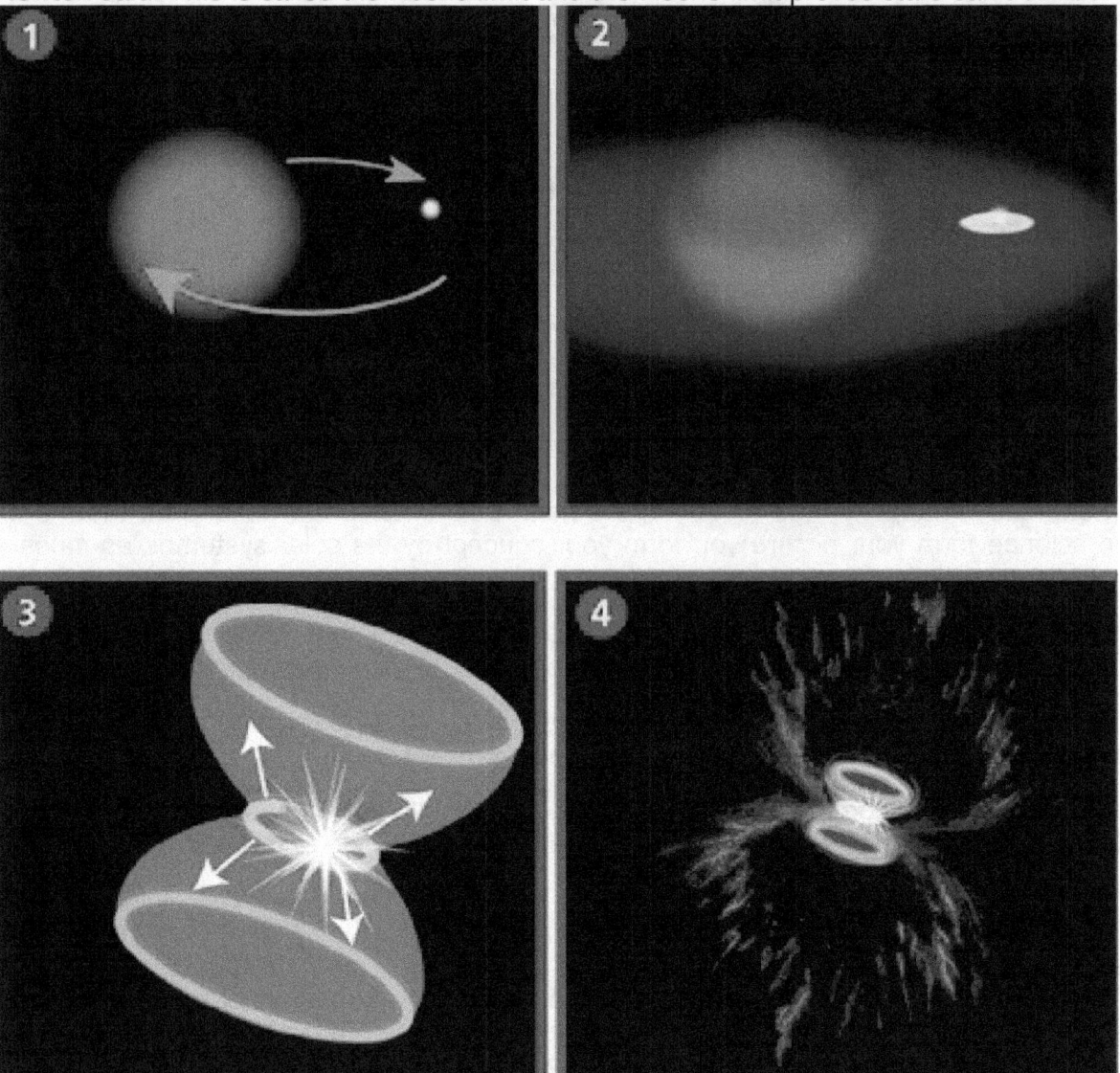

The picture explains one of the principles showing what happens when nature manages the working process of the four cosmic principles in Nature.

These are:
The Coanda effect **The Titius Bode law** **The Roche limit** **The Lagrangian points**.

There is an almost hundred percent chance that you never heard of these phenomena before notwithstanding that the entire Universe is built by these phenomena and these phenomena form the building blocks of the Universe. The reason why you never heard of it and science never mention the importance of the phenomena is because science are unable to explain it and so they rather ignore the importance than press the importance.

As I said nature applies the Titius Bode law to allocate planetary positions.

There is this very specific ratio that is in place whereby a distance ratio places plants numerically.

Titius Bode Law

The **Titius-Bode Law** is rough rule that predicts the spacing of the planets in the Solar System. The relationship was first pointed out by Johann Titius in 1766 and was formulated as a mathematical expression by J.E. Bode in 1778. It leads Bode to predict the existence of another planet between Mars and Jupiter in what we now recognize as the asteroid belt.

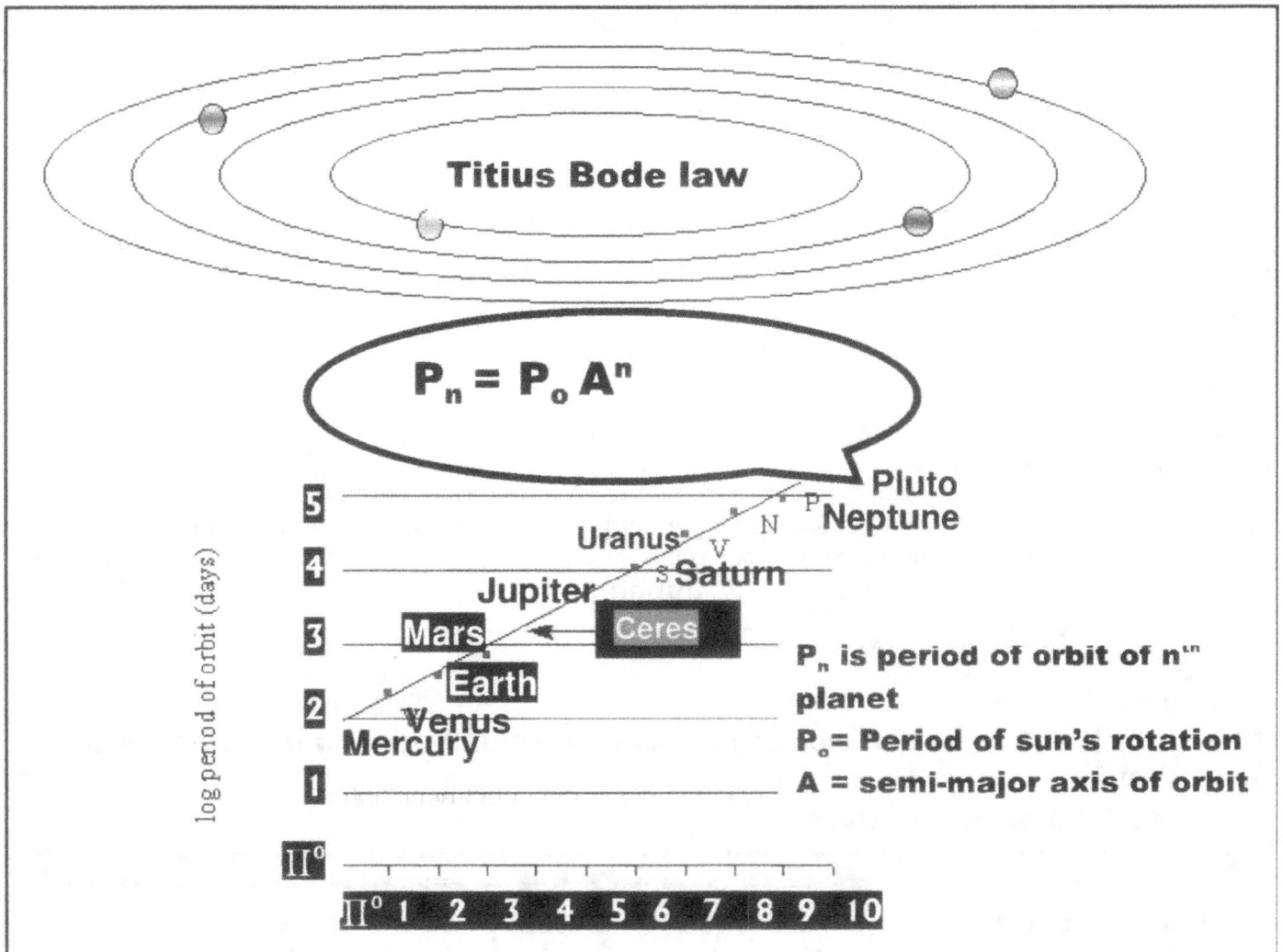

This shows a commitment to order and relevance but was never yet understood before. They could mathematically plot a ratio but never prove what is behind forming the ratio.

The law relates the mean distances of the planets from the sun to a simple mathematic progression of numbers. There is a code on which the allocation depends and this code has nothing to do with mass! In fact mass only plays a role in Newton's imagination. I dare any person to prove me wrong. Just read the book and see how science cheats.

This ratio of P_n, P_o, A shows the sequence and indicates a ration but I prove why the ratio exists and what principles in nature apply to put this ratio in place.

I show for the first time in all of history the reason why this ratio is there and why it applies as it does with this ratio of P_n P_o A. I use this to show that gravity forms not by Newton's mass but by pi.

There is a given ratio on which the allocations of planets are based and this is how nature and not Newton's imagination works.
To find the mean distances of the planets, beginning with the following simple sequence of numbers:
0 3 6 12 24 48 96 192 384

With the exception of the first two, the others are simple twice the value of the preceding number.

Add 4 to each number:
4 7 10 16 28 52 100 196 388
Then divide by 10:
0.4 0.7 1.0 1.6 2.8 5.2 10.0 19.6 38.8

The resulting sequence is very close to the distribution of mean distances of the planets from the Sun:

Body	Actual distance (A.U.)	Bode's Law <A.U.)< td>
Mercury	0.39	0.4
Venus	0.72	0.7
Earth	1.00	1.0
Mars	1.52	1.6
Asteroid Belt		2.8
Jupiter	5.20	5.2
Saturn	9.54	10.0
Uranus	19.19	19.6

It is this code of 3 and 4 that I decipher and the reasons why it structurally works by 3 and 4 and it is this growth by 3 and 4 that I unravel.

The solution to encrypt the cosmic code is in the value of 3 and four. It is because of this value that we see the sun as the sun crosses our sky while the earth spins around its axis It is a ratio that represents cosmology as we see it. We see the sun cross our sky because the sun rises in the east and sets west.

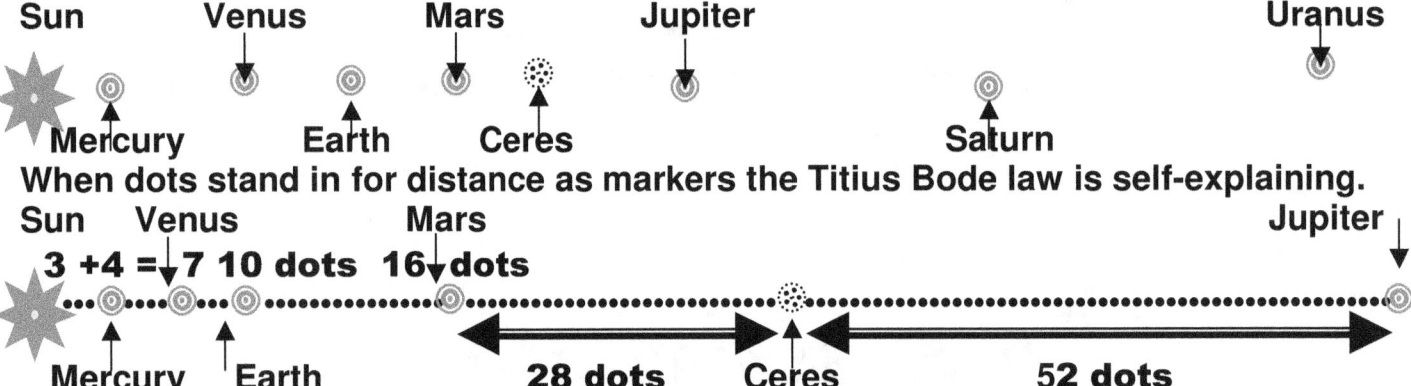

When dots stand in for distance as markers the Titius Bode law is self-explaining.

If the distance between the sun and Mercury is represented by three dots then there has to add another four to get the distance that represents four dots to allocate the position Venus has. The sun holds a value of 3 and Mercury is 5 making Venus 7 and earth 10.

Where Venus holds an allocated position of seven (3 + 4) then another three points further will we find the allocated position of the Earth. In this ratio the earth then has ten dots.

With the earth at ten dots Mercury is 16 dots further, Ceres is 28 dots and Jupiter is 52 dots. I have no more room on the page to represent a readable scale but Saturn will have another 100 dots between it and Jupiter and then Uranus world extend a further 196 dot away from the sun than Jupiter is.

Where Jupiter is 52 dots away from the asteroid belt and Ceres we have Saturn 100 dots further. On the scale only this distance between Jupiter and Saturn would cover the entire length of the page and the dots will be 100 in relation to Mercury's 3 dots and the earth's 10 dots from the sun. This is how it looks:

Sun   **Saturn**

Jupiter   **100 dots apart**

Then we have Uranus another number of 196 dots more to the outside of Saturn and then comes Pluto also almost double in distance away from this. In my book **Nature Annihilating Newton** I use rugby fields to show distances and how many rugby fields it takes to place Jupiter when the sun in ratio is the size of a soccer ball. Now put what I show in terms of this picture and see the grand scale of misrepresentation is going on. Look at the size and the distance in between and the variations and remember what I said the Newtonian scale is far more misrepresenting the truth in comparison with the Ptolemaic Universe. The truth is out there behind whatever deceit science tries to cover the truth with.

If you wish to have the Titus Bode law in a more practical sense I use a rugby field to bring ratio understanding about. Any one can walk to a soccer field / rugby field/ grind iron field because they are all very close to the same sizes.

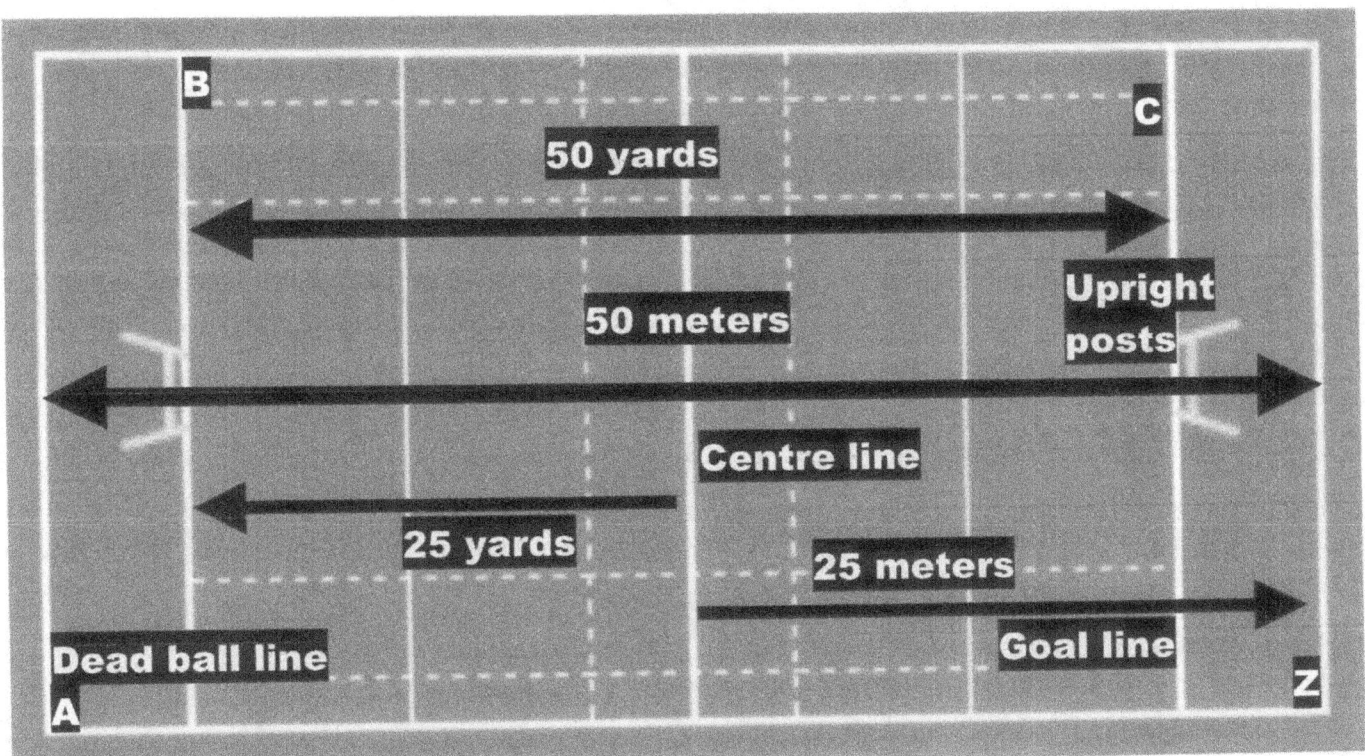

Please take note that this exercise is not about accuracy of a billionth of a micrometer in relation to a billion kilometres but it is merely to give some representation of relevant sizes of the solar system. This exercise is to bring about what the solar system in relevancies truly are and to show how misrepresenting the picture are that show the solar system as Newtonians portray it.

The sizes of material the experiment use is as follows.

Sun-any ball, diameter 21 cm inches

Mercury-a pinhead, diameter 0.075 cm
Venus-a peppercorn, diameter 0.1 cm
Earth-a second peppercorn diameter 0.1 cm
Mars-a second pinhead diameter 0.075 cm
Jupiter-a chestnut or a pecan, diameter 2.5 cm
Saturn-a hazelnut or an acorn, diameter 2 cm
Uranus-a peanut or coffee bean, diameter 1 cm
Neptune-a second peanut or coffee bean 1 cm
Pluto- a third pinhead (or smaller, since Pluto is the smallest planet)

Put a soccer ball on the dead ball line of any Rugby field. The **sun**-soccer ball is placed in the very centre of the dead ball line in the centre of the upright poles of the rugby field.

Walk 5.8 meters to the goal line and put a pinhead representing **Mercury** as a pinhead, diameter 0.075 cm in line with the post while lining up the two posts on both sides.

Place the pinhead on the grass. This will be approximately on the goal line.
This represents the 57.9 million kilometres Mercury is from the sun.

Then walk another 11 metres from the soccer ball-sun. This will be on the five-meter line placing **Venus** as a peppercorn with a diameter 0.1 cm about 11 meters away from the soccer ball-**sun**
This represents the 108.2 million kilometres Venus is from the sun.

Another way to look at it is The Earth is 12800 kilometres wide! The peppercorn is eight hundredths of 2.54 cm wide. What about the Sun? It is 1280000 kilometres wide. The ball representing it is 21 cm wide. So, 1 metre in the model represents a billion meters or a million kilometres in reality.

This means that one metre (100 cm) represents 1000 000 000 metres. Take a one metre: this distance across the grass is an enormous space-journey of one billion meters."

Now, what is the distance between the Earth and the Sun? It is 149 million kilometres. In the model, this distance will be 15 metres. This covers the first three planets. Go and stand on the pavilion and look at your soccer ball-sun with Mercury the pinhead-size and Venus and the earth two peppercorns at a distance of 11 meters and 15 meters from the sun. This is space in relation to material. That is according to Newtonians something (materials) in comparison with "nothing" which is space.

Place another pinhead 23 metres away from the soccer ball-sun. That represents Mars. Then lets skip the asteroid belt and move on to Jupiter. Jupiter is 78 metres away from the soccer ball and is the size of a Jupiter-a chestnut or a pecan, diameter 2.5 cm. This is where this model ends because the rest is totally out of even a Newtonian's imagination about reality. Go stand on the roof of the pavilion of the Rugby field and look at the solar system model that you created. Can you see the pinheads and the peppercorn seed? You can't even see the chestnut at that distance and now image this lot is going in a circle around the soccer ball.

Let's take this further but it will not be part of a sensible model on the scale we applied this far.

Saturn is a one and a half rugby field away from the soccer ball at about 143 meters Saturn being the size of a hazelnut or an acorn, diameter 2 cm. The distance Saturn is from the sun is 1427 million kilometres.

The next to follow is Uranus at a 288 meters away from the first rugby fields on the far end thereof on the dead ball line. Uranus is a peanut or coffee bean with a diameter 1 cm.

We now are one rugby field followed by a second rugby field away and then another third rugby field using almost eighty meters of the next or third rugby field, which is close to three rugby fields away from the soccer ball-sun and Uranus is the size on scale of a peanut or coffee bean with a diameter of 1 cm.

To have a view of this one would now start to think in terms of using some hot air balloon because no view from even the roof of the pavilion will allow an entire complete overall view. Uranus is a distance of 2871 million kilometres from the sun.

Try to se a soccer ball and a peanut or coffee bean with a diameter 1 cm being at a distance of 445 meters away from each other. Even from a hot air balloon this vision is impossible.

Then comes Neptune being four and a half rugby fields from the soccer ball-sun. This is as close to a half kilometre as can be and we are not at the end of the solar system.

The size of Neptune is also a second peanut or coffee bean 1 cm and at that it is at 450 meters from the soccer ball-sun. There is no way a coffee bean 1 cm can be visible at 445 in distance from soccer ball – sun just can be visible. The distance of the sun to Neptune is 4497 million kilometres.

…And then at last its Pluto. Pluto is a pinhead and at the distance it is the size of the pinhead is irrespective of anything comparable. Pluto is in distance almost six rugby fields away from the soccer ball-sun. It is 5913.5 million kilometres away from the sun. Need I say more?

There is a soccer ball over here and there are six rugby fields inbetween a small pinhead over here

That is the size and the ratio of the solar system in accordance with the Titus Bode law.

Now explain this reality that I have given as the way nature configures the solar system in relation to the way Newtonians say Newton surmising is correct.

Can you see what I call my work as reference to Newtonian madness as Corrupt Science and why the title is **A Conspiracy in Science in Progress**?

Mercury	0.06	$T^2 \div a^3 =$	0.983
Venus	0.82	$T^2 \div a^3 =$	0.992
Earth	1.000	$T^2 \div a^3 =$	1.000
Mars	0.11	$T^2 \div a^3 =$	1.000
Jupiter	317.89	$T^2 \div a^3 =$	1.000
Saturn	95.17	$T^2 \div a^3 =$	0.999
Uranus	14.53	$T^2 \div a^3 =$	1.000
Neptune	17.14	$T^2 \div a^3 =$	0.999
Pluto	0.0025	$T^2 \div a^3 =$	1.004

Sun

☼◎

Mercury $T^2 \div a^3 =$ 0.983

This is the distance from the sun to Mercury and this distance does not require even one arrow if I wish to show any ratio applying in the solar system.

Sun

☼►

Venus $T^2 \div a^3 =$ 0.992

By now at least we can introduce one arrow to show an indication of distance applying as a result of space forming ratio in the solar system.

Sun

☼►►

Earth $T^2 \div a^3 =$ 1.000

The earth do not truly double in distance however, the demonstration is not about displaying any form of total accuracy but more about leaving the impression of space becoming more in a defined ratio while time remains in the instant.

Sun

☼ ► ► ►

Mars $T^2 \div a^3 =$ 1.000

Again this ratio of growth too is not precisely representative of what is truthful

Sun

☼► ► ► ► ►**Ceres**

At this point for the first time we see a more or less accurate representation of cosmic growth.

Sun

☼► ► ► ► ► ► ► ► ► ►**Jupiter** $T^2 \div a^3 =$ 1.00

Sun

☼► ► ► ► ► ► ► ► ► ► ► ► ► ► ► ► ► ► ►

Saturn $T^2 \div a^3 =$ 0.999

That's it. That is as far as the ratio goes that this page can present as truthful. More doubling up means going so small this computer can't present that. This is the ratio applying in the Titius Bode law and that is what I will show is why this proves that gravity forms as a circle to the measured value of Π that moves as Π^2.

I have discovered the cosmic code that reveals this ratio. When applying the cosmic code all this makes sense whereas Newton's idea of mass is completely invalid.

The mass of the planets is totally at random and that is why Newton just cannot be true.

As space moves towards the sun the space becomes denser until the space forms a liquid around the sun with flames coming from the sun. Newtonians call these flames prominence because they have no idea what to call it but it is just more liquid heat. Because the sun is so large and spins so much the movement of the sun condense the space from a gas to a liquid and then freezes it to a solid all by movement of the sun That churning in rotation condenses space to freeze within the sun as a solid.

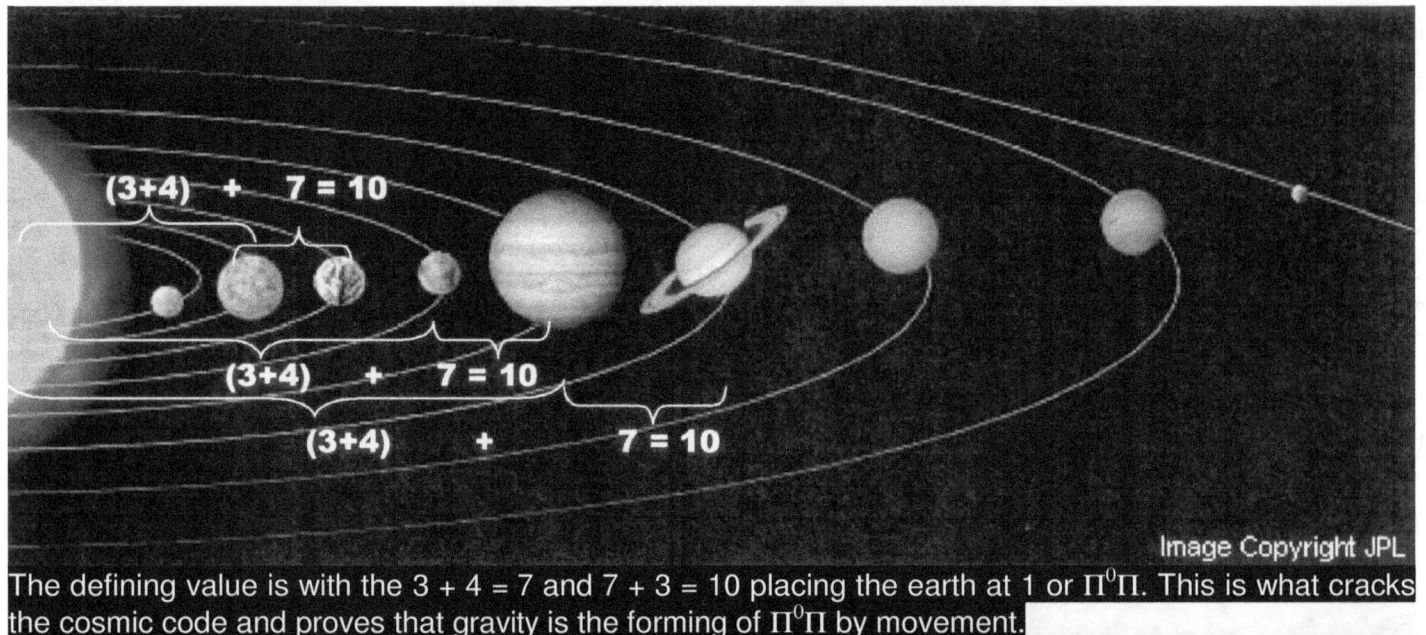

The defining value is with the 3 + 4 = 7 and 7 + 3 = 10 placing the earth at 1 or $\Pi^0\Pi$. This is what cracks the cosmic code and proves that gravity is the forming of $\Pi^0\Pi$ by movement.

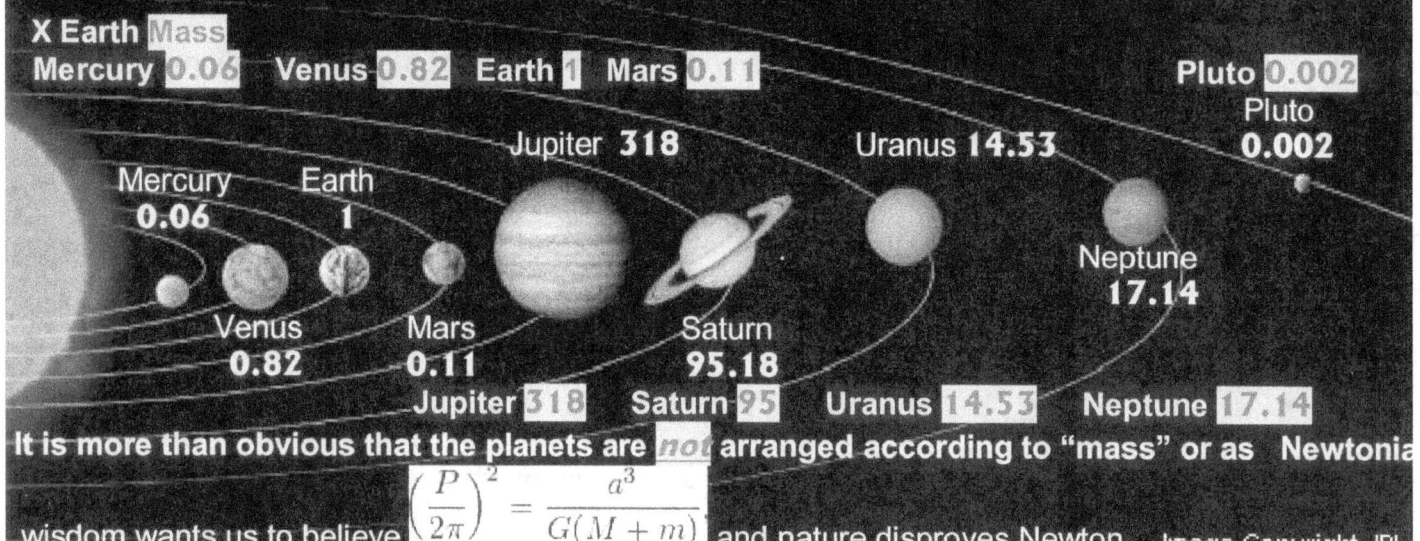

Image Copyright JPL

Look at the picture and see how mass is totally random distributed. As far as planet sizes go there is no defining order or rhythm connecting a sequence but there is a tiny one, then a small one and a another small one and an almost tiny one followed by rock debris and then all of a sudden the next one is the biggest one and that is followed by another almost big one…and so it goes. Shockingly science has been aware of this for how many hundreds of years? You think I am the first to come to this conclusion in three hundred years. Do you believe I am the first one that add 3 + 4 and get 7…I don't think so and I think those that came before me was killed like science tries to kill me. It is not the 3 + 4 forming 7 but it is what pans out from the movement this ratio of 3 + 4 forming 7 develops. This is so simple to understand and yet science has killed my work up to now by ignoring me flat. Please don't allow them to kill me; spread the word. If it is this simple to solve all along why would I be the one that cracked the code do you think? The solution is truly so simple it is adding three and four and then arrive at a number of seven. How long did science try **not** to conclude or confirm the Titius Bode if it is that simple to solve would you think?

A number of three hundred years springs to mind. …And with this picture showing the mass differentiation the same picture is also telling a storey of deceit! Do you still think I exaggerate when I say they have been conspiring to cover-up Newtonian fraud for three hundred years? If you don't smell a dead rat then you are part of the brigade hiding dead rats. The Titius Bode law develops in the manner that Π forms a value. By forming the Titius Bode law gravity forms Π and it is the forming of Π as a value and the way Π forms a value that the cosmic code becomes revealed. Gravity is about forming the value of Π and the way movement forms gravity as time forms space in the form of Π. This is the way gravity builds the Universe because it is the way that gravity builds the Universe. However science forever shy away from nature in favour of Newton and that puts a hoax in the forefront of science.

Science can never take the blame for not knowing. Never is a suggestion put forward that it might be science that holds the shortfall and it is because of science not being adequate that science cannot match the Bible. I can prove how the start came about because I decoded gravity and I did that by finding an explanation about the four cosmic principles. By deciphering the Roche limit, the Lagrangian points, the Titius Bode law and the Coanda effect I am able to show how the very first instant happened when the Universe started the very first point ever formed. These principles are in place and not the principles Newton fabricated... That this book shows. It shows that the cosmos uses other principles than what the Newtonian science promotes. What science says nature uses is not in place or does not hold evidence while what nature does use science deny by just never pressing the issue. I show what is in place and I show why it is in place but first I have to reject what science says is in place because it is not in place.

All Newtonian-thinking mechanical engineers propagate that a car lifts in the air because of the wind flowing underneath the car pushes it up and therefore it lift from the ground. This is cultured hogwash

Look at the picture of the car below. The Newtonian engineers teach that at a specific speed the air flowing underneath the car lifts the car from the ground and into the air. This is an argument made by the uninformed to bullshit the stupidity of the human race and everyone that has no thought process but still wish to come across as being wise. ...And this mindless dogma has been offered for decades and was never questioned by even one person in the world of physics.

It is said that the car lifts into the air because the wind from beneath lifts the car so much it becomes airborne. Again I reiterate this point because I know that any person reading this will be ready to defend this idea as if it is his or her personal knowledge.

Giving the situation the minimal thought the truth nails this misconception out of the park. Think of the amount of air pushing the car down versus the amount of air lifting the car up during any particular speed and then reconsider your verdict. There is a ratio of air flowing over the car compared to the ratio flowing underneath the car and the higher the speed the bigger this ratio gets.

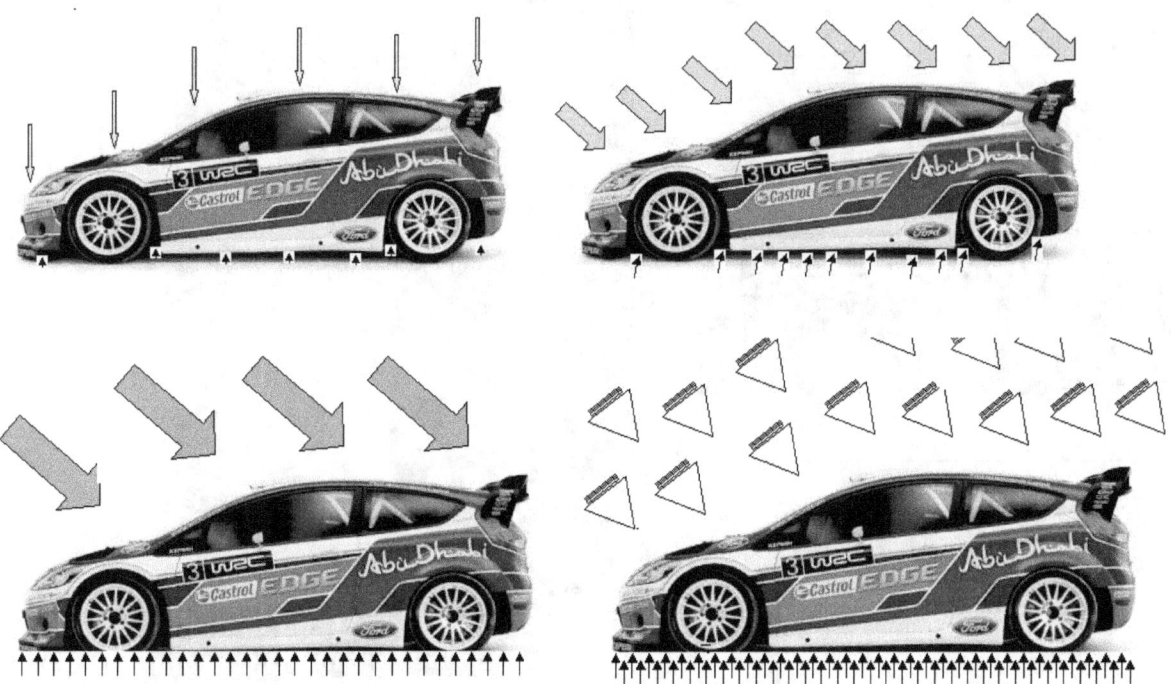

Give this a thought and guard against thought bias because of brainwashing and mind control inflicted on you under the cover of "*learning about science*". There will always be exponentially more air flowing over the car that thrusts the car onto the earth than what there can ever be pushing underneath the car enabling the car to lift into the air. What is even more a daunting fact to keep in mind is that the higher speed that the car is doing the bigger this ratio gets in favour of creating a downward movement in comparison with upward movement. Go sit back and give this a serious thought and think how can the air underneath the car overcome the air pushing the car down. The air pushing the car down will always be completely in a bigger dimension that any air can be when the car is secured to the earth. This way of

thinking clashes completely with all logical thought or intellectual mind processing.

Notwithstanding what the mass is that is required to lift but with enough movement the mass gets lifted and the mass gets air borne. Yet through many decades this got not even one engineer thinking about what applies and what rules should be discarded as rubbish.

The accepted Newtonian idea is that it is the mass of the earth that pulls the mass of the object onto the ground as the mass of the object pulls onto the mass of the earth in return and this pulling of mass results in gravity forming. This leaves us with one problem.

Take a truck of several tons and a dancer of a few kilograms next to a frog of a few grams and drop this lot simultaneously from an aeroplane while applying the same conditions to all three falling objects. It is the same temperature, the same draughts and the same identical heights applying.

Now we will have the mass of each object "pull" by "gravity" as this lot falls down

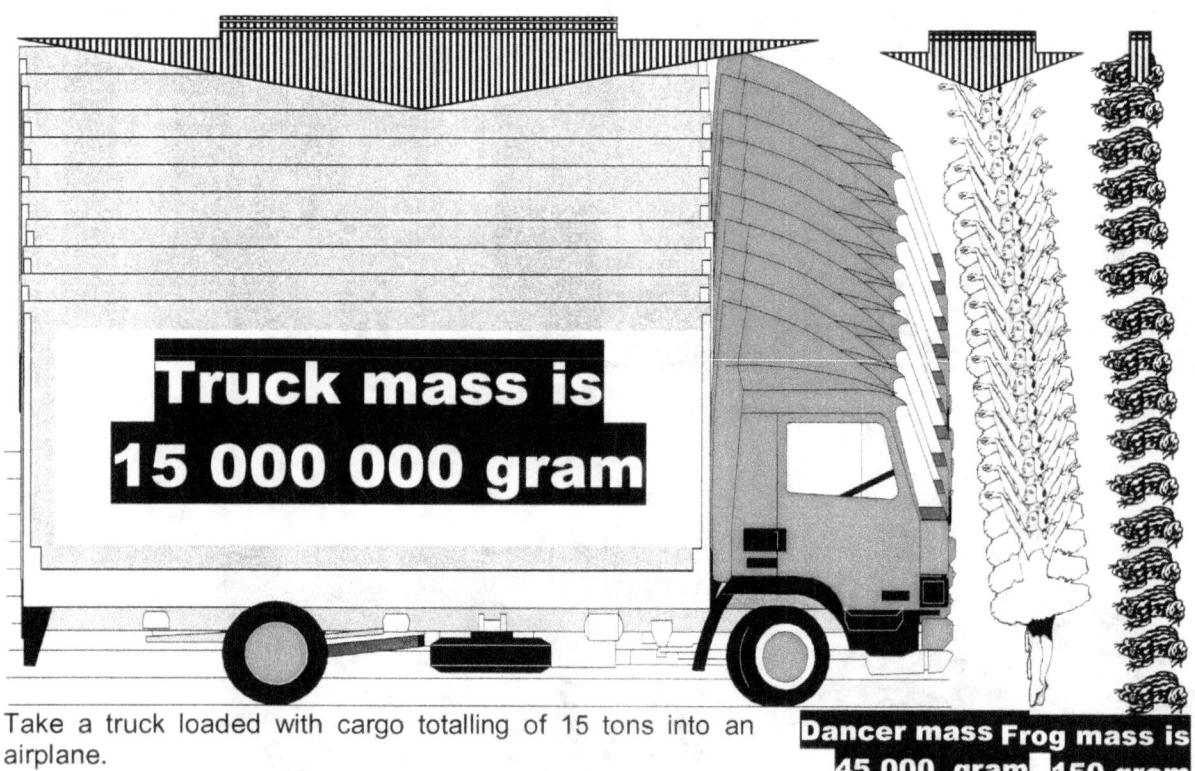

Truck mass is 15 000 000 gram

Take a truck loaded with cargo totalling of 15 tons into an airplane.

Dancer mass 45 000 gram **Frog mass is 150 gram**

Put next to the truck a petite little dancer weighing 45 kilograms.

Then to keep the dancer on her toes, put a frog of 150 g next to her.

We know all three will land at the very same instant notwithstanding mass differentiation. If that is not the case then Galileo is wrong and if this does apply as it does in nature then it is Newton that is wrong.

Newton said mass pulls mass and with the three having completely different mass values, the three still falls equally at the same time. If it is mass that pulls mass the truck must fall years before the frog and months before the dancer given the disparity in mass differentiation between the three. The earth will pull the massive truck much quicker than the small frog if it was mass pulling mass closer. This does not leave us with a problem but this becomes a problem that proves Newton's ideas about mass pulling mass as totally fictional. Now try to sell this idea to a physicist with four doctoral degrees in physics no less and see how you get nowhere in translating this into sensibility on his part. The argument never translates into him accepting this idea although it is truth.

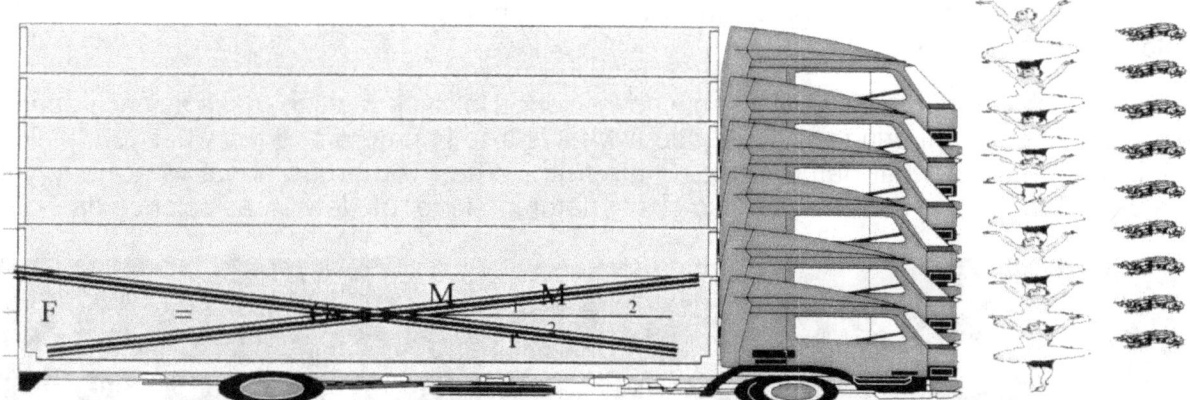

Even where nature proves Newton wrong science proclaims that it is nature that is wrong and it is Newton that is correct. They know in truth that this is what happens and yet they ignore the truth by giving some dimwit idea about a feather floating in vacuum and muddy the water so much that they then can afford to dodge the truth and hide the sensibility behind their incorrectness.

The picture above is representing the biggest normally flying structure on earth and even holding this massive structure it takes off from any surface albeit tarmac, dirt road or even sand. Notwithstanding mass or size or whatever factor that might play a role in the argument, any object takes into the air at a speed between two hundred and thirty to two hundred and sixty kilometres an hour depending on heat. Please keep in mind that there is a huge difference between the idea of getting into the air and staying in the air and that compiling factor is as important as getting airborne. Huge and small including all shapes get airborne and the reasons for that I explain later on.

If you ask what is the difference between how I see gravity works and how Newton's gravity work? Newton says objects pull while I say space compresses thereby collapses by getting reduced and everything in that space condenses. I say space reduces and Newton said objects in space pull each other. While I prove nature, Newtonian science cheats, corrupts and manipulate nature to make science work in ways nature doesn't work.

I say there is no pulling but it is space that compresses by objects producing material movement in rotating of or as gravity. The difference between my approach and Newtonian's is one Universe away from each other. I show a functional Universe and Newton show mysteries of science. I prove everything that Newtonian science this far couldn't. I prove the Universe applies four keys by which gravity works instead of unexplainable magical forces.

I show how nature works with the 4 keys while I show how science falsifies facts to make science seems to work by magical forces pulling. I bring you facts about what is true in nature and not what you think is true or science thinks is true but what nature uses as the truth. When you argue about what I say you argue with nature and what I bring you then you also falsify nature in favor of Newton as science do.

The Titius Bode law is how planets distribute places in the solar system. There is a precise sequence and order in which planets hold places and this annihilates the idea that mass plays any part in this sequence. In contrast the Newtonian system that science promote currently says that gravity "pulls" according to mass. This is fictional. The mass of the planets is totally random and Newton holds no theoretical basis that nature supports. Look at the arrangement and you can visually see using your eyes that the distribution by mass is a hoax and is an invention Newton concocted and science validates for the past 300 years.
By explaining the Titius Bode law for the first time ever this book breaks the 300-year-old myth and brings truth

The Roche limit is the law that says stars do not collide ever in spite of Newton's ridiculous idea that stars can or do collide. If two stars are in each other's atmosphere the law reads that when the minor of the two stars is closer than 2.4674 of the diameter of the major star, then the major star will liquefy the minor star into a gas plume that it then treats as more atmosphere. This is most significant in the cosmic principles I now put forward. There is not one instance or any evidence that shows where two stars do collide. When two or more stars are evenly matched by gravity the two stars do not collide either but becomes binary stars that spin around each other. Stars or planets never collide and when a meteor enters the atmosphere like at Tunguska in Russia in 1908, the earth vaporises and liquefies the meteor and the meteor becomes fragments as well as more dust clouds in the atmosphere. This totally annihilates Newton's idea that a radius between structures diminishes by the gravitational attraction of mass.

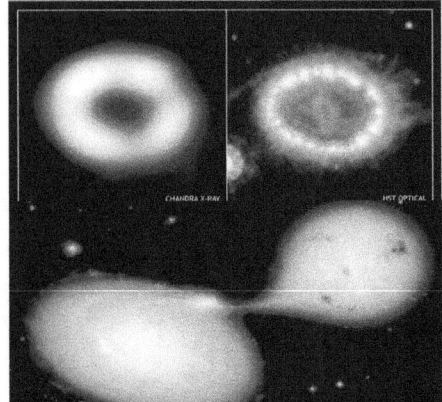

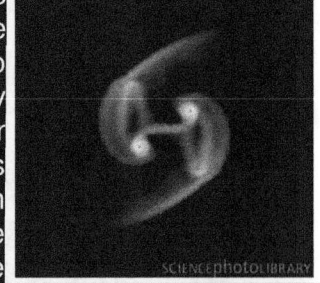

2) The Roche limit

The Roche limit is:
The region surrounding each star in a binary system, within which any material is gravitationally bound to that particular star. The boundary of the Roche lobes is an equipotential surface, and the lobes touch at the inner Lagrangian point, L_1, through which mass transfer may occur if one of the components expands to fill its lobe. It names after the French mathematician Edouard Albert Roche (1820-83).

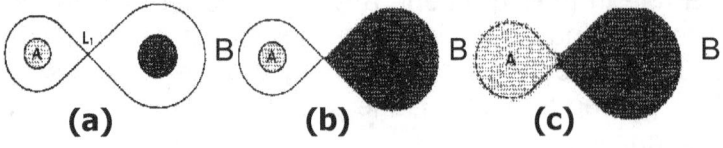

THE ROCHE LOBE: In a binary system, the Roche lobes of components A and B meet at the L_1 Lagrangian point. (a) In a detached system, neither star fills its Roche lobe. (b) In a semidetached system, one massive component, B, fills its Roche lobe. (c) In a contact binary, both components overfill their Roche lobes and share a common envelope.

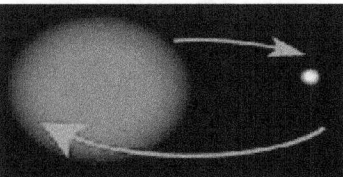 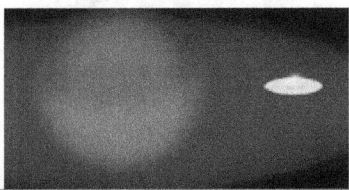

The Roche limit in the practical sense

This book shows what really happens in truth in nature.

The Coanda effect connects with the previous law. Gravity is a ratio that forms between what is solid space and what is liquid space. This applies when the movement of the solid (star or planet or atom) turns and by turning $7°$ it reduces the value of the circle of the space surrounding the solid from $\Pi = 21.991/7$ to the compressed value $\Pi = 3.142/1$ which connects to singularity. There is no pulling of material but only reducing of space when gravity contracts gas into liquid such as what the atmosphere is.

4) The Coanda affect.

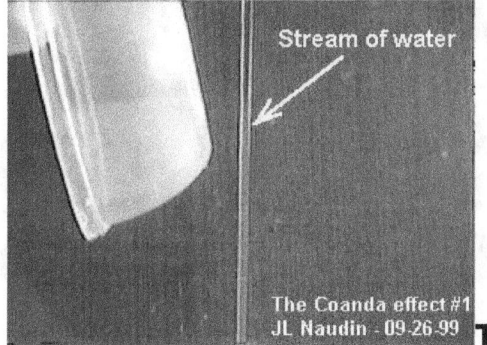

Stream of water

The Coanda effect #1
JL Naudin - 09-26-99

Stream of water

The Coanda effect #2
JL Naudin - 09-26-99

The Coanda effect

The Coanda effect applies as a gravitational phenomenon where moving liquid concentrates around the surface of round solid structures and by movement of either the liquid or the solid or both these concentrates the density of the liquid to gather and compact the flow of the liquid while remaining following the curve of the round surface. The liquid rather follows the curve of the round bowl than to fall straight to the Earth as on should expect. The liquid maintains relevance to the centre of such a round solid. I discard the idea that mass could be responsible for forming gravity because in almost four hundred years all evidence is indicating that the truth is to the contrary.

It is the Coanda effect that forms the Universe. Open your eyes and see what there is to see. The Universe holds liquids around solids by the means of circles and by the measure of gravity and any one can see that gravity is Π. In the Universe we solids, liquids and gas and elements are solids.

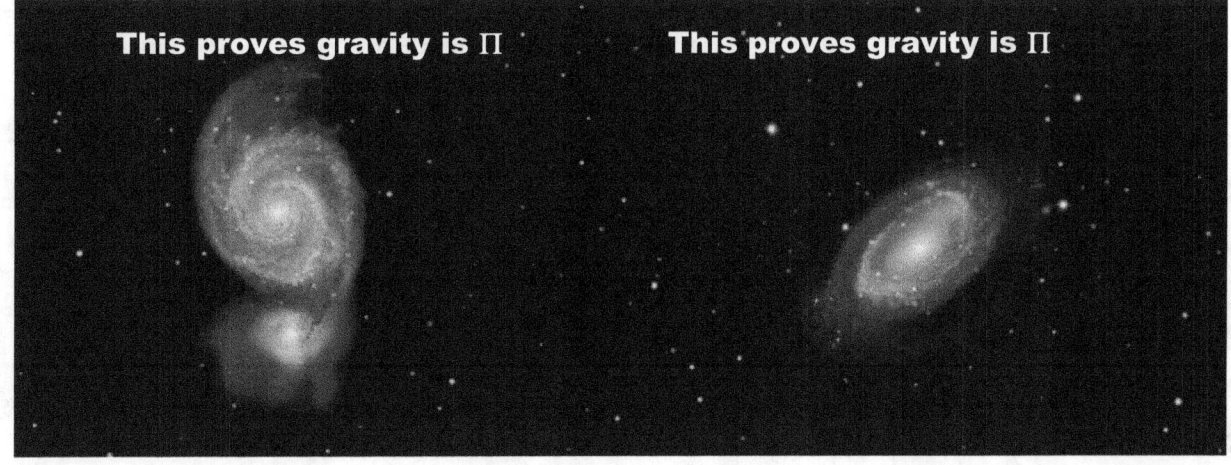

This proves gravity is Π This proves gravity is Π

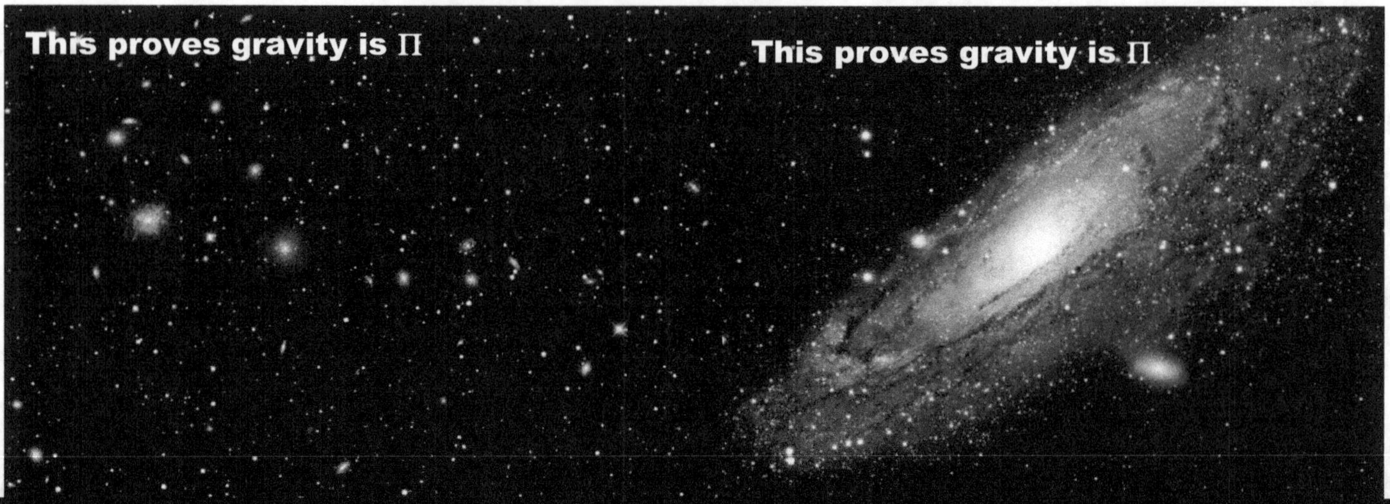

This proves gravity is Π

This proves gravity is Π

This proves gravity is Π

This proves gravity is Π

The article had the purpose whereby I aim to introduce singularity to physics in the way the cosmos uses singularity in the cosmos. Either I fail or everyone else is blind but it is clear I am not projecting my message in a manner that I am reaching the minds of many. If those practising and preaching science would only stop being blinded by Newton's brainwashing then they would see that the entirety in the Universe is formed by circles and circles are formed by Π so gravity must be Π. If you remove Π from the cosmos then the cosmos is still left with Π, as it is Π^0 located in the Black Hole. Singularity connects to Π by a means that could never be parted and the article proves that above all.

3) The Titius Bode law

The Titius Bode Law in table form:

The incorrect application of the Titus Bode law lies in subtracting the figure of 3 from 10 leaving 7. The other way of reasoning is to add four each time to the firs value of three starting with 3 and so on. The true significance of the Titus-Bode law is that it points directly to a circular growth of 7 stages. The 7 relating to 10 is a precise derogative of the Roche limit or the Roche limit is a precise derogative of the Titius Bode principle because he two systems interlink.

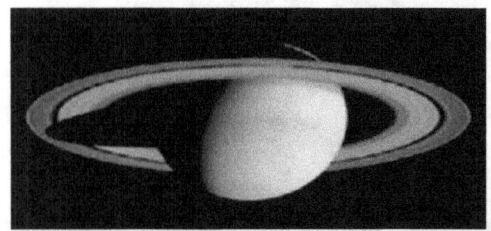

The Lagrangian Points This law proves that science is committed to deception for centuries. It is known that sattelites turn around planets in the same fashion as do planets orbit the sun. Saturn does not "pull" the satelites into its atmosphere or even "pull" the satelight closer and this knowlidge never drove science to question the validity of Newton ot his concept about "mass" pulling "mass". Even the rings formes as debris spins in a circle that always remain constant and science should have rejected newton'ss ideas on the grounds that nature never not once supports Newton and Newtonia science in any way thinkeble.

1) The Lagrangian system

LAGRANGIAN POINT:

LAGRANGIAN POINT:
*The Lagrangian points
are five equilibrium points
in the orbit of one body
around another, such
as a planet around the Sun*

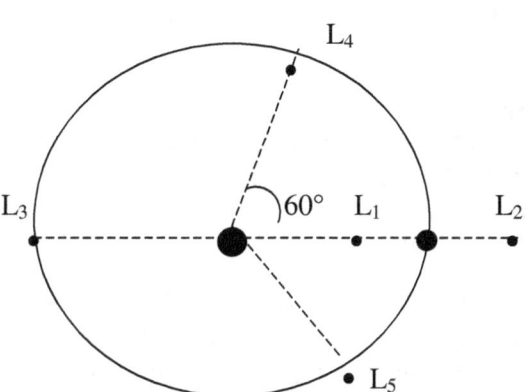

A Lagrange Point is mathematical name for an area in an orbital system between 2 bodies where another orbiting object can seemingly hold a steady position, relative to the two other bodies. In this "magic area," the combined gravity of the two larger objects provides just the right amount of force to 'cancel out' the centripetal force required to have a third object rotate "with" them.

In this sense, the Langrage point is analogous to a geosynchronous orbit, whereas the third object is "fixed" with respect to the orbits of the two larger objects. Barring other physics phenomena; there are always 5 Lagrange points in every 2 body orbital system. Three provide largely unstable fixed orbits, white the remaining 2 allow for more general, seemingly stationery fixed orbits. The 3 unstable points are often referred to as L1, L2, and L3. The 2 remaining stable points form the apex of an equilateral triangle with the large bodies as vertices. The differing in stabilities is due to the gravitational strengths of the two large objects creating the Lagrange Point. The Lagrange point can be likened to a ball at the top of a hill: if the gravity is slightly greater in one direction, the ball will roll off the top of the hill in that direction. But if the ball is in just the right spot, it won't roll in either direction.

NASA scientists take advantage of the L1 Lagrange point in the Earth-Sun orbital system to fix the SOHO satellite in position to observe the Sun. Lagrange points are ideal for satellites that require a fixed position relative to orbiting bodies, in order to make observations and calculations. L4 and L5 are areas where asteroids and other small bodies often accumulate, due to the long-term stability that is free from the Sun and Earth's gravity.

I show how Newtonians fabricate Newton's ideas about gravity. This is ongoing since the end of the dark ages and Newton. There is no mass that can pull. Most people reading this and who are schooled in physics never heard of the Roche limit, the Lagrangian points, the Titius Bode law and the Coanda effect and these principles are what builds the Universe while I am going to show that there is no factor such as mass. While it serves their purpose notwithstanding never finding evidence to the fact, still science uses *only* and *exclusively* Newton's idea of mass while the principles in place the Roche limit, the Lagrangian points, the Titius Bode law and the Coanda effect are never ever mentioned. They sometimes put referring to these principles as law in brackets to deny the status that any of the above law have. I am going to show you within the next few pages the silliness Newtonian principles hold. While I discuss the principles please see where I am incorrect or going wrong and convince you whom is wrong.

This is because the Roche limit, the Lagrangian points, the Titius Bode law and the Coanda effect disputes Newton and science would rather discard what the Universe uses than to put a question mark behind the fabrication Newton put in place. Where everyone knows the fabricated information and hiding the reality, which is in place within the cosmos, and that is the conspiracy I show to all. Science stupidity ensures they don't understand the working principles that are in place and that was known for centuries in some cases as the Roche limit, the Lagrangian points, the Titius Bode law and the Coanda effect and therefore not knowing how the principles should be interpreted they hide the concept due to not want to be seen as the ignorant fools knowing the cosmos implements the principles as reality. Science hides their limitations and incompetence behind providing the public selective of information. Take for instance the edge of the Universe they talk so much about. Reading the above will not impress you because your response will be: "So what, those laws mentioned are far away and in outer space and we are on earth". Well I put it to you with the following example.

There is no edge of the Universe because there is only an unlimited everlasting Universe out there. What the limits are that they see as the edge of the Universe is the limitation of their equipment that can't trace time back beyond what they see and that serves as their limit in understanding what the Universe offers and how the Universe unfolds.

What the article introduces is that singularity puts a value of $\Pi°\Pi$ to gravity where Π secures all the roundness we find gravity has. Look at the Universe in this way: Everything that one see within the Universe is round and everything that is round goes by circles and every circle holds Π as a value.

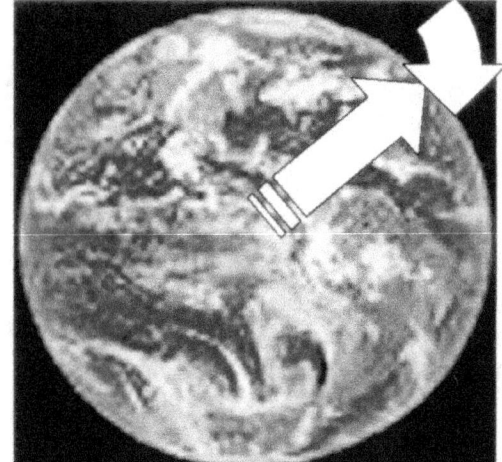

Therefore gravity has nothing to do with mass but connects totally to the value of Π. During the studies I conducted I found how to value gravity as Π by applying the law of Pythagoras. I now am able to show mathematically by applying the law of Pythagoras how gravity forms by forming a circle using Π because I have located the centre of the Universe.

In other words I say by the earth turning this turning motion of the earth compresses space and the space we call the atmosphere. As this atmosphere density increases the space it hold decreases and this motion contracts movement towards the centre of the earth. It is not material alone that falls but it is all space surrounding a cosmic structure that condenses including the space filled with material.

In the centre of every spinning object, which therefore includes the earth as also all atoms farming the earth, forms a point that has no space. The object has to spin in order to establish the centre line that is void of space. If one reduces the circle value from Πr^2 and take away **_all_** space by reducing the radius to r^0 we have left only form Π, which then is a circle. By reducing the radius to r^0 the radius then becomes singularity. Then the value of the circle is Πr^0 and only form is left. Please note that all objects being part of nature is a round circle that spins and this spinning action is crucial in forming singularity. That brings about the smallest space there ever could be because at Πr^0 is where all space starts and this then is where the Universe starts. By reducing the circle even further we have to reduce the value of Πr^0 to $\Pi^0 r^0$. With a space value of $\Pi^0 r^0$ the numerical value is 1 x 1 = 1. Having a value of 1 means we have located singularity because singularity presents a value of no more than 1 and $\Pi^0 r^0$ is 1. Because every atom and every sub-atomic particle also holds singularity at a value of $\Pi^0 r^0 = 1$ this line in singularity forms as a straight line all to the value of 1. This brings about a line running from $\Pi^0 r^0$ all through singularity forming Π^0 to a point where the curve of the earth forms Π. Since this line has Π spinning by $7°$ the moving value of Π goes square in movement $\Pi \times \Pi = \Pi^2$. This fact brings the equal value of singularity in the centre of the earth from $\Pi^0 r^0 = 1$ to where the earth meets air or liquid having a value of movement at $7x\Pi$ x (Π x Π =) Π^2 or $7\Pi\Pi^2$. This then forms the value of singularity that is forming gravity being $7\Pi\Pi^2$. Because space includes the measure of time where one cubic meter is per second or degrees the value of $7\Pi\Pi^2$ also refers to space (meters) time (per second or degrees) and this then becomes $7\Pi\Pi^2 = 217$ km / hour. Once an object gets airborne this value of $7\Pi\Pi^2$ changes too $73\Pi^2$ for reasons I explain elsewhere.

Gravity forms by the dual directional movement of objects going according to the speed of movement. An astronomer will hang suspended in space and high above the earth if the circular movement is high enough to keep him there. If he spins slower he will start to drop. It is not gravitons grabbing him by pulling him closer to the earth. If his circular momentum can sustain the rotation his linear distance will maintain his orbit but as soon as the circular rotation become insufficient to maintain the orbit the linear distance will reduce and then the "gravitons" are called into action. How ridiculous can they get!

Gravity is the sound barrier because the sound barrier forms by four cosmic principles that form gravity and therefore gravity and the sound barrier are the same principles in conflict because of movement differences. The sound barrier applies by movement in space in space shared.

Everything I am about to show Newtonians say is nothing new. Newtonians say they are aware of everything that I show and then they turn around and simplify physics by putting the entirety of their physics down to the pulling power of mass that forms gravity. The sound barrier is gravity and to understand gravity is to understand the sound barrier. They say that I bring nothing new to the table and they say they apply every aspect I indicate that forms gravity. When I ask them to explain the sound barrier they use the Mach principle and Doppler's effect, which both dates from a time when no one was aware anything man-made could fly let alone go faster than sound can. Doppler made his contribution to science at a time when the train he measured was slower than a horse. The statistics he left to science applies to going as fast as a man on a horse could ride. That is a far cry from a jet breaking the cosmic boom. They can claim what they like because they never have to prove anything and they never have to listen since they know everything while their shortcomings makes science more the jesting of a buffoon. That annoys me to my guts because all of that comes down to the conspiracy whereby they whitewash their stupidity from the brainwashing they inherited from their predecessors and pass the brainwashing on to their students. Their ignominious indifference makes me want to shout to the mountains in agony and unbelievable frustration.

In the earth's atmosphere no object can move slower than the earth does. If it moves slower than the earth does, it moves as fast as the earth does by receiving mass and being pinned onto the earth as being part of the earth. There is always movement because of the earth moves and all other movement goes in anticipation of moving above what the earth does. Movement is space that is duplicating filled space at a pace and in relation to other space within other unfilled space. That is why using parachutes slows the falling process of falling objects down because falling is a ratio between solids and liquids. This is proven in these photos about the sound barrier where the movement of the solid jet contracts the space of the liquid cloud. What happens in the photo is vivid proof of my theory and it proves the Coanda effect is what is forming gravity. The Coanda effect is a relation between material spinning and air or liquid compressing and that is what happens between the Earth and the atmosphere. This picture shows objects are always moving extraordinarily as they move in relation to the earth's atmosphere that always moves in relation to the earth's gravity compressing space into forming the atmosphere. But it also shows the movement of the aircraft compressing space in relation to the aircraft moving as well. That shows that all movement is gravity or time related. This shows objects moving is extraordinarily because everyone always forgets about that it is the earth that normally applies all of the movement while all else stands still in relevancy. There is always movement because of the earth moving and when anything moves above and beyond the movement the earth provides that then forms the <mark>sound barrier</mark> using a modified version of the <mark>Titius Bode law</mark>

I have explained these 4 laws and why only these 4 laws apply.

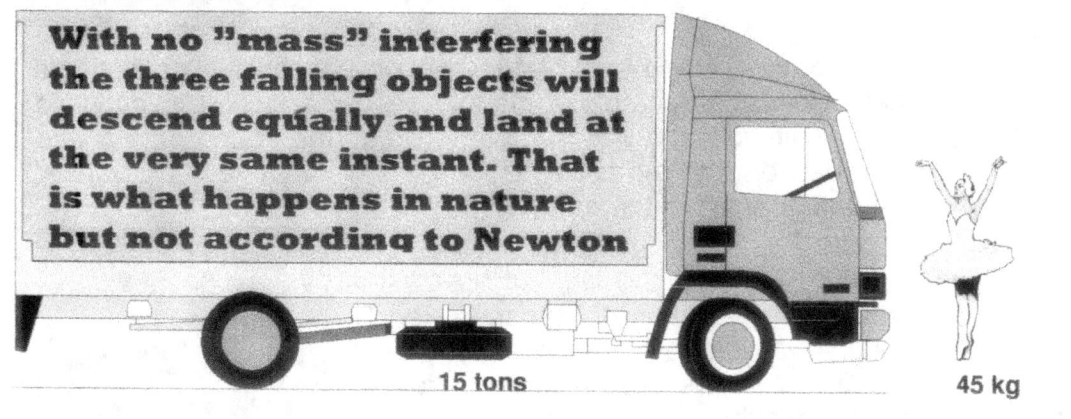

With no "mass" interfering the three falling objects will descend equally and land at the very same instant. That is what happens in nature but not according to Newton

15 tons   45 kg   150 g

All things fall equal. This is what Galileo said. Notwithstanding mass, all object under the precise same conditions such as heat, wind, height and so on will fall through the sky at precise equal speed. …And that evidence we see on TV daily.

This is so clear as daylight and it is so blatantly obvious that it cannot be mass that produces the gravity and still every Physics academic that I presented my work to in the past simply ignores these facts where some even said in writing that my arguments on this matter is incoherent. The answer is simple. If it is mass that does the pulling then the more mass must produce more pulling and a little mass must produce much less pulling. This then has to translate to massive objects falling faster and little mass falling slow. We know that this is not the position but everything falls equal under the same conditions.

Every layer formed as the atmosphere of the earth holds a relevancy inclining towards singularity.

To the above the gravity forming Π holds a value of $\frac{21.991}{7}$ but as the earth turns, in the instant that the earth turns by turning in that instant of time seven becomes singular and in concentrating space it then divides space to the value of 21.991 to form $\frac{3.1412}{1}$.

Then at the bottom gravity is $\frac{3.1412}{1}$ and at the top gravity is $\frac{21.991}{7}$ and that is the relevancy called gravity. There is no pulling. There is no graviton. There are no magical forces of any sorts as science wishes to use.

The entire idea of the four forces comes from the spinning of the earth or the atom or the electron or the proton spinning and in every instant the dynamic changes and that forms a new criteria what then becomes the intensity re-applying. However as big as you wish to take the concept or as small as you want to go it is a changing of Π where Π by spinning revalue $\frac{21.991}{7}$ to interact with singularity to then form $\frac{3.1412}{1}$. That is gravity and now it is no mystery but just part of God's Creation explained. It becomes nature without mass.

I also have already explained why all these gravitational tendencies apply. Because of the four cosmic pillars we can fly, we can fly at altitude and we are able to break the sound barrier and break free from the earth gravitational control that keeps us under arrest and chained to the earth.

The value of Π forming is what drives the Universe the "ability" to "expand". By Π forming a value above the axis value of 3 this addition shows growth coming from a future in time where it first forms .991 which is less than one in a form where one is the epitome of what can be in the entirety we call the Universe. By coming from .991 going to 1 and then when spinning through the double square of seven becomes ten on both sides of the divide we see how time in growth drives space grow from .991 to 1 to ten on either side of the seven square that spins.

In these pictures it shows clearly that movement of a body compacts the liquid (atmosphere) surrounding the moving body to the point (in these cases) where it con tracts the moisture so much the moisture in the air becomes a cloud.

I explained why gravity takes this form and how the laws come into play.
Mass has no function in gravity applying less the object connects to the earth. Gravity applies by placing

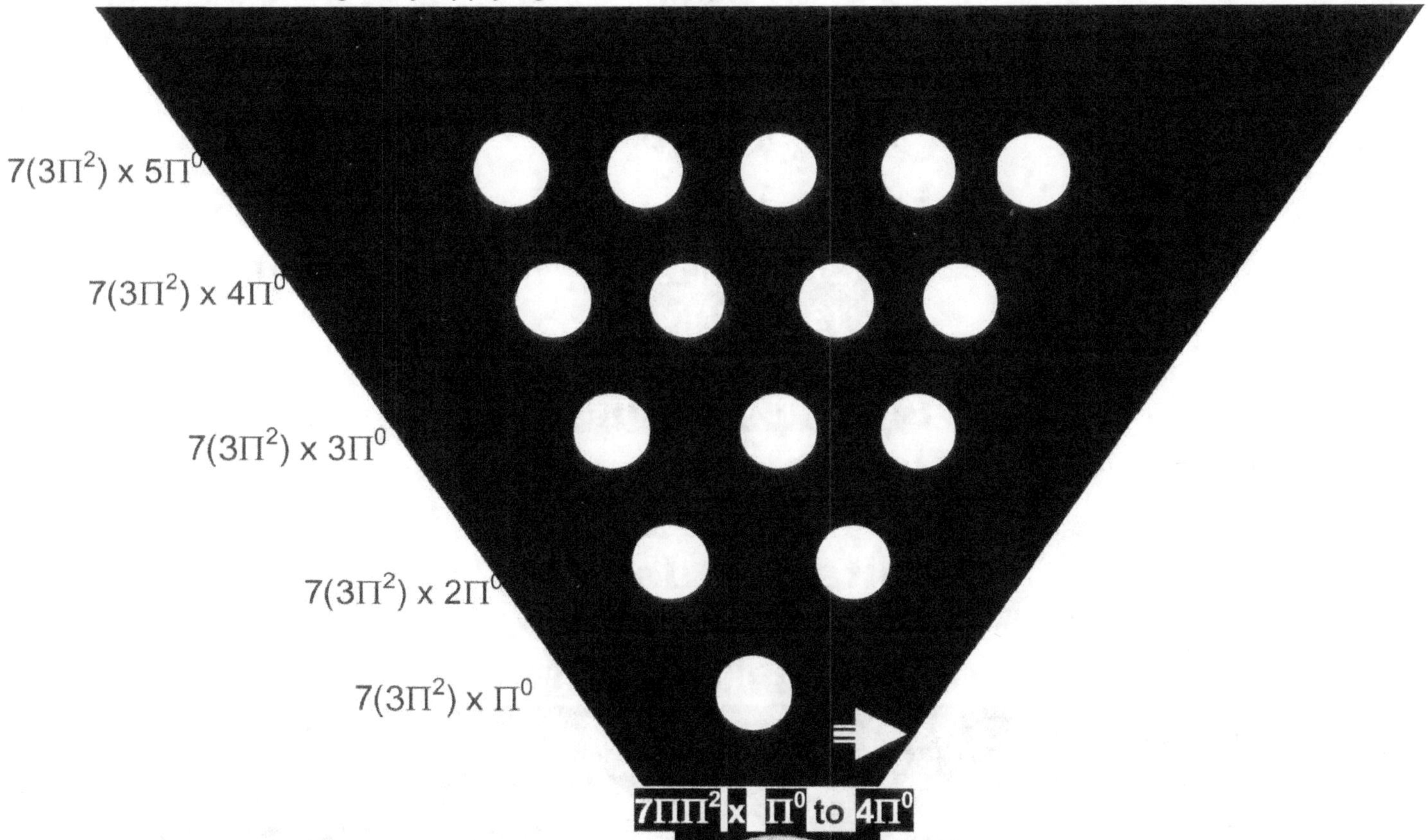

$7(3\Pi^2) \times 5\Pi^0$

$7(3\Pi^2) \times 4\Pi^0$

$7(3\Pi^2) \times 3\Pi^0$

$7(3\Pi^2) \times 2\Pi^0$

$7(3\Pi^2) \times \Pi^0$

$7\Pi\Pi^2 \times \Pi^0$ to $4\Pi^0$

relevancy in movement between what is cosmic liquid and what are cosmic solids. There are two forms holding space where one is frozen heat, which is what we call solids, and the other is liquid or gas heat, which we call outer space or light. All solids notwithstanding can become liquids and also gas depending on the heat mixture between solid space and heat blending to produce the mixture. Outer space is the hottest place in the Universe since outer space expands to the very limit. The sun and other stars produce cold by movement that condenses space and the flames we associate with heat is liquid heat escaping the bitter cold of the sun. The sun is not a giant steam pressure cooker like they thought in Newton's time but it is one giant air conditioner. The sun works with gas and not flames boiling water to steam, as a steam train would do. It works with gas pumping heat as it removes compressed overheating space form the inside of the sun. Gravity works by compressing heat or expanding heat and containing heat by movement in very specific density compartments. All these sectors in gravitational division depends on heat levels producing density within the space and that density sets requirements for heat constricted by movement of objects within that specific space boundary. In that light I set out to name the boundaries of density forming layers.

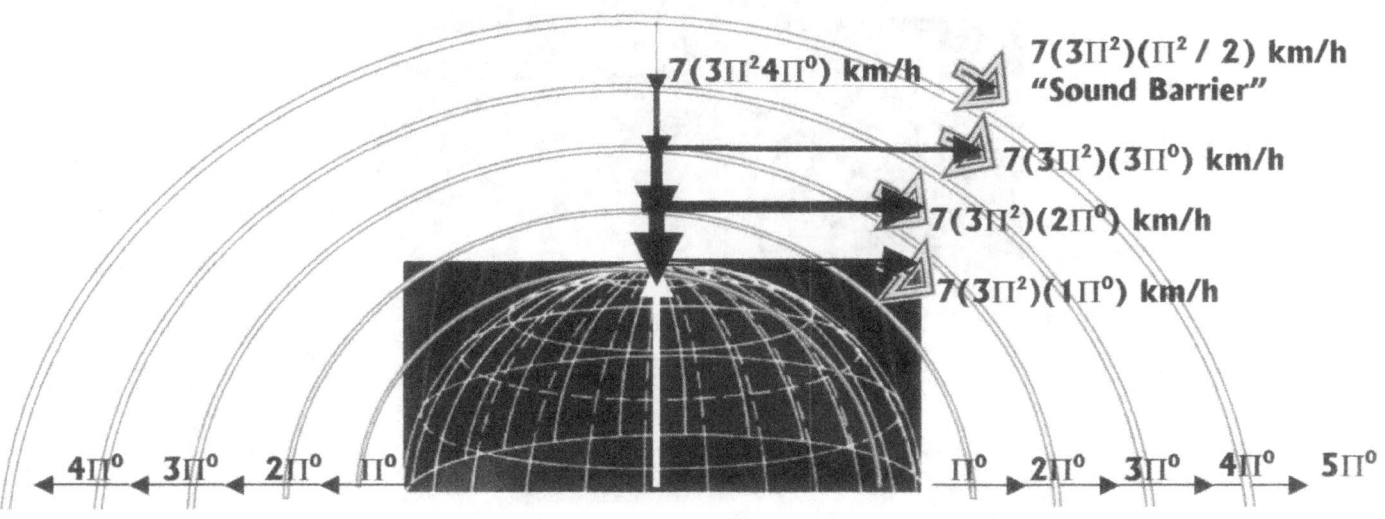

$7(3\Pi^2 4\Pi^0)$ km/h

$7(3\Pi^2)(\Pi^2 / 2)$ km/h "Sound Barrier"

$7(3\Pi^2)(3\Pi^0)$ km/h

$7(3\Pi^2)(2\Pi^0)$ km/h

$7(3\Pi^2)(1\Pi^0)$ km/h

$4\Pi^0$ $3\Pi^0$ $2\Pi^0$ Π^0 Π^0 $2\Pi^0$ $3\Pi^0$ $4\Pi^0$ $5\Pi^0$

Specific density
This is $7(3\Pi^2)$ km / h

Static density
This is $7(\Pi\Pi^2)$ km / h

Virtual density
$7^0(3\Pi^0)(\Pi^2)(1.1\Pi^0)$
$7^0(3\Pi^0)(\Pi^2)(4\Pi^0)$
$7^0(3\Pi^0)(\Pi^2)(\Pi^2/2)$
$7^0(3\Pi^0)(\Pi^2)(\Pi^2/2)(\Pi^2/2)$

Variable density
From $(1.1\Pi^0)$ to $(5\Pi^0)$

Limitation density
$\Pi(7^0(3\Pi^0)(\Pi^2)(\Pi^2/2))$
$7^0(3\Pi^0)(\Pi^2)(\Pi^2/2)(\Pi^2/4)$

At $(\Pi^2/2)$ $(\Pi^2/4)$ this is $7(3\Pi^2)$
At $(\Pi^2/2)$ this is $7(3\Pi^2)$
At $(4\Pi^0)$ this is $7(3\Pi^2)$ km / h Virtual density
At $(3\Pi^0)$ this is $7(3\Pi^2)$ km / h Virtual density
At $(2\Pi^0)$ this is $7(3\Pi^2)$ km / h Virtual density
This is $7(3\Pi^2)$ km / h Specific density
This is $7(\Pi\Pi^2)$ km / h Static density

$7(3\Pi^2)7(\Pi^0)(\Pi^2/2)$

$7(3\Pi^2)7(4\Pi^0)$

$7(3\Pi^2)7(3\Pi^0)$

$7(3\Pi^2)7(2\Pi^0)$

$7(3\Pi^2)$ km / h

$7(\Pi\Pi^2)$ km / h

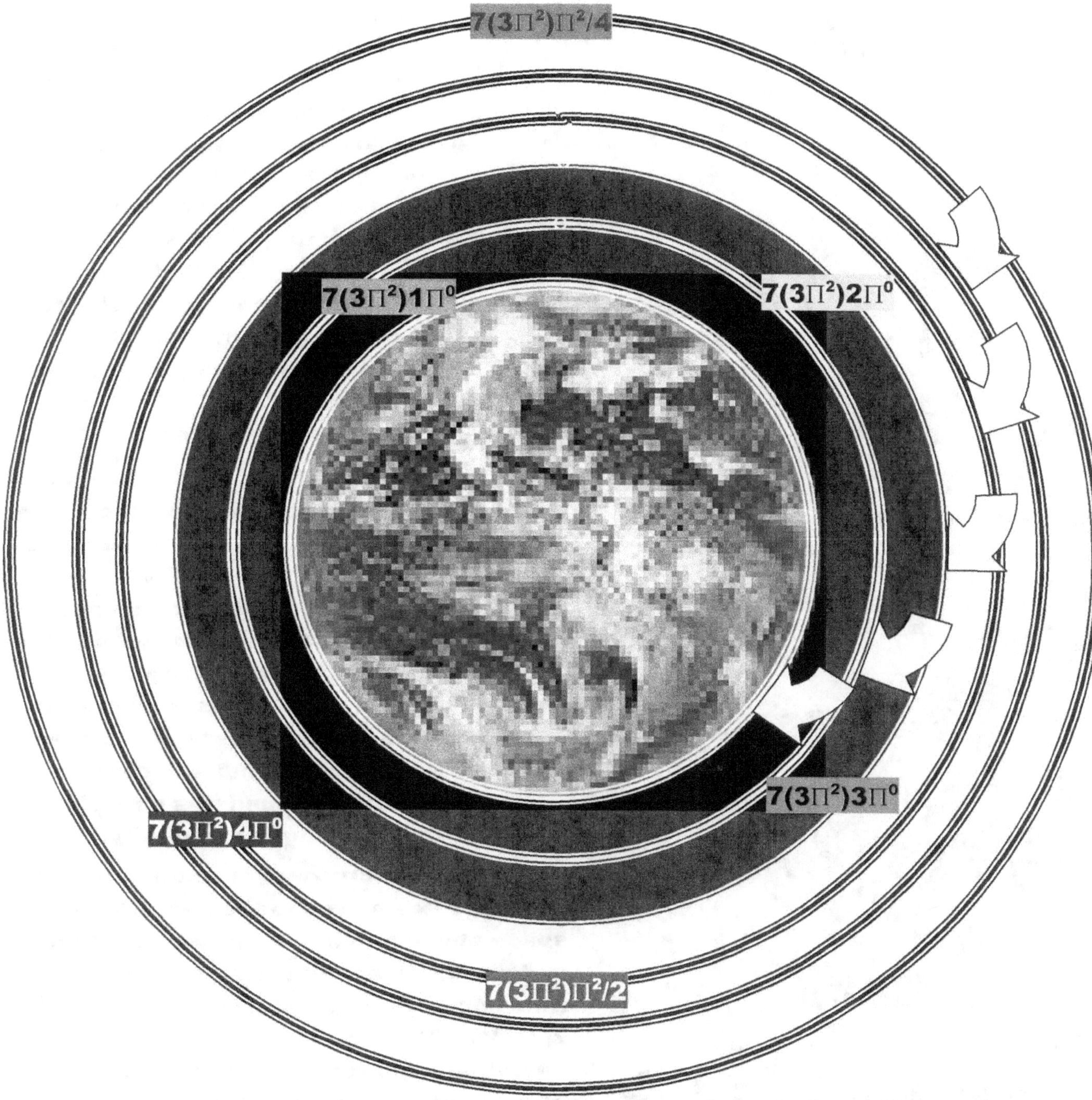

$7(3\Pi^2)\Pi^2/4$

$7(3\Pi^2)1\Pi^0$ $7(3\Pi^2)2\Pi^0$

$7(3\Pi^2)3\Pi^0$

$7(3\Pi^2)4\Pi^0$

$7(3\Pi^2)\Pi^2/2$

With all these numbers and relevancies I prove that there is a distinct and undeniable relevancy between movement and air in which the object moves. This forms as a result of the culmination between all four laws that forms gravity. The faster the object moves the more does the dynamics of the moving body change and this is due to the circumstances in the liquid (air) that allows different conditions to apply different relevancies and these relevancies change as movement increases.

As the speed increases the air to solid ratio changes because the solid body comes into contact with more air, which allows the air to stretch the body that moves. The air to body ratio increases and such increases demands a higher ratio of air to pass over the body structure. Therefore in terms of the body the movement forces the ratio of the body versus the air to compensate and the body becomes Longer as a result of more air flowing.

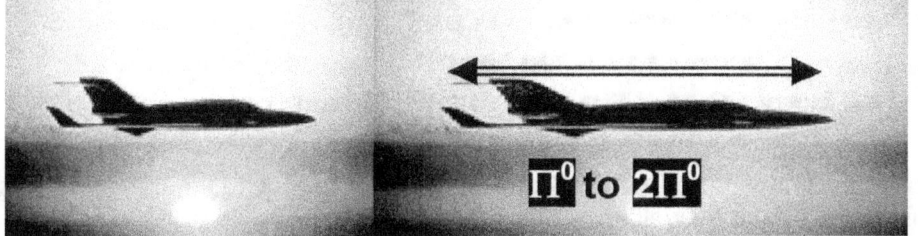

Π^0 to $2\Pi^0$

At $7(3\Pi^3)2\Pi^0$

At $7(3\Pi^2)3\Pi^0$

At $7(3\Pi^2)4\Pi^0$

At $7(3\Pi^2)\Pi^2/2$

At $7(\Pi\Pi^2)\Pi^0$ the designated space in relation with the aircraft will not be sufficient to lift the aircraf of the ground. It would then require huge wings to increase the space the airplane holds in relation to the space the airplane requires to fly. The speed must increase to $7(3\Pi^2)1.25\Pi^0$ at least to become airborne and lift from the earth. The moment the aircraft is not touching the groun it loses the "mass" component science holds so dear.

At a speed ratio of $7(\Pi\Pi^2)1.25\Pi^0$ the plain exceeds the size to space ratio which the plain requires to remain at a level of $7(\Pi\Pi^2)1\Pi^0$ and it will according to the increase in space ratio lift from the ground and into the air. This is when the aircraft gets airborne. However this is where the craft loses the "mass" it had where "mass" is just an indication that by touching the ground it is part of the earth according to singularity.

The aircraft "increases" in space required and therefore moves to the next level of space above ground. Being above ground and not touching the earth changes the relevancy in movement from $7(\Pi\Pi^2)1.25\Pi^0$ to $7(3\Pi^2)1.25\Pi^0$ which places the aircraft in the liquid zone and forms no longer part of the earth's solid factor. Now the aircraft becomes space, which moves towards the earth. This linear movement has exceeded the downward gravity of $7(\Pi\Pi^2)$ or the craft will fall to the earth. It is the need for space that increases. Depending how you look at it the space the movement goes through increases and the size of the plain increases or the space remains the same and then the plain shrinks in size. This relevancy between that which moves and that which it moves through becomes so big that the space increases at low levels that it "breaks" the space or change the space to being so big sound does not exist

Also I explained how every stage of flight occurs and what effect the Titius Bode law, the Roche limit, the Coanda effect and the Lagrangian points play in this event above as well as al other forms of gravity.

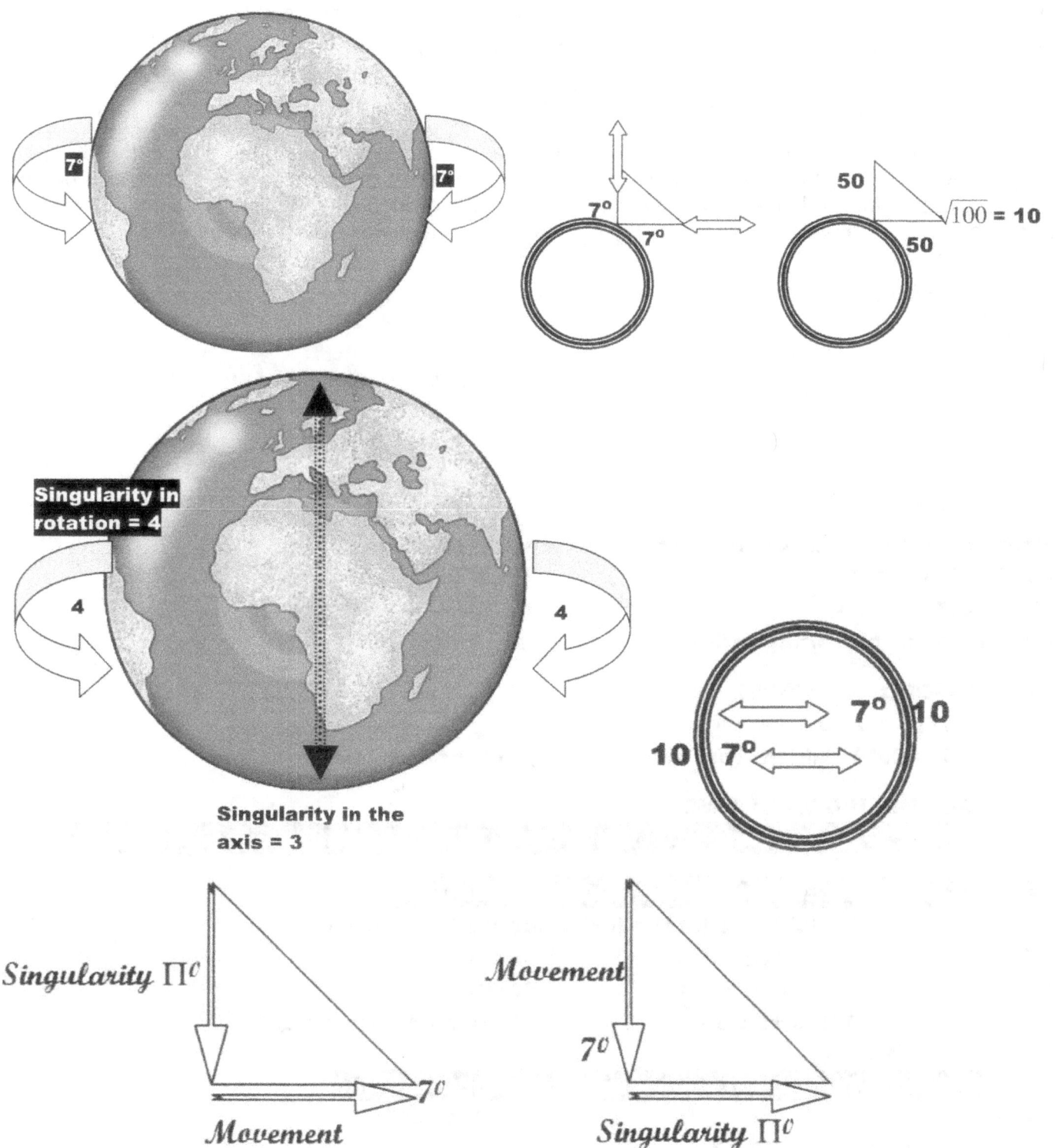

We find the earth or any or any other rotating object turns by $7°$. However as ever this observation is not quite that simple. As Kepler proved gravity is $a^3 = T^2 k$ and that says $T^2 = \dfrac{a^3}{k}$, $k = \dfrac{a^3}{T^2}$ $k^0 = \dfrac{a^3}{kT^2}$, which proves where to find singularity and the reason why singularity controls the Universe. If $a^3 = T^2 k$

Then movement has to be in $T^2 = \dfrac{a^3}{k}$ as well as $k = \dfrac{a^3}{T^2}$, a fact that science ignores in favour of Newton's idea of $a^3 = T^2$. This means that as much as it turns by $7°$ it circles forward by $7°$. That gives a triangle in singularity where in singularity the straight line is equal to the half circle that is equal to the triangle. Therefore the triangular movement converts that same triangle into two triangles all working with the law of Pythagoras. In the one triangle the movement runs down and the other the movement is sideways. Therefore the triangle holds ($7^2 + 1^2 = 50$) moving in one direction and ($7^2 + 1^2 = 50$) moving in a second direction. That leaves the total of movement in singularity ($50 + 50 = \sqrt{100}$ **= 10)**

This is how Π forms an expanded value $\Pi = \dfrac{21.991}{7}$ and the compresses value is $\Pi = \dfrac{3.1416}{\Pi^0}$.

Gravity forms by singularity forming a positional dimension in relation to every other point in the Universe holding a position. The time is the change in position of every point that relocates to a new point putting the entire Universe in a new relation to every other point holding singularity. Whilst the object turns a new position co-ordinates every point in singularity in relation to what it was, what it is and what it is going to be.

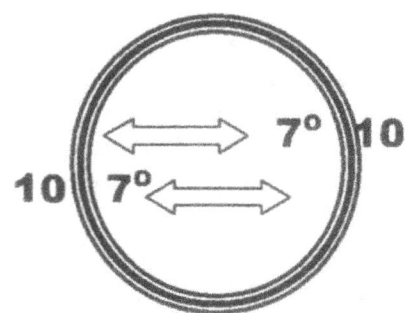 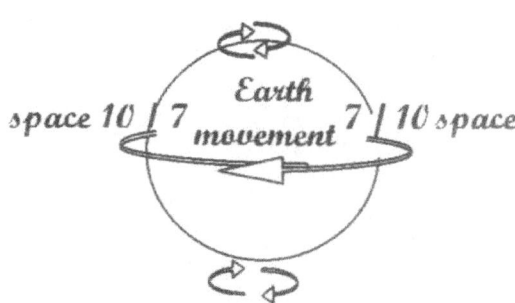

Matter in relation (part of) to the total dimension of space.
(10 / 7) \ (7/ 10) = 2.04
1.4285 / 0.7 = 2.04
Taking from both orbiting influences

SPACE DIVIDED INTO TIME
(7/10) / (10/7) = 0.49
.7 / 1.4285 = 0.49 Taking from both orbiting influences

SPACE MULTIPLIED WITH TIME
7/10 / 7/10 = 1 and   10 / 7 X 7/10 =1 Therefore not influencing change

THE PROCESS PARTED USING THE ROCHE PRINCIPLE

10 / 7	$(\Pi/2)^2$ The Roche influence on Titius Bode	
7/10	2.04 x $(\Pi/2)^2$ =	5.033
$(\Pi/2)^2$	2.04 x $(\Pi/2)^2$ =	5.033
10 / 7	5.033 +5.033 =	10.066 from both objects

           7/10
7/10 / 10 / 7= 0.49 + 0.49   on both sides of the divide

   10 / 7   **0.49 X 2 =.98**

     10 / 7   10 / 7
     7/10 =.49   7/10 = .49
          .49   +   .49   = .98
     .98 X 10.066 = 9.86 =Π^2 **TIME SPACE = Π^2= 9.86 TIME**

This calculation mathematically confirms how gravity forms by the interaction of material (7) and space (10)
Remember I try to facilitate as much information in as little space in this introduction and it might come across as compact but with 500 or 750 pages everything becomes clear.
This book and all my other books introduce a new concept about science. Science at present is flawed and before throwing this introduction aside in disgust then first go and look at the planets varying in size. The distribution of mass is completely random and there is no mathematical or theoretical or any other complying order to the way planets arrange by mass. Therefore it is madness to promote any idea that

mass can place planets in located positions. Nature and Newton is not compatible even in the least. We have to distrust science or we have to distrust nature. Put in the correct sense we have to trust nature or we have to trust Newton because what Newton says is incompatible with nature and this is true above and beyond the brainwashing mainstream science inflict on us all. You now can for the first time in human history choose to read the truth for the first time or you can remain brainwashed forever believing Newton. I deciphered the following cosmic laws that totally annihilated Newtonian wisdom.

These laws are what nature put in place instead of mass and moreover being in place of what Newtonian fantasy makes every person believe is science. Using these laws enabled me to achieve what science considers is impossible. I use these principles to prove that what the Bible says happened is exactly how it started…and I have five books written to prove that. This means I had achieve what no one in science this far could or would or even tried to achieve in science. By writing these books I had to break a code of silence. Before I prove how the cosmos started I first have to prove how nature forms the Universe. This is how nature forms the Universe and these laws form the Universe. From deciphering how these laws apply I could decipher how the Universe started and guess what; I then managed to use mathematics to support what the Bible says how this Universe came about.

This is the Titius Bode law

The Titius Bode law proves that mass has no place in science. See in the picture how random mass is and with such randomness, how can mass place planets in the positions they hold? By my effort to solve the mystery of the Titius Bode Law, I prove that gravity forms not by mass but gravity forms by π forming in movement π². Solving the Titius Bode Law and proving from that how gravity works opens up a new view on the cosmos.

From mathematically explaining how the planet formation forms I could gather the way the cosmos formed right in the very beginning when the Bible said it started the way the Bible said it started. Mathematically proven it is precisely as the Bible says the cosmos started.

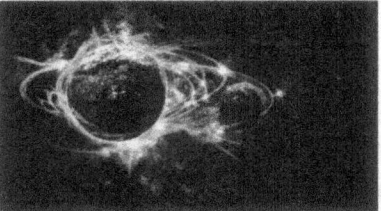

This is The Roche limit

The Roche limit has been around for centuries and with all the mathematical splendour available to apply in order to fathom concepts behind this phenomenon, still with all the computing ability of a machine all those physicists with all the mathematical superiority could not touch any understanding about the concept forming the background. Yet when using the truth about gravity in physics the answer is simple; it is that gravity is Π.

This is the Lagrangian points

The Lagrangian points have been known to science for centuries and with all the mathematical splendour available not one calculation could ever explain why this event is taking place. The satellites form precise locations positioned around the major planet and never comes closer while remaining in their positions.

I introduce these four laws as I introduce you to science in the manner as nature forms the Universe. These four laws shows how and even why the Universe started because these four laws are the cosmos and the way the cosmos started. There is no Newtonian trickery. The four cosmic laws hold the basis of what forms the Universe where every one of these laws form part of what forms Π and together these laws form as a group the value of $Π^2$ and $Π^2$ is gravity. That I prove mathematically. The Universe forms the value of 1. No more, no less because that is the value of singularity. In the beginning there was possibilities before material formed groups that formed material in groups that formed groups of arterial, which we then gave individual names. But before that there were spots and dots as there now are spots and dots forming a Universe. To understand how this lot formed we have to go back and find out how the spots and dots formed before the Universe formed our Universe we give so many names to. Essentially only singularity forms a Universe so small it is outside the Universe we recognise.

This is the Coanda effect

The Coanda effect has powered turbine engines and aeroplanes in flight for almost a century and with all the mathematical splendour available to design the most terrific aircraft, not one engineer could mathematically compute one fact to show understanding why this takes place. How sad it is that those claiming of much superior intellect in physics remain just no more than having computing power. The understanding is not complex. I have to warn the readers that the topics are showing a very new approach with no quick answers. Understanding is in the proof and that does not come by reading just a few lines and then forming conclusions. The information is new but not hard to grasp. I did not put these phenomena in place and these phenomena nullifies Newton's correctness and the proof I bring goes beyond any doubt. I prove the Titius Bode law. Go to the internet and see how science doubt the Titius Bode Law and the correctness thereof while to solve the problem you add 3 plus 4 to get 7. That is if you want to find a solution. I have published the Titius Bode Law in four already published books but in this one I go deeper than the four already published. In each of the books I present I disclose how the Titius Bode Law forms gravity. These books are:

Everyone is in agreement with Albert Einstein that the Universe started with one spot, a point we call singularity. Now that you saw the laws I wish to give a glimpse about how the Universe started. As I said, the Universe started with 1 and this fact even Newtonian science accepts. Singularity started as the first spot and not with a massive already formed Universe that only afterwards grew in size. Since the Big Bang event the Universe only grew but everything that is in the Universe already was in the Universe. What was in place at the Big Bang only then grouped and became material but not more of what already was.

Science is forever looking for the point the Universe started but the Universe never started. The Universe ended and from that ending the Universe flowed into a change. The spot holding eternity and infinity were there because both these values are everlasting. Why would they end if they were everlasting in one unit? The spot never ended and is continuing on even in this day. At first the spot was perfect. The spot was invisible and yet present as it is today. The spot ruled everything that was not yet in place because within the spot was all possibilities the Universe holds. The Universe was and is everything and anything but never was it and never is it null, zero or nothing. Nothing excludes all possibilities and the spot includes all possibilities.

The spot was there in an everlasting capacity where eternity latched on infinity and within the realms of the combination the perfect point formed. The spot was precise and the very same. The spot flowed from one instant in the future to the same instant in the present forming the same instant in the past. Since the spot was everlasting correct there was no change in appearance from moment to moment. Since there was no change in the flow of time and the past was never changed from what it was in the present where the present was an exact repeat that duplicated what came from the future there was no division parting time. The dot was never visible while it held everything within. It is exactly as it is at this day and so the proof is in the Universe as the Universe at present functions.

Then the Universe ended. It did not begin because where would it begin and what will begin and wherefrom would such a beginning originate? The Unversed in its present form still is perfect but for one aspect, it changes all the time to such extent it is change that produces the flow of time. The Universe parted infinity from eternity by producing definite in between where nothing can start and on the other side nothing can end. This placed a difference in what was and what is compared to what will be.

Hot
Cold
Elements locking heat into matter by excluding and exclusive space.

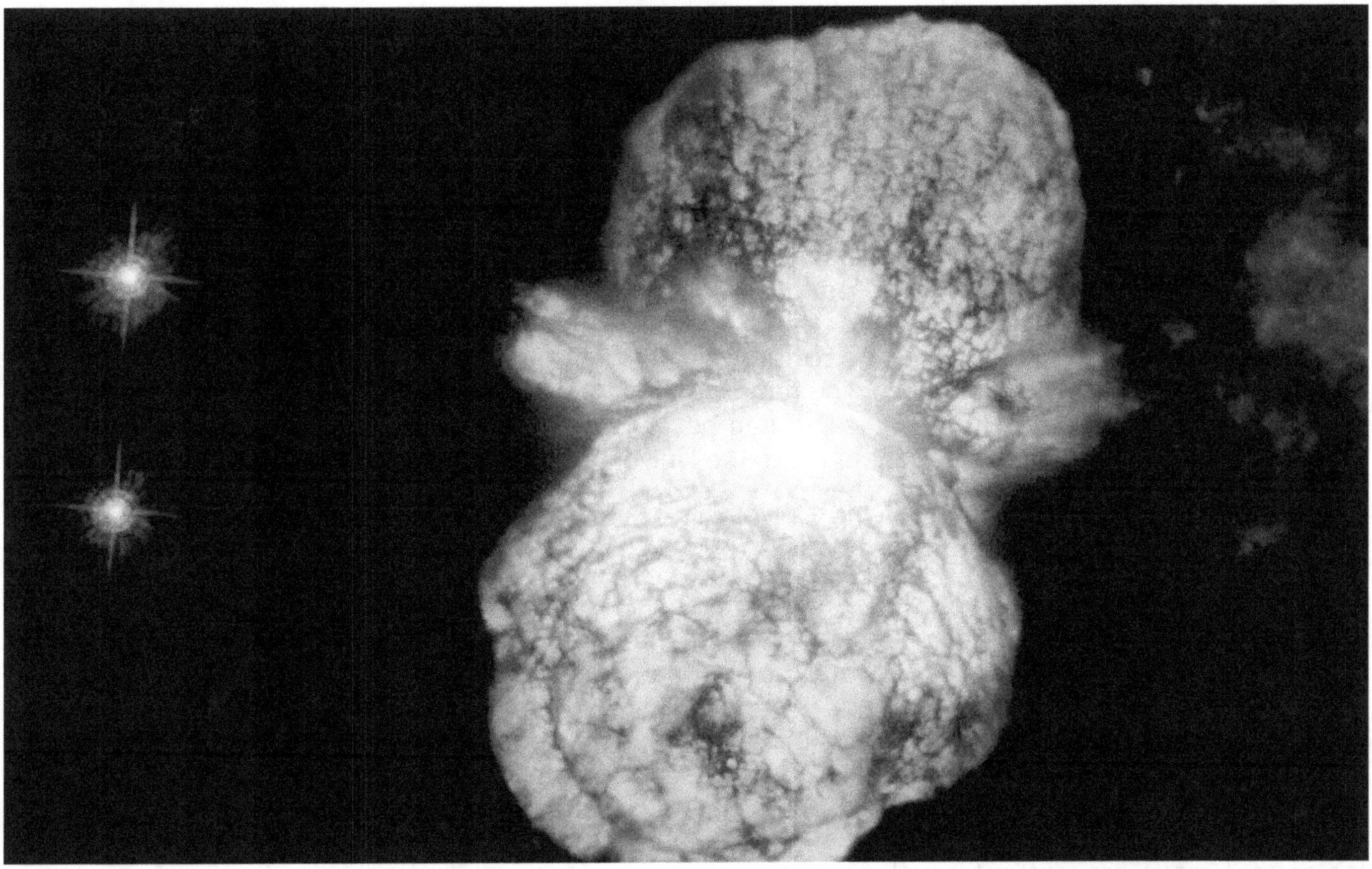

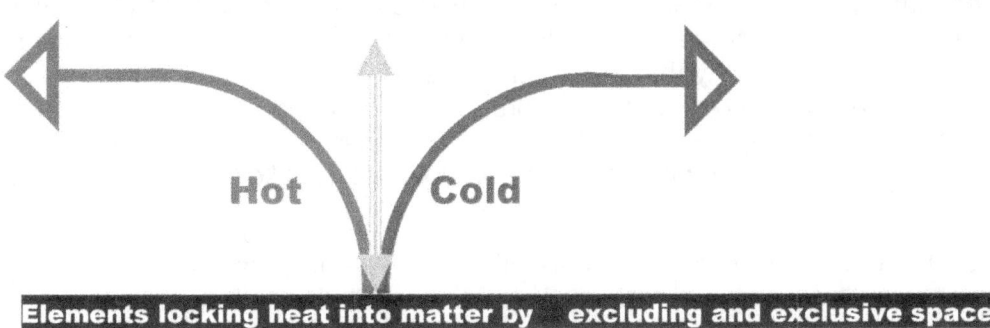

Then came the biggest outburst that ever could be and ever would be. Heat created space as the Universe overheated. Then at one point heat parted from cold and a difference came about in the Universe. That which is hot moved away from that which is cold in relevance by expanding while the cold is contracting. This remains the cardinal rule in the cosmos.

CHANDRA X-RAY
HST OPTICAL

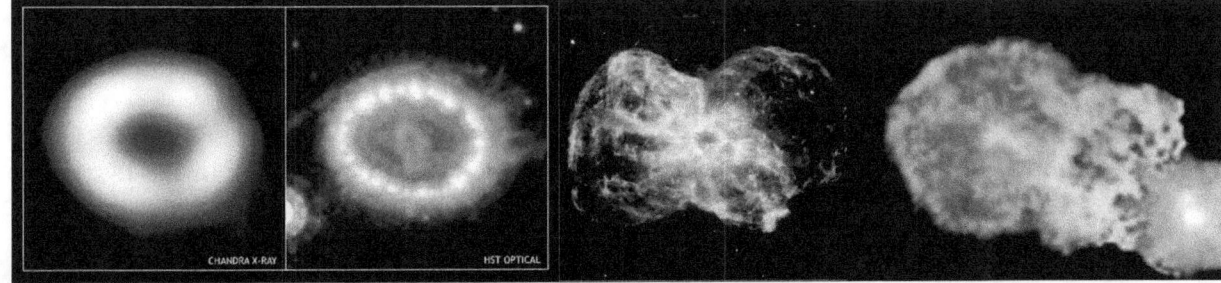

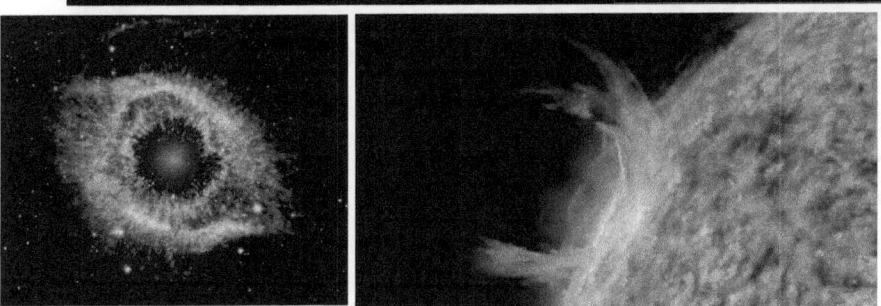

Every picture I show is heat expanding into space. Wherever we look at and whatever we view, we find that the Universe is the interaction between what is cold versus what is hot being in that space. Everything in the Universe is heat that produces different levels of density and intensity in movement. Movement cools down and non-movement produces heat. Material contracts heat and overheating expands heat into space and that is what gravity is. Gravity is control by heat levels.

When the spot overheated and grew into space forming the first dot the very first moment of imperfection arrived. The spot did not start the Universe and the dot did not begin the Universe but the difference that came about from this moment-alpha brought in time differentiation. Then the difference of movement brought changes in time formed as space and that changing brought about time segments we now think of as moments. This was where infinity intervened with eternity to change space. That is why infinity is twice as long as eternity.

Think of it this way. The spot overheats and expands into the dot as the movement comes by the spot enlarging to form the dot and by growing into the dot the movement enlarges and the movement as well as the increase in size reduces the heat the dot accumulates. The growth then by reducing cools the spot off and the cooling removes the structural size of the dot as it returns to form the next spot. However what is in the Universe can never leave the Universe once it becomes part of the Universe but has to remain a component within the Universe as long as the Universe exists.

However before heat brought about the Universe we know time was one continuing everlasting spot, which is a line that never went further than one spot.

The spot formed the future. The spot formed the present. The spot formed the past. Since the spot in the future was an exact image of the spot in the present and that was identical to the spot in the past the new spot had no identifiable difference between the future and the one in the past. It was a repeat of what was being identical to what was coming and therefore it stayed the same. The spot was so small it was invisible and unnoticeable and yet it is so big that at present time it holds the entirety of what is within. It is as big as nothing could ever be and it was so small as nothing that could ever again be. This was the only time nothing had a validation because as soon as the Universe came nothing disappeared and became something. One should be very clear about understanding this aspect. However the "nothing" became a factor of 1 as it brought not a value but the absence of a next dimension as in 1^0 The spot then represented nothing in as far as dimensions $1^0 = 1^1 = 1^2$ and so on. It was consistent with nothing since only nothing existed at the time that produced space. The Universe was in singularity and in singularity space as a factor is absent or nothing. It was infinity that which can never start united with eternity, which is that which can never end. To start that which can never end first had to end and to end it first had to wait for that which can never start to find a way to start to lead to the process of ending. Since neither process could apply due to unfavourable circumstances making it impossible to follow on the other the process was eternally infinite.

To start with giant formulas as Newtonians do only impresses other dim wits that can't think and only shows incompetence and ignorance as is the case with Newtonian science. Mathematics came about as the Universe formed and as the Universe formed it formed mathematics during the process. I show

exactly how did the first spot arrive and what was in place before the first spot that became one. $\overset{\bullet}{\Pi^0}$ I show why the first spot came into place and what made it be.

$\overset{\bullet}{1^0} \qquad \overset{\bullet}{1^1}$ I show what made 1 grow into two and why did one grow into two. This is the most fundamental reason why the Universe started. It started with the fact that one came about that became

two. $\overset{\bullet}{1^0} \qquad \overset{\bullet}{1^1} \qquad \overset{\bullet}{1^2}$ What changes took place to allow two that was one to become three and why did this process begin to form a Universe. That I show in a very detailed explanation.

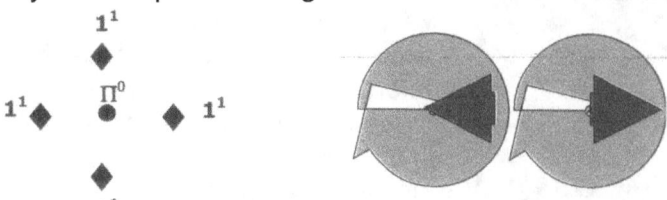

Then also why did four become in the place of three without replacing one or two, as a number and I will give a hint; it was because 2 + 2 = 4 then became 2 x 2 =4. This brought the Universe into a new era such as was never seen before.

Then that which can have no inside parted from that which can have no outside. $\Pi^0 = 1^0$ parted from 1^1. The spot became dot and then relevant as the Universe transformed. Another hint is that this is what the Universe are made of and this is how the Universe formed and transformed into what it became. To claim proof I have to show that this indicated that the motion produces the space and the space finds limits in the motion confirming the space while the space is conforming the motion. Singularity is where and in that

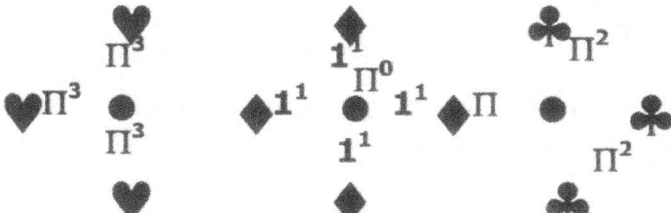

the triangle and the half circle and a straight line is equal in 180°

Realizing why these three forms are equal unlocks the information in Genesis 1 verse 1. This was when singularity as 1^0 parted from 1^1 that the motion then came about as 1^2. That in reality left little consolation because with $k^0 = 1^1$ that left the space formed by the motion way outside the realms of the emerging Universe. I decided to replace the symbol Kepler used of $k^0 = 1^1$ to a more appropriate Π^0. However I am not going to go into detail before I go into detail about what the four cosmic laws are about and why they in movement control the Universe. The concept is confirmed in the fact that by using a pendulum in a time device such as a clock then it proves that time is movement and gravity is movement and that is why gravity moving the pendulum arm is able with that movement to measure time.

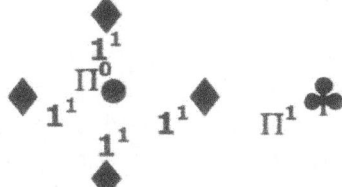

        Then the Lagrangian law came in place and this allowed four to advance to five. This principle, as is the case with all the principles are still applicable in nature. However how did the development of gravity take one dot to four and then shifted it to five. This is where the Universe formed $\Pi^0\Pi\Pi^2$ for the very first time.

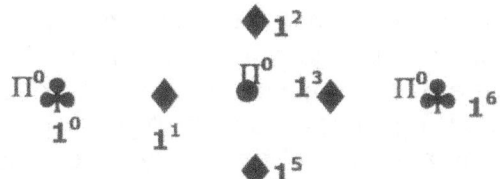

        Then as a result using the reason why two became three then applying the very same reason, it came about to bring five to form six.

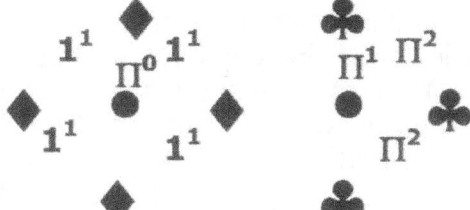

        However the reason why four became five is the same reason that brought seven into place because the one is the result of the other forming and the two being five and seven are so well interlinked it had to become in place simultaneously. Therefore this then brought about that gravity formed space as $\Pi^0\Pi\Pi^2$

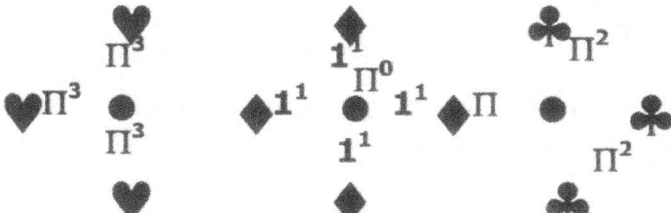

        By forming seven it then formed ten and that was where that which was eternal and perfect became temporary and imperfect. Singularity distorted into a concept outside the eternal. The truth is at that point all there was, was an inside and an outside still to form and a promise of what might come.

The mighty wise can push their present into forming the past by clinging onto the worthless they represent as worthwhile that will become the past when these books becomes the present. However, that choice will doom them into the past along with all other things and thoughts not worth the burden to take into the present and onto the future from the past. Their adopting the worthless and not adapting to the truth moves them to the past only worth to be forgotten as the worthless part of the past. The choice is theirs to make.

Again I repeat: You are going to read some <u>mathematics in equations</u> and expressions in <u>mathematical formulas</u> placed to defend my position but if you <u>don't like it</u> then just <u>skip</u> the mathematics because the content and grounds the mathematics proves or disproves is not important in the arguments and it is there for physicist to hide behind. I don't have to hide behind mathematics to make others feel inferior because my arguments make people understand physics and make people feel empowered and superior. The mathematics I include is to show what mindless clots those Superior Humans are that portray their position as superior in mathematical ability and it is there to disprove the Members of the Physics establishment that advocates the necessity of bringing mathematical proof to prove? It is there not to scare readers away but to silence the Brainy Bunch critics by showing them the foolishness of their arguments. By using mathematics the Brainy Bunch have been cheating the public and have been brainwashing students for centuries. That cheating is how they do it and I have to show and uncover the dishonesty in mathematics.

The Universe consists of gravity that forms by the working of the four phenomena never mentioned. That does not say much for the bountiful prestige that mathematician's claim as their lawful bragging rights in areas where true human intellect is called on. Is it not high time to begin to admit you are playing the game of fools with you arrogance about your achievements using mathematics when designing space whirls and travelling to galactica while not even understanding what movement asks for? You do not even understand the neutron and the neutron is compressing density increasing, which is what gravity is, which is what time is, which is what all movement is…that is why the neutron has no mass because mass is the principle coming about where independent movement ends.

You're mathematics could not get you any closer than playing games in a fairy tale Universe using misguided presumptions about mass forming gravity and living the Universal farce which Newton created because that fairy land is what all the Kings clever heroes and all the King's splendid wise could never prove in hundreds of years. If you feel superior as a scientist practising physics on the highest level having a gloating hail of superior mental capability covering you like an aura, then I have very saddening news for you.

If you have the ability to compute and calculate at the highest level, then look at your computer and see one that machine has abilities as a machine which is equal to you, but it's a manmade machine. Stop playing games by creating fairy worlds making up fairy tales about fairies and little people, mass that can create forces, four of them no less, and come and join the rest of us living in reality that does not need to compute forces to be able to not understand what it is that you compute, but to use human intelligence and in that way to understand what only human intellect could ever understand.

Then what in the present is not worth carrying into the future as the past being worthwhile?

Notwithstanding your mathematical brilliance you completely lack any understanding of mathematics or of physics.

The spot was one perfect spot that overheated and parted into a dot.

•• Then by overheating the spot split into two being a spot and a dot.

••• The spot shifted to the past leaving the dot in the present while the dot cooled of and formed another spot one in the past and one in the present and one coming from the future.

The spot   expanded into becoming a dot going to the value of Π.

While the spot expanded into becoming a dot it   had nowhere to go because it was still inventing space, which was a concept that did not exist yet.

This was when 1^0 became 1^1

From the four cosmic laws we can see how the Universe started. I will allow a glimpse into the process, which I wrote several books about and still I know I have not scratched the surface. It started with one spot that became a dot through overheating on the one side and cooling on the other side. The dot became two, the next spot and the previous dot, because space interrupted time and in between two instances of time landed one speck of space forming a dot on the one side and a spot on the other side. Remember once anything even a process forms part of the Universe it ha no place to go but to remain part of the Universe.

From this the Universe started exactly as we find the Bible says it started. It started with Π^0 or then singularity or if you wish, then from a void where there is ns space and there is only time in the instant. The entirety we think of non- material space is filled with this. It is a spot that is not and while it is not it still fills and maintains one entire Universe. Wherever you look you will find this spot that holds space and forms space without being able to claim space. Material fills as a solid because movement compresses this dot so tightly it forms a solid construction of compressed material while the material forms of a substance that claims no space. Non-material is this same heat but is much less compressed and so because of a reduced density because of the flow of time from space to material we can see the density increasing on the one side and decreasing on the other side. The proof of this we see in Kepler's tables where the Titius Bode law process that material is regularly spaced in accordance with the growth formula that the Titius Bode law is indicative of.

Also the Roche limit at $\Pi^2/4$ shows this limit between material forming and the Coanda effect shows a clear growth of density developing around spinning materials as well as the satellite positions of materials around structures we think of as planets. There is a definite distance maintained between matter and therefore it must be non-material the compresses around the sun and while spinning the sun compresses the non-material because this maintains the heat balance within the sun.

When this heat balance goes array the structure overheats and we then find what is thought of as a Super Nova or an exploding star. Everything in the Universe is reliant on density caused by movement, which causes movement, allowing density to development specific time or specific gravitational ratios.

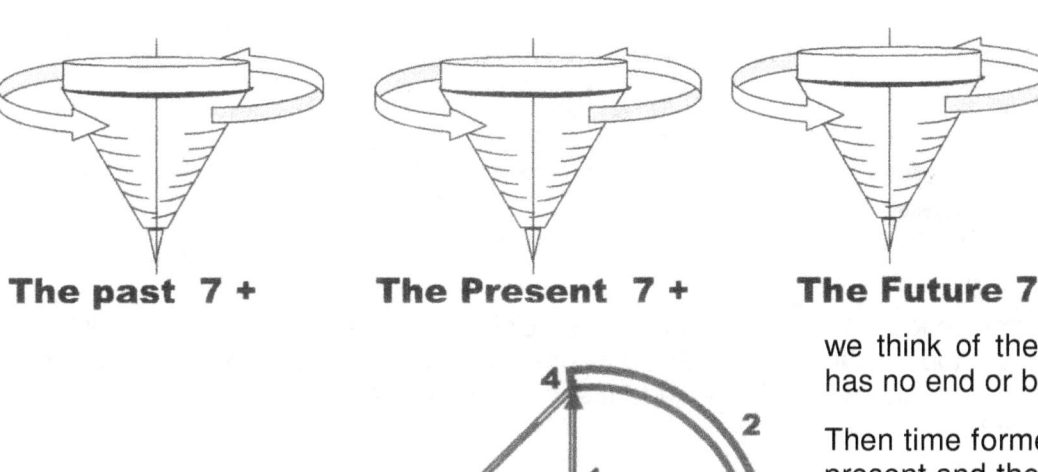

The past 7 +   **The Present 7 +**   **The Future 7**

This left one dot and one spot that developed into the next dot. Where is the first dot, the mother dot, and the original dot from where everything came. We are within that first spot that became the mother dot and that dot we think of the entirety as the Universe that has no end or beginning.

Then time formed a sequence of the past, the present and the future leaving three dots lined up in a line we think of as time.

Afterwards space becomes a factor because we see the second dimension develop, as time becomes 2 + 2 = 4 and 2 x 2 = 4. This takes the Universe into a complete new level.

There fore the one line forming time became three and the line becoming space became the line (2 + 2 = 4) as well as the circle (2 x 2 = 4).

1 = past   2 = present   3 = future

In this we find that the triangle, the half circle and the straight line all three has equal value of 180°. Please keep in mind that we are within the limits of singularity forming 1^0 going 1^1 going 1^2 going 1^3 going 1^4, which all is still Π^0.

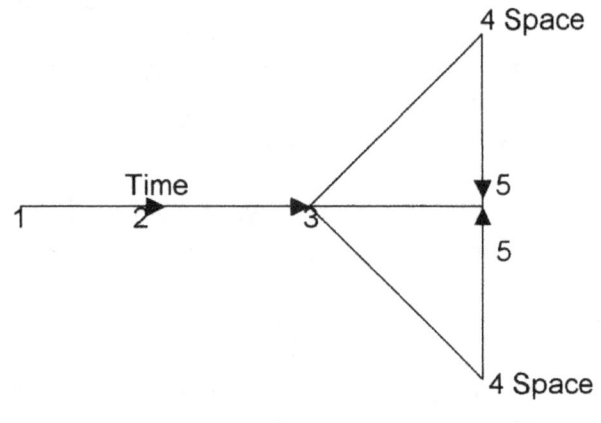

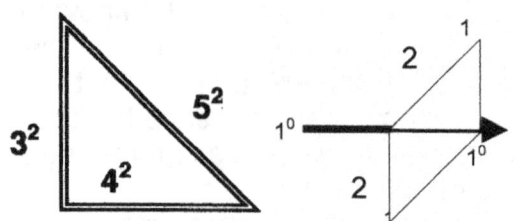

In that we find the value of the n ext spot forming at 5, which is the Lagrangian number for the next dot.

At the point where the four cosmic pillars form 7 as a relevance we have also two values 7 + 7 + 7 = 21, which is time (3) times space (7) = 21 and in the space sector there is 5 times the four sides forming space (4 X 5 = 20) plus singularity 1 = 21.

I shall get to the other part forming the value of Π, which is $\Pi = \dfrac{21.991}{7}$ and when in relation to Π^0 it is $\Pi = \dfrac{3.1416}{\Pi^0}$. The Universe we know begins where the two values forming Π begins because less that Π we have singularity. This is extremely important to realise because realising this confirms the Bible and the way the Bible states that the Universe started. In the Universe there is infinity within the centre forming time and there is eternity forming a circle where both infinity and eternity forms singularity.

$Space = \dfrac{\Pi r^2}{\Pi^o}$ To get to form cancel space: $\Pi = \dfrac{\Pi r^2}{\Pi^o r^2} = \Pi$. To get to singularity cancel form: $\Pi^0 = \dfrac{\Pi}{\Pi} = \Pi^0$ and Π^0 is singularity. The value of Π forms as $(3 + 4)^2$ and $(3^2 + 4^2)$.

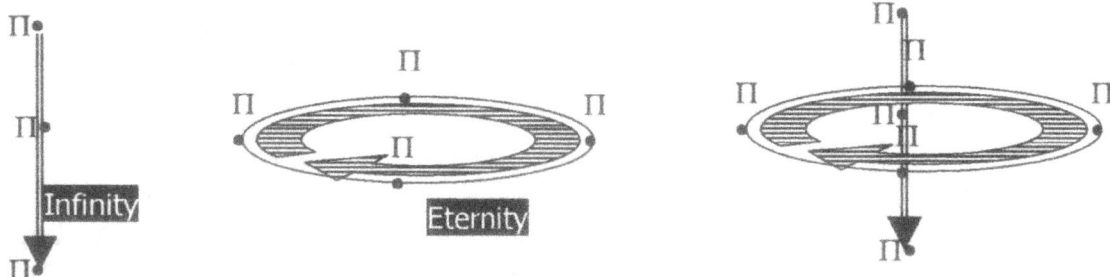

It is through the spinning top that we learn the process of how the Universe started from one dot to what we now have. This is because the top teaches us how the Universe started. Time that can't move parted from time that can never stop moving and within the boundaries that time formed space came about and that space we know as the Universe.

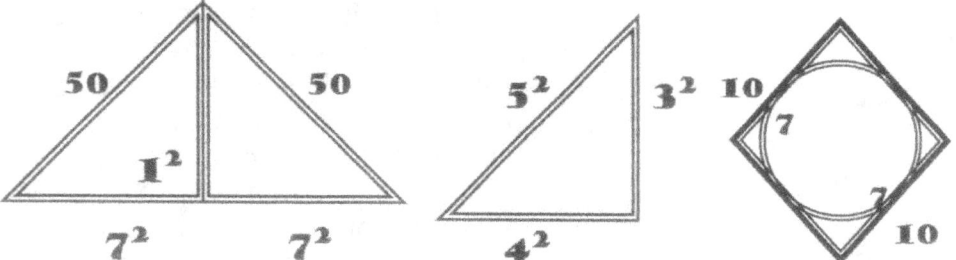

It is by 7° that the Universe turns. In turning it diverts direction by 7°. This application still moves the universe ever instant of movement or time.

The divide is $(3 + 4)^2 = (7)^2 + (1)^2 = 50$ and forming the next point is $3^2 + 4^2 = 5^2$ where both values is the result of the Pythagoras triangle.

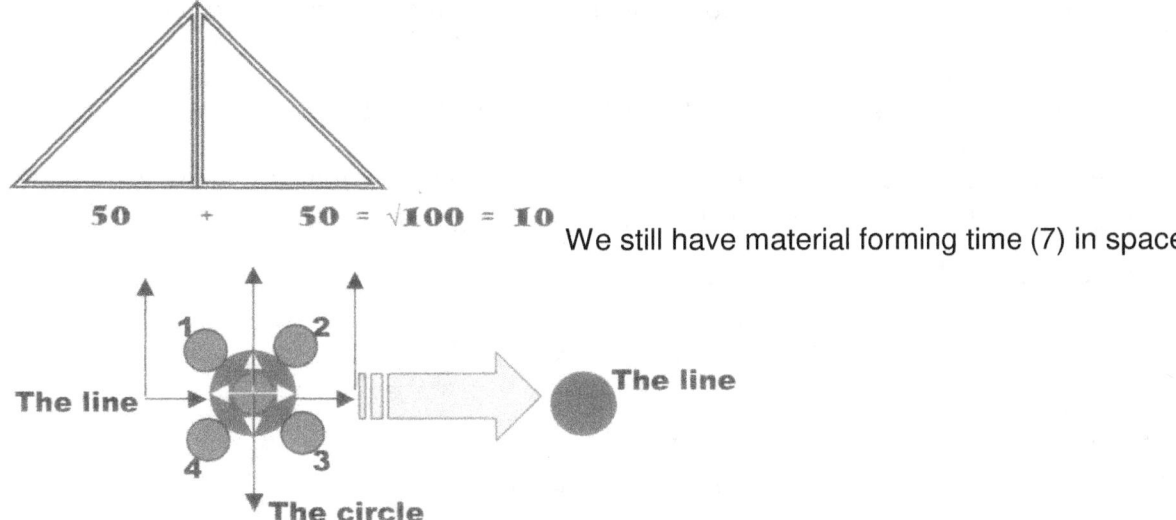

50   +   50 = √100 = 10

We still have material forming time (7) in space (10).

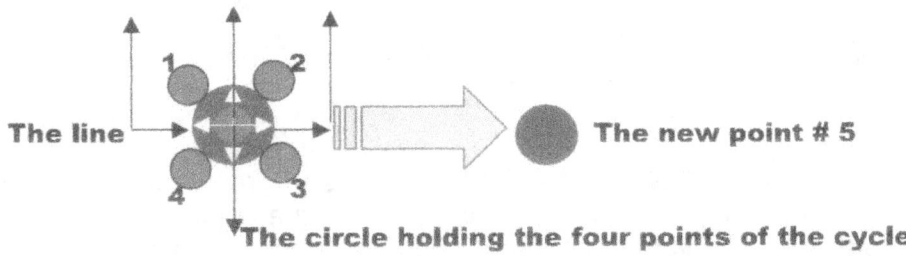

The line — The circle

In the one $\{(3 + 4)^2 = (7)^2 + (1)^2 = 50\}$ a line forms a double value and in the other triangle a new point forms the next point in the line.

Both values end up as 10 where the one is $\sqrt{100} = 10$ and the other is $5 + 5 = 10$

The line — The new point # 5

The circle holding the four points of the cycle

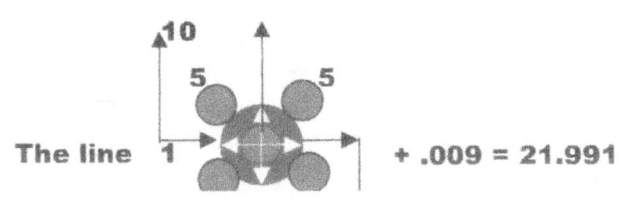

10
5   5
The line   1   + .009 = 21.991

However we know that singularity as 1^0 is the smallest point any point can be in our Universe because that point in value falls outside any space within our Universe.

Forming a value as Π of 3.1416 indicates a value of less that 1 which indicates that the line forms a future that comes as the line continues from far outside singularity. In the black hole we can see this clearly where the future absorb and consumes the past. The past is the darkness and the future is the line holding singularity.

1   +
5   5
$(1/ 10 \times 3/10 \times 3/10) = 0.009$
Π found a new value of 21.991 / 7
The line
5   5
The total points worth became 4 x 5 is 20

The process in which the circle that holds the five points form a triangle in reference of the circle of four points forming a new cycle.

This is how the Universe began. It began simple as it formed numerical order in numbers that brought about order to spots that became dots. It is so simple that the complicated mindset of the Newtonian physicist rejects this as being too simple for their liking.

How does the world of Newtonian science accept this reality and how does science respond to this being the way that the Universe formed. In twelve articles to Annalen Der Physics I tried to explain the process that I name the Absolute Relevancy of Singularity and I went about to insist on the simplicity of the Universe such as you could see in the manner in which I did. I introduced a concept that was never penned and was new as a concept to everyone.

It is forever that Newtonians try to bullshit me by implying that I am too stupid to understand Newton. Never, not in over 40 years was there even one that explained to me what I did not understand about Newton. Whenever I disagree with science and Newton I am forever and always brushed off by suggesting without suggesting that I lack the mental facilities to follow reason and in that I did not have sufficient mental intellect to understand Newton.

I sent twelve academic articles in which I explain the four cosmic laws and moreover the influence that this had on the outlook we have on cosmic science. I explained in the articles how singularity forms and this is far from how Newton works.

I now wish to present you the reply exactly as I received it using the precise words as it came via e-mail. I was (again) rejected because I was not Newtonian. This came back…

To: Mr. Peet Schutte

27 March 2008

Dear Mr. Schutte:

I am sorry but it is apparent from your letter that you are missing the basics of mathematics and classical mechanics.

Sincerely, U. Eckern
Editor in Chief
Annalen der Physik

Prof. Dr. Ulrich Eckern

Theoretische Physik II
Institut für Physik

Universität Augsburg
D-86135 Augsburg

Telefon +49-8 21-598-3236
Telefax +49-8 21-598-3262
eckern@physik.uni-augsburg.de
www.physik.uni-augsburg.de/theo2

This how gravity forms:

In short I said: The curve of the earth is $7°$ on both sides ($7° + 7°$) but because $7°$ represents the earth turning in movement it is also ($7^2 + 7^2$). By turning it crosses singularity (1^2) in the centre of the turning circle in rotation and then according to the law of Pythagoras the circle in movement becomes a triangle moving in singularity. Numerically ($7^2 + 1^2$) = 50. So $2(7^2 + 1^2)$ = (49 + 1) + (49 + 1) = 100 on both sides of a circle.

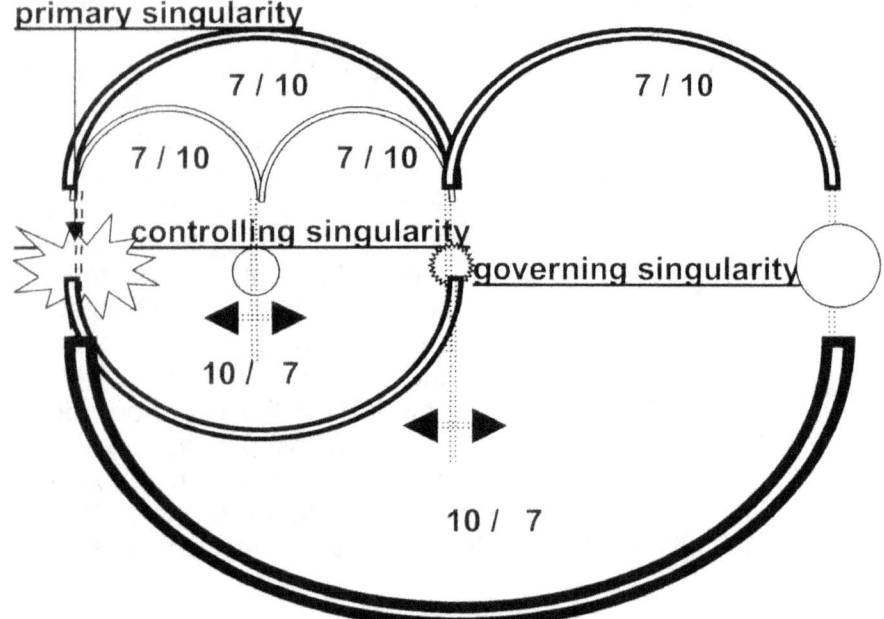

A triangle then forms turning the direction = 50 + 50 = 100.

Therefore the space in which the circle turns is $100^{1/2}$ to the root thereof = 10 and therefore the Titius Bode law shows the inside of the circle factors forming Π as gravity where 7 goes related to twice times 10.

That is Π and that is why 7 goes double squared in a circle adding the centre part of the circle, which is 1.991 divided by the space in which the planet orbits around the sun that forms singularity at 1.991. This adds to the 10 +10 placing each planet in the allocated singularity position, which is then derived. It is implementing Π as gravity. The Titius Bode law shows that gravity is when (10 + 10 + 1.991) / 7 = Π.

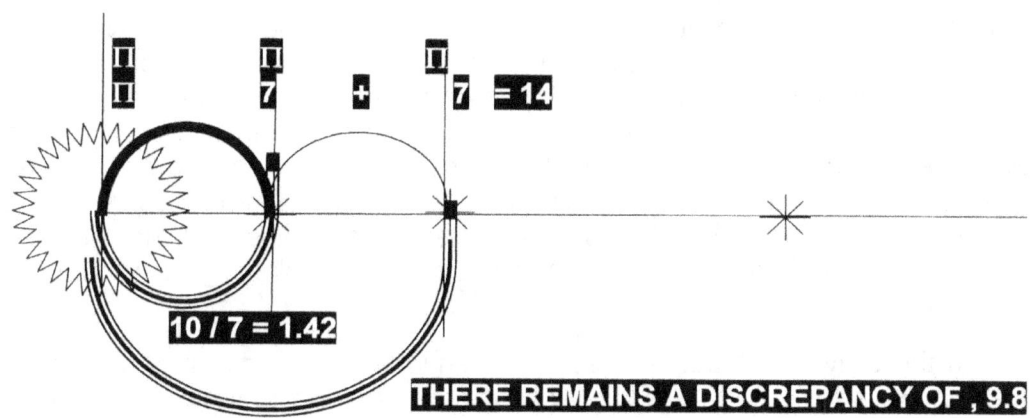

10 / 7 = 1.42

THERE REMAINS A DISCREPANCY OF , 9.8

14 / 1.42 = 9.86 and the value of 9.86 is Π^2

Sound complicated.. not bloody likely when I use twelve articles to prove this fact...
And the e-mail sent to me only again confirmed yet another Newtonian rejection...

Dear Dr. Schutte,
You submitted an article of 15 pages to the Annalen. The content of this paper doesn't constitute a theory in physics. With a lot of words and some simple algebraic relations, there is no way to "explain" the world of physics. You seem to be out of touch with modern developments. This is also shown by the fact that you don't quote any relevant literature. I am sorry to say, but the Annalen is not able to publish your work. I am sorry for having no better news for your.
Best regards, Friedrich Hehl
Co-Editor Annalen der Physik (Berlin)
--Friedrich W. Hehl, Inst. Theor. Physics
* University of Cologne, 50923 Koeln _____/_____ Germany
fon +49-221-470-4200 or -4306, fax -5159
hehl@thp.uni-koeln.de, http://www.thp.uni-koeln.de/gravitation
* Univ. of Missouri, Dept. Phys. & Astr., Columbia, MO, USA

In the macro the four phenomena or the four cosmic principles form the following structures in the greater cosmos. However these four cosmic laws are the pre-micro that project a pattern onto the macro cosmos but forms the smallest particles one can find. In truth these four principles take the Universe into the abyss where space does not yet form. My discovery of how to decipher the laws took me to an understanding of how the Universe began and that understanding brought me to where I can mathematically explain the Bible according to Genesis 1 verse 1. I was able to achieve this because I discovered the manner in which the smallest particle forms and then I retraced how everything started. I repeat: these four pillars forming everything. These four principles begin the Universe before the Universe begins.

I am very aware of the fact that most everybody including physicist that is practising physics never heard of these laws and are unaware of the existence of these law. These laws are what nature applies instead of Newton and that is why these laws are kept under cover and in the dark as far as the significance it represent. Nature and Newton don't even share a Universe and because Nature nullifies Newton science hide nature under a blanket and out of the sunlight. If you wish to remain in the dark then don't purchase this book but it you wish to read about nature's reality you can't afford not to read. Without recognising these four laws science does not make sense unless they cheat facts to make it seem to be sensible just as is the case presently.

Science forever hides the truth of any subject behind this veneer that the public is too stupid to understand and therefore they disclose their opinions about whatever field they wish to promote and not telling is because they know all. It is the information about the unbelievable oversight of Newtonian mistakes I disclose that proves how they keep the oversight of Newtonian mistakes silent and why they don't divulge that which they keep silent about. It is about them never committing to the entire story by giving an all-round presentation of everything anyone would require to know to be in a position to evaluate.

Science went corrupt in 1705 when Edmund Halley told the world he used his friend Isaac Newton's physics formula to calculate the route and time that the comet that was named after him would arrive. This was where science went crooked and started to corrupt science, a position that went on ever since because that same dishonesty is still present in Newtonians science. Halley calculated the time periods since 1066 at the battle of Hastings and found a comet was mentioned every seventy-six years. This was very ordinary for a man of his class so he had to get far cleverer than backdate history to get a time frame. So he really got clever and conspired with the biggest fraud in science ever since; the man hat stole all the formulated physics Doctor Hook invented, the man that even got Kepler' figures wrong by cheating it to support his ideas, the man called Isaac Newton.

If Halley was honest about tracing the arrival of a comet that was mentioned ever seventy – six years then Halley was no more than bloody ordinary and hat Halley could never be. So to look smart Halley said he used the formula of Newton to calculate the rout the comet took. This says he used mass to calculate how the comet came to the sun. That says the sun's mass pulled the comet and the comet's mass pulled right back and this way the comet came to the sun. I don't go into the comet that much in this book but I do in other books. In this book I show how Newtonian science started to go corrupt in 1705 with one conspiracy to cheat and became the corrupt myth it now developed into. How do I know Halley did not use the mass pull mass idea because if he did then how did he calculate that the comet was cyclic or that it returns every seventy-six years. If mass pulled the comet to the sun what then pushed to comet back into outer space? Halley's big ambition was to prove the comet comes and goes but if mass makes the comet come what pushes the comet back. You know what is the biggest fraud that came to be called Newtonian science? The most brilliant minds on earth this past three hundred years failed to asks this simple question: if mass pulls the comet closer what pushes the comet away? If mass forms the force of pulling and pulled the comet closer than what pushed the comet back into the darkness of the beyond. How did he know the mass of Halley's comet? Nobody then asked questions.

No one asks uneasy question…except I. I show the fake science we have by just questioning science in search of the truth. Newton and Halley got away with corrupt science. Today Newtonians get away with corrupt science. Then those in science question my integrity because I question the integrity of those in physics and in this book you read how a bag of stinking shit flies into the faces of the most holly, the most intellectual mind the world ever produced.

I deciphered the cosmic keys by which the Universe is formed. These keys are so small it can only be mathematically equated and yet it forms everything within the Universe. These keys are applied to form all forms of material and non-material of which the Universe comprises. By deciphering the keys I also deciphered the manner in which the cosmos started LONG before the Big Bang started. Using the way the keys form material and space made me see how the Universe began before zero was in place. If you purchase and read this book you will find that the Bible is correct about Creation and I challenge any **mindless atheist** to show me **MATHEMATICALLY** where I am wrong with my **mathematical interpretation** of the Bible on the first page of the **Bible in Geneses 1 Verse 1**.

However if you purchase this book AND read the content you will see how I use MATHEMATICAL numerical equations with which I show how the Bible is correct when presenting the Biblical version of how the Universe was created. I use science and mathematics that is a numerical understanding to convey what the concept constitutes of. When the cosmos came about words were not yet a reality and therefore using words brings a shortfall when trying to understand a process that came about before the Universe came about or the spoken word was a reality. I DO NOT use theology or any religious doctrine or any form of evangelistic approach in any way but conduct mathematical science to show the Biblical version of Creation is the only correct interpretation of how the cosmos came to be. This is the first time ever that this achievement was successfully conducted and I did it with the grace and mercy of the Almighty Creator that Created a Universe by applying mathematics. It was not mathematics being God that formed the Universe but it was the Universe by Creation that brought about mathematics! It started when one started...

Take note that the Almighty God does not use words or language to write His laws applying as nature because word and language is human methods to write what applies to humans and manmade thoughts. In nature where it really matters we see the Almighty God uses cosmic laws written as cosmic structure using mathematics to write on a cosmos as a canvas using light as ink and time as pigments.

I have cracked it...I have finally cracked the cosmic code. The Universe is built on four cosmic principle laws and the four laws form gravity as a unit. However this proves Newton wrong and in proving Newton is a hoax I am a cast out and rejected.

These four principles I mention is the four building blocks nature uses to form our Universe. Why don't science use it...because science does not understand it. Science denounces its importance. Why would science not support what nature uses...because it destroys Newtonian mythology of mass. I am the first person in human memory to figure out how the four cosmic principles, which in fact are the four laws nature uses apply. Science would not even recognize these laws that are nature in space in place and used by nature. Science rejects these laws formed by no less than nature because through these laws nature condemns Newton and what Newton said as being all fraud. I challenge any of the super brain cheats to PROVE Newton. Using these laws I prove nature and from these four cosmic laws I prove gravity forms by rotating movement that by turning forms pi to form pi square. Newtonian science would not recognize my work because this destroys anything science brought about since the hoax started in 1705 and these gangsters calling themselves physicists protect a fable and a fantasy and the mess they make of science where they hide the truth under fabricated cover-ups. Now I call on science to prove Newton correct. Everything in the Universe is round. Anything that is round has to apply the value of Π.

This is a fact of mathematics but while Newtonian science forever tells the Universe to have "mass" and to use "mass" nowhere in science would one find Π used in prominence. Whatever you may study in astrophysics, go where you wish but never would you find Newtonian science taking the fact of Π into any prominence. When you read any of my books you will see that gravity forms by movement applying Π as a value. I have found the four phenomena that put Π in astrophysics. By valuing gravity as Π therefore the Universe consists of gravity that forms by the working of the four phenomena that Newtonian science hardly ever mention but dubbed "a freak of nature".

The **Titius Bode law** has been around for centuries and with all the mathematical splendour available there for all to use, all the brilliant mathematicians could never come close to show any ability of understanding any of this very important phenomena. They could mathematically equate the formula the sequence applying as the formula, but then after that their superior human intellect dries up as they hide behind worthless equations.

The **Roche limit** has been around for centuries and with all the mathematical splendour available to apply in order to fathom concepts behind this phenomenon, still with all the computing ability of a machine all those physicists with all the mathematical superiority could not touch any understanding about the concept forming the background. Yet when using the truth about gravity in physics the answer is simple; it is that gravity is Π.

The **Lagrangian points** have been known to science with all the mathematical splendour available not one ever explain why this event is taking place. The form around stars that have satellites and rings and pattern applying. Just the fact that the satellites

for centuries and calculation could Lagrangian points this shows a definite

around the planets are not drawn to the planets red flags Newton as a hoax.

The **Coanda effect** has powered turbine engines and aeroplanes in flight for almost a century and with all the mathematical splendour available to design the most terrific aircraft, not one engineer could mathematically compute one fact to show understanding why this takes place. How sad it is that those claiming of much superior intellect in physics remain just no more than having computing power. The understanding is not complex. I have to warn the readers that the topics are showing a very new approach with no quick answers. Understanding is in the proof and that does not come by reading just a few lines and then forming conclusions. The information is new but not hard to grasp. I did not put these phenomena in place and these phenomena nullifies Newton's correctness, therefore don't blame me because falsified Newtonian astrophysics claims on correctness never ever existed but in Newton's imagination. Now to set the record straight it is time we wash dirty laundry in public. Since I present the truth I can call their conduct criminal.

This is what this book reveals for the first time in human history, it proves the law locating planets being the **Titius Bode law.** The **Titius Bode law** is the law showing the existence of relations between the mean distances of the planets from the sun to form the measured value of Π and that is what I prove and that is how I prove gravity is Π. Science can't support me … I destroy what they say we must believe.

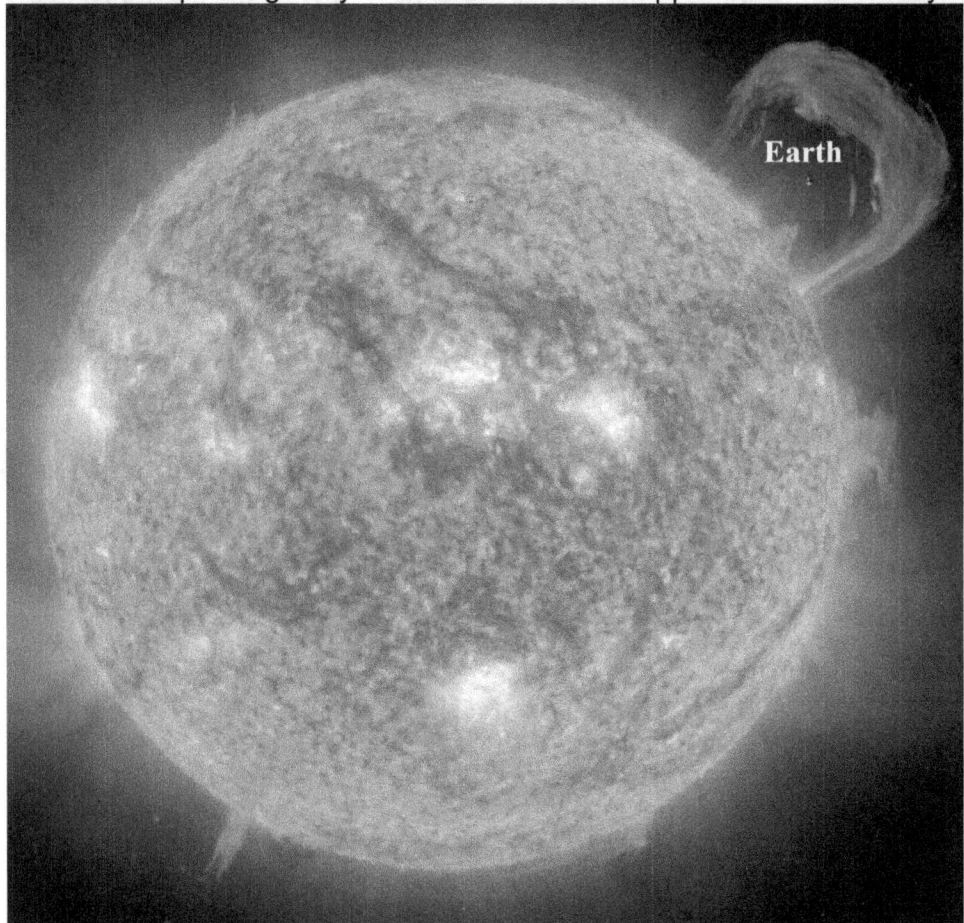

Look at the size of the sun in relation to the size of the earth. The sun freezes space into a liquid just by turning and by that turning it uses the law of Pythagoras to condense the space it contracts.

It is no coincidence that the sun by turning controls the space one third all the way to the next star system. It is that massive and the influence reaches that far.

However in this giant's turning forms the movement that is so veracious it turns outer space which is what I call cosmic gas and which is what fills all the dark space we see into cosmic liquid which is the flames you see and that is named prominence.

Whatever you think of the Newtonian terminology that Newtonians think up and use, in most cases it is to hide their inability to understand what they try to say. If it is prominence then what is it but if it is liquid heat or flames we all know what that is and you can see with your eyes that it is flames. That is all part of the conspiracy to hide reality away from the truth. Look at what you see and you see one giant gas pump spinning gas that freezes heat into liquid flames. Again I repeat that it is the space that moves and that Kepler and the Titius Bode law proves. The material holds position in the flow of cosmic space that condenses and change density as it flows to and towards the sun

There are four cosmic laws that form gravity and I wish to introduce a small part of how we are affected by the working of gravity as it forms the cosmic code.

This is the Titius Bode law which nature uses to allocate planetary positioning in the solar system. You might say that is far away and in the cosmos but this is how the Titius Bode law apply the very same principles that form our gravity on earth. This following explanation presents the truth about gravity as much as it explains what the Titius Bode law is all about and how the Titius Bode effect movement on earth while material is displacing space. On earth the following applies in terms of gravitational density or gravity, as we all know it in science. Should you find the following information slightly bewildering it would all be explained and understood when you arrive at the end of this book. I put this in to set a trend to prove I can substantiate every claim I make where I dismiss current mainstream science. As you should notice I am the first one in history that made an in-depth study on the figures of Kepler.

The way science at present view gravity or the movement of space within space is that objects draw as much as it is been drawn to other objects by going down in a linear fashion. This is true as much as this is only half of the story.

I maintain that gravity is the space surrounding the earth that becomes more compact and therefore becomes denser as the volume of the space reduces. It is space that moves taking material with it. However all space move as space becomes denser.

When any object falls, it travels downward by a certain distance in a certain time period, thus providing space-time. In other words it displace a certain space through a specific time period and this proves that space displacement in time period forms the time applying in that gravitational density.

That places gravitational density in relation to applying time. An object falling will fall while also being restrained and that is why we see the object travel through space while we see space restrict the fall. It seems as if while the object is in free fall the object has a lot of wind blowing through the object. This is the restriction of gravity. However where will this restriction that seems like air blowing through the falling object come from when the object falls with space?

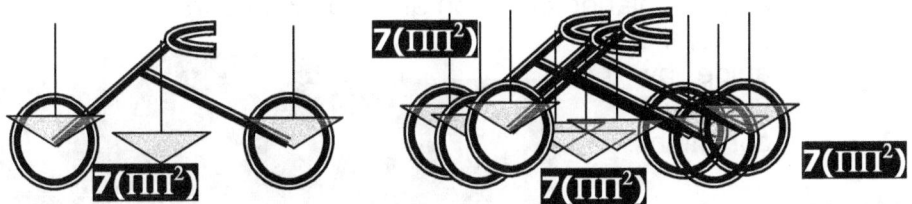

Science only acknowledges the moving downward part as that which forms naturally. They ignore the other part of gravity because they concentrate on one aspect only and that is the moving down and towards the earth aspect of gravity. This is because science goes about resolving the question of gravity by "mass pulling mass" which then only provides for the downward thrust aspect of gravity. However if we take Einstein's view on gravity we find another aspect applying, which is $E = mC^2$. In the $E = mC^2$ scenario shows that gravity has two aspects or time factors and not just one as Newton said. Since this statement is cloned directly from Kepler's formula $a^3 = T^2k$ this formula is the most correct or only correct gravity formula applying. To dissect this formula we have to look at the two (T^2k) time issues to valuate the space a^3 that forms. This forms a grid made up of lines formed by singularity criss-crossing as these line form the gravitational density that makes up space.

$$k = \frac{a^3}{T^2}$$

We all are aware of the line factor forming the line or **k**. But what forms the "wind" part.

However trying to "balance" the bicycle in an upright position is the tricky part and all of us that mastered this technique of bicycle riding know about this all too well. The "wind" you feel is the horizontal lines forming the resistance in the movement or restriction in the grid.

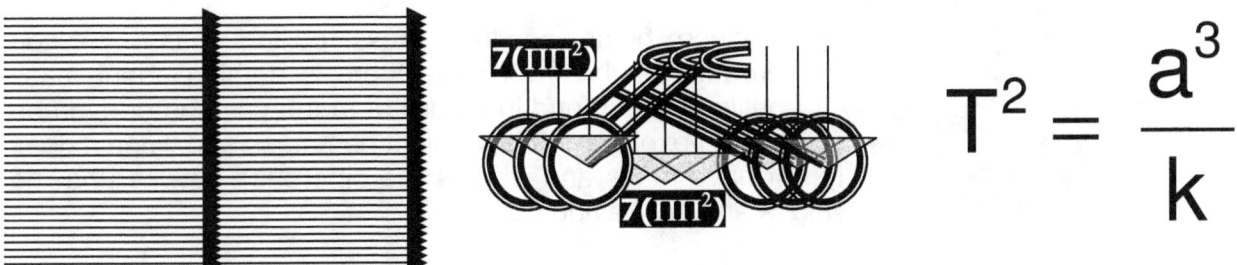

$$T^2 = \frac{a^3}{k}$$

The countering the downward thrust there is the curve factor of space or T^2.

This is the part that stays unacknowledged and this is the most important issue of gravity at this point. The "wind" you feel is the vertical lines forming the resistance in the movement or restriction in the grid.

When $k = \dfrac{a^3}{T^2}$ = $T^2 = \dfrac{a^3}{k}$ to form $k^0 = \dfrac{a^3}{kT^2}$ gravity is in harmony but when the circular T^2 exceeds the k factor, then T^2 will compromise its allocation in relation to k in order to bring about the correct ratio again. This is why cars and planes get airborne when the value exceeds $7(\Pi\Pi^2)$

The following figures I derived from Kepler's tables where Kepler proved space – time mathematically and showed precisely where to find singularity. Kepler said the following; $a^3 = T^2 k$, which then is

$$T^2 = \frac{a^3}{k} \text{ as it is } k = \frac{a^3}{T^2} \text{ because singularity is } k^0 = \frac{a^3}{kT^2} \text{ where } k^0\text{=1 which is singularity or 1.}$$

$$7(\Pi\Pi^2)\Pi^0$$

Standing still the car is moving down and moving along with the earth at a relevancy rate of $7(\Pi\Pi^2)\Pi^0$

In gravity all movement is as much going down vertically as it goes in circles around the earth and there are density levels forming movement requirements much the very same as the Titius Bode law applies between planets in orbit around the sun and the sun. The very same laws that apply in planet formation also apply on earth to gravity and the lot there is abide by the same conditions.

$7(3\Pi^2)7(\Pi^0)\ (\Pi^2/2)$

$7(3\Pi^2)7(4\Pi^0)$

$7(3\Pi^2)7(3\Pi^0)$

$7(3\Pi^2)7(2\Pi^0)$

$7(3\Pi^2)\ km\ /\ h$

$7(\Pi\Pi^2)\ km\ /\ h$

These 4 laws nature applies prove that gravity is the result of density requisitioning different speeds in movement at certain levels in space. These 4 laws in nature proves that gravity has nothing to do with mass forming gravity but movement bringing about space either expanding or contracting and this is resulted by movement. At every level in space a certain value in relevancy between space displaced and time requires the movement to maintain that density. This puts everything about gravity in a very new light. These values provide the gravitational density applying at each level of space density around the earth.

At $(\Pi^2/2)\ (\Pi^2/4)$

At $(\Pi^2/2)$

At $(4\Pi^0)$

At $(3\Pi^0)$

At $(2\Pi^0)$

This is $7(3\Pi^2)$

this is $7(3\Pi^2)$

this is $7(3\Pi^2)\ km\ /\ h$

this is $7(3\Pi^2)\ km\ /\ h$

this is $7(3\Pi^2)\ km\ /\ h$

This is $7(3\Pi^2)\ km\ /\ h$

This is $7(\Pi\Pi^2)\ km\ /\ h$

From the centre of the earth or the sun singularity compresses space and it is not material that draws closer to the centre, which is what holds singularity but it is the space holding material or not holding material that moves towards the object and towards the centre of the turning object. This movement compresses the space and as the space is further away from such a centre the density reduces as the volume of space increase. This puts a density relevancy on all space all space from the centre.

Science holds on the "mass pulling mass" Newton created as the truth while there is nothing about mass pulling to be found in nature or in the cosmos or even on earth. I explain why any object has mass and when it does not have mass and what is the difference between having mass and not having mass.

Gravity is the movement of objects in space moving through space.

Gravity is the gravitational density of movement of object in space and the difference the object move through space, Everything relies on singularity and everything depends on singularity and how movement effects the relocation position of al moving things in the cosmos in relation to where it was, where it is and where it will be the next moment. In this gravity forms by gravitational density. Science in the Newtonian version ignores these cosmic laws but that does not change the prominence they hold. Nature uses them notwithstanding that science ignore them. Science ignores them because they destroy Newton and that science can't understand.

CURCULAR ATMOSPHERIC DISPLACEMENT Π^2 Π

Everything forming gravity applies the Titius Bode law. When an object stands on earth holding "mass" the object serves a position as a unit of the earth while it moves with the earth. This is part of the four cosmic principles forming a unit that forms gravity I named on the previous page.

$\Pi \setminus \Pi^0$
LINEAR DISPLACEMENT

$\Pi^3 \setminus \Pi^2$
AIRCRAFT
TOTAL DISPLACE-
MENT

The reason why an atmosphere forms is the Lagrangian points forming atmospheric boundaries with density limitations applying.

When the objects starts to move in a horizontal fashion that exceeds the gravity going in a vertical fashion the Coanda Effect bring about that the object then forms a part of the atmosphere where it releases from the earth and becomes part of the atmosphere. We call this release flying.

BETWEEN MACH ONE AND THREE Π^3 / Π^2

ATMOSPHERIC CIRCULER DISPLACEMENT $\Pi^2 \setminus \Pi^0$

ADAPTED   CIRCULER DISPLACEMENT

$\Pi^2 \setminus \Pi^0$

$\Pi \setminus \Pi^0$

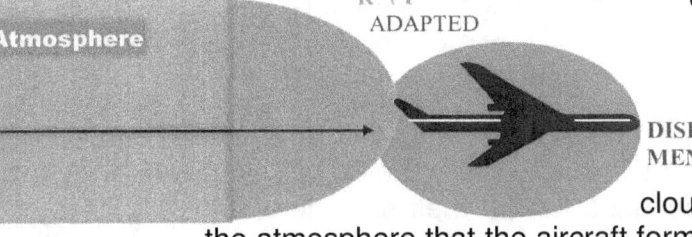

Above the Roche limit of $(\Pi/2)^2$.

CIRCULER DISPLACEMENT
$R^2 \setminus T$
ADAPTED

Then with much more movement applying the aircraft breaks the sound barrier. This is actually where the movement exceeds the Roche limit and in that movement creates that the moving object forms an independent atmosphere.

R/T LINEAR
DISPLACE-
MENT

This is when as the Roche limit the aircraft by moving concentrates so much independence it captures a cloud of vapour by concentrating the density of the atmosphere that the aircraft forms as a unit to become a unit within the earth's unit but an identifiable part that is no longer part of the earth's atmosphere.

That is why at that point it carries its own rules on sound outside the borders of sound on earth. Very briefly this is the information about physics that the four cosmic principles or laws provide us on gravity. Gravity is immensely more than simply "mass" pulling "mass" that then bring a number of complex formulas and equations about.

What you are about to read is a new approach to science as never seen before and it introduces nature as science for the first time in history. Gravity depends on four laws by which the Universe started from the point being one single dot to what we now have as a Universe.

What would this about understanding of the Titius Bode and the other three laws hold for science would you ask? How does this improve our understanding of science? Well for one fact: I can prove there just can't be aliens because the Titius Bode law will not permit such travel! Our understanding of gravity becomes much clearer because we can see the four laws working within what we call the sound barrier.

There are density changes that enforce different speeds at different altitudes and the aircraft has to travel at certain speed to fly at certain altitudes. It is the density at that altitude that forces the aircraft to maintain the speed. At the speed of sound the aircraft enlists a cosmic border that breaks the earth's atmosphere, where the differentiation will bring about a shudder, that changes the time dimension with re-adjusting space. The sound barrier is the dimensional gravitational difference and the aircraft's structure releases from the earth's atmosphere valued at 7 // 10 and by such release gives the aircraft a value of Π^2 /2 in relation to 3. Since the aircraft is going in a direction the air that it goes thought holds a value of Π^2, which then takes on a double value of the Lagrangian law, which is 5.

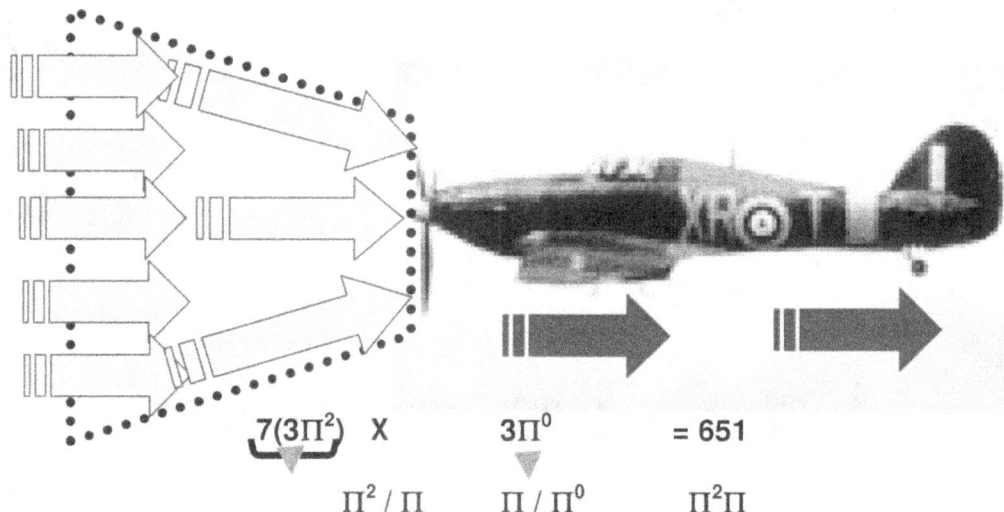

$$\underbrace{7(3\Pi^2)}\ \ X \qquad 3\Pi^0 \qquad = 651$$

$$\Pi^2 / \Pi \qquad \Pi / \Pi^0 \qquad \Pi^2\Pi$$

A propeller driven aircraft can divert from singularity with a maximum diverting of $3\Pi^0$.

The aircraft holds a relation to heat in dimensional change provided by the earths "gravity" at a value of $7(3\Pi^2)\ 3\Pi^0$ = 651 km / h. Any more speed needed the aircraft must introduce additional heat from own supply.

This picture alone must revolutionise the engineering's understanding of movement in as far as gravity goes and yet not one person shows interest. There is a cosmic code written in space by time where everything in nature in the cosmos uses one value to form whatever space. The code is $\Pi^0\Pi\Pi^2$ and this formulates gravity. Now why won't the world of physics not even read my endeavour or challenge my arguments or contest my correctness. I shall say why; it is because I show their science is rubbish. If they could shred me to bits, boy they would have enjoyed doing just that. Now on reading this remark you will immediately climb on your high horse and get aroused and become very opinionated!

When flying at specific altitudes it requires not only displacing the air but also contributing to more air or heat forming. This extra heat only a jet engine can deliver.

$7(3\Pi^2)$ X $2\Pi^0$ ⇔ $7(3\Pi^2)$ and $3\Pi^0$
Staying inline with the earths atmosphere can take the craft to a maximum value of
$7(3\Pi^2)$ X 5Π =1036.3 km. / h. After that point the Titius Bode law will stretch no further and the "SOUND BARRIER" becomes compromised

$7(3\Pi^2)$ X $2\Pi^0$ ⇔ $7(3\Pi^2)$ and $3\Pi^0$ = 621.78 this is only achievable with jet power.

$7(3\Pi^2) \times 2\Pi^0 \Leftrightarrow 7(3\Pi^2) \times 4\Pi^0 = 829$

The maximum toleration the earth's singularity will allow
$7(3\Pi^2) \times 5\Pi^0 = 1036.3$ and the entry of the Titius Bode law is the sound barrier at $7(3\Pi^2)(\Pi^2/2) = 1022.795$.
This is how the sound barrier works. In the centre of the earth a pointy forms holding singularity at a value of 1 or as I like to call it Π^0.

The shuttering that pilots report is the barrier between the maximum point holding the earth's singularity and the craft obtaining cosmic singularity.

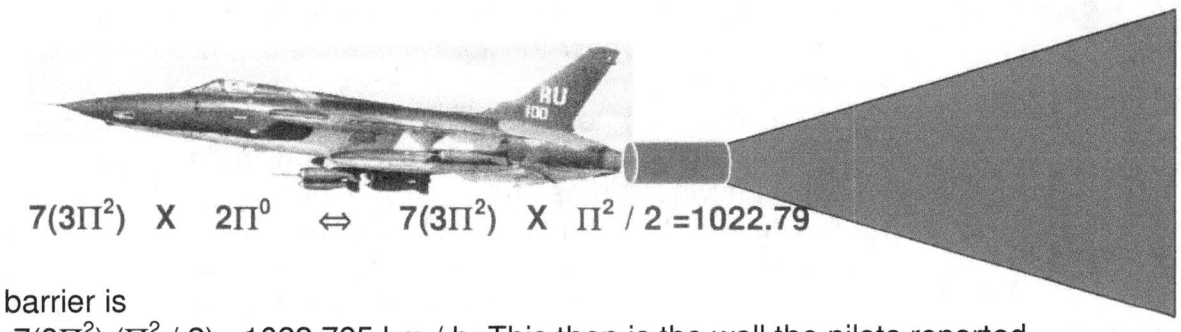

The velocity the aircraft reaches before the sound barrier is
$7(3\Pi^2) \times 4.5\,\Pi^0 = 932.67$ km / h and after the sound

$7(3\Pi^2) \times 2\Pi^0 \Leftrightarrow 7(3\Pi^2) \times \Pi^2/2 = 1022.79$

barrier is
$7(3\Pi^2)(\Pi^2/2) = 1022.795$ km / h. This then is the wall the pilots reported.

THE OBJECT IN FLIGHT MAINTAINS ITS POSITION AS PART OF THE EARTHS STRUCTURE 7 /Λ\ 10 but going in one direction

$7(3\Pi^2) \times \Pi^1 = 651.1$ km. / h

...And what have I accomplished up to this point with all this information I provided?
Not only have I scared away all my critics but also I have chased off any support I might have gained because I show what my approach to science is about and in that alienated any readers that were interested before. Please take note that everything said this far is so simple that even I understand the concept and science thinks of me as being so stupid that I don't even understand anything about Newton...but I am getting to that!

For those that should show some degree of interest in what I say, I have and later show a letter with which my work and I am brushed off by the great minds in physics...

Everything that I show is how nature forms gravity and how nature builds the Universe because I am the first to study and understand Kepler and with that understand nature.

Understanding this about nature and how nature applies science I should think is most important and yet no one in science in 17 (seventeen) years could be bothered one bit to read my work. I can't think how arrogant science can be to brush nature off by favouring Newton!

Remember; <u>if you don't like the mathematics ignore it</u> because it doesn't contribute to anything. With it I dispute Newtonian thinking and it <u>is only put there to shut my critics up and show them what true physics are but if you don't need proof</u> but just want explaining then ignore the maths and read the articles. That way you will understand the entire concept. It is so simple that even I understand it and even a simple person such as I could work it out and realise how it works. I never tried to be clever by telling the Universe what Newton wanted it to be but I studied nature in the cosmos to see what nature applied in the Universe.

Singularity started as the first spot and not with a massive already formed Universe that only afterwards grew in size. Since the Big Bang event the Universe only grew but everything that is in the Universe already was in the Universe. What was in place at the Big Bang only then grouped and became material but not more of what already was. The division was brought about by differences in density in accordance with movement.

Think of it this way. The spot overheats and expands into the dot as the movement comes by the spot enlarging to form the dot and by growing into the dot the movement enlarges and the movement as well as the increase in size reduces the heat the dot accumulates. The growth then by reducing cools off and the cooling removes the structural size of the dot as it returns to form the next spot. However what is in

the Universe can never leave the Universe but has to remain a component within the Universe as long as the Universe exists.

However before heat difference brought about the Universe we know time was one continuing everlasting spot, which is a line that never went further than one spot. This I realised when I studied how the four phenomena worked and how nature applies gravity. Please note that I have no opinion because it is not testing Newton compared to my view but it is how nature works against the Newtonian hoax that never apply anywhere in nature.

The fact that we have an atmosphere and the moon does not have an atmosphere has everything to do with the fact that the moon uses the axis of the earth to turn, which puts the moon in place as an extension of the earth's gravity. The atmosphere has everything to do with the four cosmic pillars and most of all due to the Coanda effect that turns the gas in outer space to a liquid we call the atmosphere.

$4\Pi^2/(6^2 \times 10^2)$

$\Pi^2/2$

$7 \times (3\Pi^2)$

To leave the atmosphere one has to displace $7 \times (3\Pi^2)(\Pi^2/2)(4\Pi^2)/(6^2 \times 10^2) = 11$ km / sec

In order to fly at maximum altitude of 31000 meters then the slowest displacement is $7(3\Pi^2)(\Pi^2/2)(\Pi^2/4) = 2523.64$ km/h.
This displacement is the very same speed as the displacement at Π^0, which is $7 \times (3\Pi^2) = 207.261$ km/hour. If going slower the plane will fall out of the sky at $7 \times (3\Pi^2)\Pi^0$ and will bring the same result as going slower than $7 \times (3\Pi^2)(\Pi^2/2)(\Pi^2/4)$ at 31000 meters. This shows it is a density difference applying at various levels.

These values form part of a cosmic code I developed with which and also how the relevancy of the cosmic order can be determined.

It was developed by studying the four cosmic laws or as I named it the four cosmic pillars.

In twelve articles I went out of my way to explain this and many more and I was informed as follows:

In order to fly ant the edge of the atmosphere one needs to develop a relevancy between liquid (air) and solid (aircraft) of at least $7 \times (3\Pi^2)(\Pi^2/2)(\Pi^2/4) = 2523.64$ km / h. Should the relevancy be any slower the aircraft will fall from the sky just as it will do at $7 \times (3\Pi^2) = 207$ km / h at ground level. These two speeds are exactly alike but each serves at a greater altitude in order to comply with the gravitational density laws.

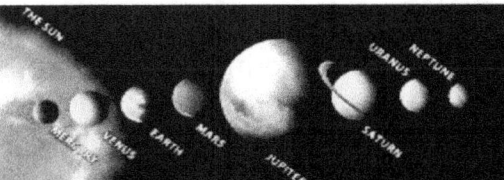

The Titius Bode law is how planets distribute places in the solar system. There is a precise sequence and order in which planets hold places and this annihilates the idea that mass plays any part in this sequence. In contrast the Newtonian system that science promote currently says that gravity "pulls" according to mass. This is fictional. The mass of the planets is totally random and Newton holds no theoretical basis that nature supports. Look at the arrangement and you can visually see using your eyes that the distribution by mass is a hoax and is an invention Newton concocted and science validates for the past 300 years. By explaining the Titius Bode law for the first time ever this book breaks the 300-year-old myth and brings truth

The Roche limit is the law that says stars do not collide ever in spite of Newton's

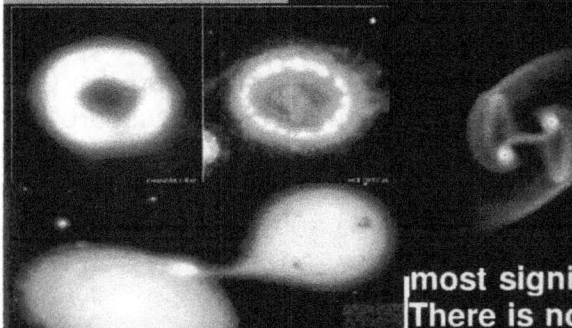

ridiculous idea that stars can or do collide. If two stars are in each other's atmosphere the law reads that when the minor of the two stars is closer than 2.4674 of the diameter of the major star, then the major star will liquefy the minor star into a gas plume that it then treats as more atmosphere. This is most significant in the cosmic principles I now put forward. There is not one instance or any evidence that shows where two stars do collide. When two or more stars are evenly matched by gravity the two stars do not collide either but becomes binary stars that spin around each other. Stars or planets never collide and when a meteor enters the atmosphere like at Tunguska in Russia in 1908, the earth vaporises and liquefies the meteor and the meteor becomes fragments as well as more dust clouds in the atmosphere. This totally annihilates Newton's idea that a radius between structures diminishes by the gravitational attraction of mass.

The Coanda effect connects with the previous law. Gravity is a ratio that forms

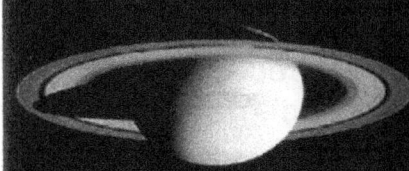

between what is solid space and what is liquid space. This applies when the movement of the solid (star or planet or atom) turns and by turning 7° it reduces the value of the circle of the space surrounding the solid from Π= 21.991/7 to the compressed value Π = 3.142/1 which connects to singularity. There is no pulling of material but only reducing of space when gravity contracts gas into liquid such as what the atmosphere is.

The Lagrangian Points This law proves that science is committed to deception for centuries because the sattelites spinning around the big planets never come closer. It is known that sattelites turn around planets in the same fashion as do planets orbit the sun. Saturn does not "pull" the satelites into its atmosphere or even "pull" the satelight closer and this knowlidge never drove science to question the validity of Newton ot his concept about "mass" pulling "mass". Even the rings formes as debris spins in a circle that always remain constant and science should have rejected newton'ss ideas on the grounds that nature never not once supports Newton and Newtonian science in any way thinkeble.

This is Nature's physics and this is how nature's physics works. Now read what nature is and how nature forms the Universe.

Everyone is in agreement with Albert Einstein that the Universe started with one spot, a point we call singularity. Now that you saw the laws I wish to give a glimpse about how the Universe started. As I said, the Universe started with 1 and this fact even Newtonian science accepts.

For the first time in human history you can read how the Solar system forms. It forms by the Titius Bode law and not Newton's mass pulling. I discovered for the first time how to interpret the Titius Bode law and decipher the ratio that forms the Titius Bode law. Does science applaud me? No because I trash Newton…and if you thought Newton explained it then know that is the Corrupt Science I Reveal. Nature uses the Titius Bode law (go look it up) and I am the first to manage to explain that. Because the Titius Bode law trashes Newton Newtonians keep quiet about the Titius Bode law. I found the four building blocks that form the Universe and from that found gravity forms by Π. I show what nature uses and Newton and nature share nothing. If you disagree with me you disagree with nature because I present what nature uses in contrast to what Newton said is physics. If you read my work you will find that I explain nature because I discard Newton and the idea that mass has anything to do with forming forces.

That above statement I prove and I prove undoubtedly by example. There is no pulling of any object on any other object as Newton said. Gravity does not work like a magnetic field but works in principle by the object rotating… But it is space that compresses as rotation reduces in a principle called the Coanda effect. It works by material objects producing movement in rotating that condense space around the object to liquid. Gravity turns space surrounding the rotating material from gas to freeze it into liquid. Outer space is a gas that turns to atmosphere that is a liquid. This compressing of space is the Coanda effect. The difference between what my approach must be when using nature and Newtonian's approach that is unsupported by nature is one Universe away from each other. I show a functional Universe and Newton show what they call the mysteries of science in the Universe. I prove everything that Newtonian science this far could never prove. I prove the Universe applies four keys by which gravity works instead of unexplainable magical forces that pull each other and this no person (not even Einstein) this far could ever explain. There are 4 principles applying gravity and forms a value of Π as a circle rotates.

What makes the difference between reality and science is Nature does not use Newton in any way or form. Newton says objects pull by an unexplainable magical force called gravity while I say it is round objects that rotate. As it turns it forms Π and by collapsing Π the space around the star / planet that space forming an atmosphere compresses and thereby collapses from 21.991 / 7 to 3.1416 / 1. As it turns it divides 7 from 21.991 / 7 by 7 to form 7/7 = 1. As the direction rotates Π changes the travel by 7° and that 7° is 7 / 7 = 1 or Π = 3.1416 / 1. Then by getting reduced it compresses everything in that space as the entire space the object holds and claims condenses by reducing Π from 21.991 / 7 to 3.1416 / 1. Space reduces and Newton said objects in space pull each other. While I prove nature, Newtonian science cheats, corrupts and manipulate nature to make science work in ways nature doesn't work. There is a link between the space that compresses and material that spins and condenses space.

This is the first time ever in the history of human intellect that this has been attempted or achieved.
Why does science not go on a song and dance all night long…because this totally rubbishes Newton and therefore also mainstream science.

These four books prove gravity is not about mass as Newton said but it is movement of the turning that forms gravity as pi going pi square.
These books give answers about questions no one asks such as why does the sun cross our blue sky every day?

Have you ever thought why the sun crosses our sky?

These books raise questions and give answers but if you wish to have proof you should read the other six.

This first table of Kepler proves how nature forms the Universe because this is the way it forms the solar system. It shows space expands with the adding of every planet doubling the distance from the previous planet to the next planet, just as the Titius Bode law indicates.

Planet	Mass per Earth unit	$k^{-1} = T^2 \div a^3$ Movement	a^3 of space volume	T^2 During time units
Mercury	0.06	$T^2 \div a^3 =$ 0.983	$(a^3)=$ 0.059	$(T^2)=$ 0.058
Venus	0.82	$T^2 \div a^3 =$ 0.992	$(a^3)=$ 0.381	$(T^2)=$ 0.378
Earth	1.000	$T^2 \div a^3 =$ 1.000	$(a^3)=$ 1.000	$(T^2)=$ 1.000
Mars	0.11	$T^2 \div a^3 =$ 1.000	$(a^3)=$ 3.54	$(T^2)=$ 3.54
Jupiter	317.89	$T^2 \div a^3 =$ 1.000	$(a^3)=$ 140.6	$(T^2)=$ 140.66
Saturn	95.17	$T^2 \div a^3 =$ 0.999	$(a^3)=$ 868.25	$(T^2)=$ 67.9
Uranus	14.53	$T^2 \div a^3 =$ 1.000	$(a^3)=$ 7067	$(T^2)=$ 7069
Neptune	17.14	$T^2 \div a^3 =$ 0.999	$(a^3)=$ 27189	$(T^2)=$27159
Pluto	0.0025	$T^2 \div a^3 =$ 1.004	$(a^3)=$ 61443	$(T^2)=$61703

This second table disproves Newton completely because it shows that "mass" play no part or has no influence in the location of planets. The mass is completely randomly distributed.

If Newton said Πr^2 but $\quad F \;=\; G\,\dfrac{M_1 M_2}{r^2}$ __is totally ridiculous__

PLANET	Mean Distance from the Sun (AU)	Equatorial Radius (km)	Mass of planet (Earth=1)	Mean density (grams/centimeter³)	
Mercury	0.3871	2439	0.06	5.43	
Venus	0.7233	6052	0.82	5.25	
Earth	1.000	6378	1.000	5.52	
Mars	1.524	3397	0.11	3.95	
Jupiter	5.203	71490	317.89	1.33	
Saturn	9.539	60268	95.18	0.69	
Uranus	19.19	25559	14.53	1.29	
Neptune	30.06	25269	17.14	1.64	
Pluto	39.48	1160	0.002	2.03	

There are the numbers and there are the facts. <u>No one can deny the facts as the truth.</u> **The tables speak for itself and by ignoring it science commit fraud. It is no use to shoot the messenger because science wants to hide the truth. The truth is out and someday some one will read it and that day everyone in science will regret that science ignored the truth to protect a lie.**
Yours truly
Peet (P.S.J.) Schutte

You and everyone else can contact me at __peet@naturescosmicconcept.co.za__ or at __mail.naturescosmicconcept.co.za__ if you feel you don't agree don't waste your time with this then go to a book that holds this theme at a higher level.

www.ingramcontent.com/pod-product-compliance
Lightning Source LLC
Chambersburg PA
CBHW080240180526
45167CB00006B/2359